# Soil and Groundwater Pollution

# Soil and Groundwater Pollution

Edited by **Sheryl McMillan**

SYRAWOOD
PUBLISHING HOUSE

New York

Published by Syrawood Publishing House,
750 Third Avenue, 9th Floor,
New York, NY 10017, USA
www.syrawoodpublishinghouse.com

**Soil and Groundwater Pollution**
Edited by Sheryl McMillan

International Standard Book Number: 978-1-68286-050-2 (Hardback)

# Contents

# Preface

Soil and groundwater are two resources of paramount importance to all living organisms. Due to the increasing levels of pollution worldwide, soil and groundwater have also been adversely affected. This book unravels the recent studies in the field of soil and groundwater pollution. Assessment of risks from water, soil and air pollution, effective and viable remedies, waste disposal strategies, techniques and methods for protection of soil and groundwater, etc., are some of the areas that have been discussed in the text. Comprising of detailed analyses and data, this book will prove immensely beneficial to professionals and students involved in the study of environment at various levels.

The information shared in this book is based on empirical researches made by veterans in this field of study. The elaborative information provided in this book will help the readers further their scope of knowledge leading to advancements in this field.

Finally, I would like to thank my fellow researchers who gave constructive feedback and my family members who supported me at every step of my research.

**Editor**

# Probabilistic Risk Assessment for Six Vapour Intrusion Algorithms

Jeroen Provoost[1], Lucas Reijnders[2], Jan Bronders[3], Ilse Van Keer[3] & Steven Govaerts[3]

[1] Independent researcher, Finland

[2] Open University Netherlands (OUNL), Department of Science, Valkenburgerweg 177, 6419 AT Heerlen, Netherlands

[3] Flemish Institute for Technological Research (VITO), Boeretang 200, 2400 Mol, Belgium

Correspondence: Jeroen Provoost, Independent researcher, Finland. E-mail: Jeroen.Provoost@yahoo.co.uk

**Abstract**

A probabilistic assessment with sensitivity analysis using Monte Carlo simulation for six vapour intrusion algorithms, used in various regulatory frameworks for contaminated land management, is presented here. In addition a deterministic approach with default parameter sets is evaluated against observed concentrations for benzene, ethylbenzene and trichloroethylene. The screening-level algorithms are ranked according to accuracy and conservatism in predicting observed soil air and indoor air concentrations at two contaminated sites to determine their suitability for regulatory purposes and the possible occurrence of false-negative errors. Dominant parameters that drive the predictions are grouped by either physico-chemical, soil or building parameters, and also by parameters that are either uncertain or variable, to determine the prioritisation for further research actions such as additional measurements. The findings from this study suggest that the screening-level algorithms that have a higher degree of conservatism for their default parameter set are the Johnson and Ettinger model, Dilution Factor algorithm from Sweden, Vlier-Humaan and VolaSoil. From these four algorithms the Johnson and Ettinger model and VolaSoil have a relative high accuracy (discriminative power). For the latter two algorithms different parameters, that are variable and uncertain, contribute to the variation in indoor air concentration, and differences were observed between the aromatic and chlorinated hydrocarbons. For the chlorinated hydrocarbon trichloroethylene, the default parameter set of Vlier-Humaan, CSoil and Dilution Factor algorithm of Sweden might be adapted to arrive at a higher deterministically predicted indoor air concentration if more conservatism is required. The deterministically predicted air concentrations for aromatic hydrocarbons seem to be sufficiently conservative. It is shown that the probabilistic approach allows for an improved insight into the relative importance of parameters in the risk estimates.

**Keywords:** deterministic, probabilistic, risk assessment, vapour intrusion, benzene, ethylbenzene, trichloroethylene

## 1. Introduction

Soil contamination has become an important issue, especially in industrialized countries. Against this background governments in industrialized countries have developed contaminated land management policies to reduce risks to humans and ecosystems originating from soil pollution. Such policies often include legislation and soil screening values, which trigger further actions when exceeded (Provoost et al., 2008a). The exposure to soil contamination and the exceedance of soil screening values are established by multi-media exposure algorithms (Carlon, D'Alessandro, & Swartjes, 2007). One of the major pathways of exposure is inhalation of indoor air as a result of sub-surface contamination with volatile chemicals. "Soil vapour can become contaminated when volatile chemicals evaporate from subsurface sources such as (DOH, 2006):

a. groundwater or soil that contains volatile chemicals,

b. non-aqueous phase liquid, which exists as liquid volatile chemicals,

c. buried waste,

d. underground storage tanks or drums".

Soil vapour can enter a building and affect the indoor air quality (Kaplan, Brandt-Rauf, Axley, Shen, & Sewell, 1993; Fugler & Adomait, 1997). When contaminated vapours are present nearby the foundation of a dwelling, vapour intrusion can occur. Soil vapour can enter a building irrespective of the age of the dwelling and the type of basement (DOH, 2006).

Predicting the soil air and indoor air concentration, and the related human exposure, is complex and is affected by numerous factors. Factors can be divided generally in three categories: environmental, building and physico-chemical (McAlary, Provoost, Dawson, & Swartjes, 2011; Provoost et al., 2010; DOH, 2006), which are subject to variability. Examples of environmental factors are: soil conditions, concentrations, source location, groundwater conditions, weather conditions and biodegradation. Examples of building factors are: mechanical ventilation and heating systems, air exchange rates, foundation type and surface features. Physico-chemical factors are for example: the Henry constant (Provoost, Ottoy, Reijnders, Bronders, Van Keer, Swartjes, Wilczek, & Poelmans, 2011), solubility and vapour pressure. These categories need to be taken into account when conducting a soil investigation and when evaluating air measurements (observations) and modelling (predictions) with screening level algorithms (DOH 2006). The latter output is subject to two sources of variation: uncertainty and variability. Variability regards variation that can be naturally expected, while uncertainty regards precision by which a quantity is measured (Van Belle, 2008). Parameters from algorithms can be uncertain because there is insufficient information about a true, but unknown value, for example the fraction of organic matter in the soil. One can describe an uncertainty parameter with a probability distribution. Theoretically, it is possible to reduce uncertainty by gathering more information on the site. Practically, information can be missing because it was not gathered or it is too costly or technically difficult to gather. Variability is inherent in the system, and therefore cannot be eliminated by gathering more information (Cullen, & Frey, 1999; Finley & Paustenbach, 1994; McKone & Bogen, 1990; Ragas, Brouwer, Buchner, Hendriks, & Huijbregts, 2009; Filipsson, 2011).

Most of the present algorithms for indoor air vapour intrusion calculate point estimates based on a set of default parameter values and therefore give no indication of the variation of the output. Within this so called deterministic framework, the value of each algorithm's parameter is chosen in such a way that a worst-case or conservative point estimate is obtained for indoor air exposure. Due to this way of setting parameters the variation as a result of uncertainty and variability of the predicted soil air and indoor air concentration remains unclear (Ragas et al., 2009). It needs also to be recognised that vapour intrusion screening algorithms need to be further advanced (Fisher, Ireland, Boylandb, & Critten, 2002). Algorithms applied with a deterministic approach could give a false sense of security to decision makers and the public (Krupnick et al., 2006) as research has shown that deterministic risk assessments of contaminated land may underestimate and overestimate risk depending on the algorithm used and default parameter settings (Oberg & Bergback, 2005; Ririe, Sweeyey, & Daugherty, 2002; Kuusisto & Tuhkanen, 2001; Nassar, Ukrainczyk, & Horton, 1999; DeVaull, Ettinger, & Gustafson, 2002; McHugh, Connor, & Ahmad, 2004; Oberg & Bergback, 2005; Johnston & MacDonald, 2010; Johnson, Kemblowski, & Johnson, 1999; Provoost, Cornelius, & Seuntjens, 2002; Provoost et al., 2008b; Provoost et al., 2009; Provoost et al., 2010). However there is published literature available containing conceptual frameworks that allow regulators to deal with variation as a result of uncertainty and variability (EPA, 2001; Fisher et al. 2002; van der Sluijs, Janssen et al., 2004; ITRC, 2008; Warmink, Jansen, Booij, & Knol, 2010; Little, 2013; Provoost, Reijnders, & Bronders, 2013).

As opposed to the deterministic approach, the probabilistic approach includes variation in the algorithms parameters. This involves the computation of the probability distribution function for the relevant algorithm parameters from the data obtained. Followed by a selection of values from the probability distribution by using a sampling method and displaying the results as sensitivity charts (i.e. ranking of parameters according to the correlation between parameter and algorithms output) and frequency distributions (i.e. graphs displaying the distribution of soil air or indoor air concentrations). For a probabilistic approach, Monte Carlo simulation has been introduced (Hammersley & Hanscomb, 1964; McKone & Ryan, 1989). This allows the user to define probability distribution for uncertain parameters and frequency distributions for the variability of parameters, which propagates these input distributions into an output distribution of the predicted air concentration. "Monte Carlo simulation provides insight in how specified variability and uncertainty in inputs propagates through an algorithm. It forces analysts to explicitly consider uncertainty and interdependencies among different inputs and it can cope with any conceivable shape of probability distribution function and can account for correlations" (van der Sluijs et al., 2004). A potential advantage of probabilistic indoor air vapour intrusion calculations is that they may give insight into the major determinants of calculated outcomes, and the effect of variation therein.

Several studies are available that evaluate a particular algorithm probabilistically or assess its sensitivity (Seuntjens, Provoost, & Cornelis, 2001; Provoost et al., 2002; Swartjes, 2002; Webster & Mackay, 2003; Johnson, 2005; Oberg & Bergback, 2005; Tillman & Weaver, 2006; van Wijnen & Lijzen, 2006; Swartjes, 2007; Tillman & Weaver, 2007; Johnson & Macdonald, 2010). These studies do not allow an inter-algorithm comparison for the same site or input parameters and thereof few are published in peer reviewed literature. Thus the present study considers simultaneously six frequently used vapour intrusion algorithms (see chapter 2.2) and the objective of this study is to determine the variation in the soil air and indoor air concentration for these algorithms as a result of uncertainty and variability in their parameters. In addition, a sensitivity analysis for each of the algorithms reveals which parameters contribute most to the variation in the predicted air concentration (Swartjes, 2009). The soil and indoor air concentration were calculated as a result of soil contamination from two well documented sites where benzene, ethylbenzene or trichloroethylene was predominantly present in the soil. The study does not investigate uncertainty related to the algorithms itself or scenario chosen (Filipsson, 2011; Huijbreghts, Gilijamse, Ragas, & Reijnders, 2003), but rather focuses on parameter uncertainty and variability. The results presented in this paper can contribute to the evaluation of the suitability of screening-level algorithms for regulatory purposes and the possible occurrence of false-negative errors. A probabilistic approach might thus increase the robustness without decreasing the credibility of the vapour intrusion predictions, as it presents results as a frequency or probability distribution. A probabilistic approach may also help in prioritising further actions.

## 2. Materials and Methods

### 2.1 Description Sites

Measured data from two well documented sites were used to contrast predictions and observations.

The site Astral is situated in Vilvoorde (Belgium) and was used for over 30 years as an industrial plant for the production of paints and varnishes and became contaminated with volatile aromatic hydrocarbons, such as benzene and ethylbenzene. At the source zone both the vadose zone and the groundwater are contaminated. A site investigation revealed a three dimensional migration of the contamination in the soil and deeper groundwater. The conceptual site model showed that the contaminants of concern were the aromatic hydrocarbons benzene and ethylbenzene in the source zone and close proximity in the soil. According to the source concentrations, transport mechanisms, exposure routes and receptors on the industrial site, it is apparent that the migration of vapour to the indoor air as a result of soil (vadose zone) contamination is the dominant route of human exposure. Hence, humans were most likely to be exposed through the route "inhalation of indoor air" (Bronders et al., 2000). The vadose zone has an average thickness of 150 cm, with sandy-loam in the sub-surface and a loamy soil close to the groundwater level. The building above the contaminant source has a concrete floor with a thickness varying between 25 to 50 cm with some cracks and gaps.

The site from the Colorado Department of Transport (CDOT) is situated in Denver (USA), and was used from 1957 to test paints and materials used for pavements and on roadways. Chlorinated hydrocarbons, such as 1,1,1-trichloroethane, trichloroethylene and dichloromethane, were used and stored on the site in two underground storage tanks. In 1970 a leakage in both tanks resulted in a spill which contaminated the surrounding soil. Site investigations revealed that chlorinated hydrocarbons were present in the groundwater, soil and indoor air (Johnson, Ettinger, Kurtz, Bryan, & Kester, 2002; Bryan, 2000). The contaminant of concern was trichloroethene, and the dominant route of human exposure is inhalation of indoor air as a result of the soil (vadose zone) contamination. The vadose zone has an average thickness of 460 cm, with a predominant sandy-loam soil profile. The enclosed space concrete floor thickness was about 15 cm thick.

### 2.2 Selection of the Algorithms

A variety of screening algorithms are available that predict the migration of chemicals from the groundwater to the indoor air. Many of these algorithms are applied for site specific human health risk assessment, and for deriving soil screening values for volatile organic compounds (Provoost et al., 2008a; McAlary et al., 2011). The selection criteria for the algorithms on which a probabilistic assessment is performed, is done based on the following assumption:

▪ Public or commercial availability: the full algorithm with standard parameter set is (commercially) available and published in the public domain or can be obtained easily from the author.

▪ Intellectual property rights: no infringements on the intellectual property rights or copy rights are invoked by implementing the algorithm in a Microsoft Excel® based spreadsheet.

▪ The algorithm is used within a regulatory framework for contaminated land management in one or more of the EU countries.

This resulted in a selection of the following algorithms:

▪ S-EPA dilution factor algorithm (Sweden) (S-EPA, 1996), hereafter DF SE

- SFT dilution factor (Norway) (SFT, 1999), hereafter DF NO

- Johnson and Ettinger model (United States) (Johnson, 2005), hereafter JEM

- CSoil algorithm (Netherlands) (Brand, Otte, & Lijzen, 2007; Otte, Lijzen, Otte, Swartjes, & Versluijs, 2001)

- VolaSoil algorithm (Netherlands) (Waitz, Freijer, Kreule, & Swartjes, 1996; van Wijnen & Lijzen, 2006; Bakker, Lijzen, & van Wijnen, 2008)

- Vlier-Humaan algorithm (region Flanders in Belgium) (OVAM, 2004), hereafter Vl-H

The background of each of the selected algorithms is described in Provoost et al. (2010), and in more detail in Provoost et al. (2008b, 2009) and McAlary et al. (2011). An overview of the processes included in these algorithms is provided in Table 1.

Table 1. Overview of the selected vapour intrusion algorithms

| Algorithm | Diffusive transport | Advective transport | Advective transport | Biodegradation | Source-depletion |
|---|---|---|---|---|---|
| Location | Soil | Soil | Foundation | Soil | Soil |
| DF SE* | yes | yes | yes | yes | yes |
| DF NO | yes | no | no | no | no |
| JEM | yes | no | yes | no | yes |
| CSoil | yes | yes | no | no | no |
| VolaSoil | yes | yes | yes | no | no |
| Vl-H | yes | yes | no | no | no |

DF SE: Dilution Factor algorithm from Sweden, DF NO: Dilution Factor algorithm from Norway, JEM: Johnson and Ettinger model, Vl-H: Vlier-Humaan, * empirically derived from multiple sites, which can include all processes described.

### 2.3 Deterministic Analysis

Six vapour intrusion algorithms are used to predict air concentrations with the default parameters set to obtain conservative point estimates. For each of the algorithms the deterministic parameter set is available as supplementary material. Where possible, algorithm parameters were adapted to the site specific conditions such as soil (e.g. mean initial soil concentration or average depth of the soil contaminant) and building properties. Other parameters (e.g. perimeter seem crack, pressure difference soil-building) were put on the algorithm specific default parameter value. The deterministic predictions are displayed together with the box-and-whisker plots from the probabilistic predictions.

### 2.4 Probability Distributions of Algorithms Inputs

Probability distribution functions (PDF) were derived from the data gathered for the three groups of parameters (soil, building, and physico-chemical properties of the three chemicals) and fed into the six algorithms. Data were obtained from site measurements or literature. The PDF were calculated by performing a Chi-Square goodness-of-fit test where the Chi-square gauges the general accuracy of the fit and breaks down the distribution into areas of equal probability and compares the data points within each area to the number of expected data points. Generally a p-value greater than 0.5 indicates a close PDF fit. The goodness-of-fit performes a set of mathematical tests to find the best fit between a standard probability distribution and a data set's distribution (Crystal Ball, 2000). The tests resulted in 4 types of distributions: normal, log normal, uniform and triangular distributions. For each of the algorithms the probabilistic parameter set is available as supplementary material.

### 2.5 Probabilistic and Sensitivity Analysis

The probability distribution in algorithms inputs were propagated into an output distribution of the predicted air concentration in the soil air and indoor air for each of the selected algorithms (probabilistic analysis), and on the contribution of each parameter to the variation (sensitivity analysis). Predicted soil air and indoor air concentration for benzene, ethylbenzene and trichloroethylene, were calculated for 5000 combinations of parameter values using Monte-Carlo simulation. The results are displayed as frequency distributions (box-and-whisker plots displaying the distribution of soil air and indoor air concentrations) and sensitivity stacked bar charts (ranking of parameters according to the correlation between parameter and algorithm output). "Sensitivity is calculated by computing ranked correlation coefficients between every algorithm parameter and the soil air or indoor air concentrations while the simulation is running. Rank correlation is a method whereby the parameter values are replaced with their ranking from lowest value to highest value using the integers 1 to N prior to computing the correlation coefficient. Correlation coefficients provide a meaningful measure of the degree to which algorithms parameters and predicted air concentrations change together. If a parameter and predicted (soil or indoor) air concentration have a high correlation coefficient, it means that the parameter has a significant impact on the prediction (both through its uncertainty and algorithm sensitivity). Positive coefficients indicate that an increase in the parameter is associated with an increase in the prediction. Negative coefficients imply the reverse situation. The larger the absolute value of the correlation coefficient, the stronger the relationship. The sensitivity can be also expressed in percentages of the contribution to the variance of the predicted air concentration. The contribution to the variance is calculated by squaring the rank correlation coefficients and normalising them to 100%" (Crystal Ball, 2000). The probabilistic and sensitivity analysis were performed by using Crystal Ball®.

The box-and-whiskers plots for the probabilistic analysis also include the tolerable concentration in air as a reference to inhalation risks of indoor air. Tolerable concentrations were obtained the World Health Organization (WHO, 1996, 2010). For benzene, ethylbenzene and trichloroethylene the tolerable concentration in air are respectively 1.7 µg/m³, 22,000 µg/m³ and 23 µg/m³. "The concentrations of airborne benzene associated with an excess lifetime risk of 1:10,000, 1:100,000 and 1:1,000,000 are assumed to be 17, 1.7 and 0.17 µg/m³", respectively (WHO, 2010) and the tolerable concentration in air related to an excess lifetime risk of 1:1,000,000 was selected. With respect to the general population, a tentative guidance value of 22 mg/m³ (5 ppm) for ethylbenzene in inhaled air has been reported by WHO (1996). The concentrations of trichloroethylene assumed to be related to an excess lifetime cancer risk of 1:10,000, 1:100,000 and 1:1,000,000 are in that order 230, 23 and 2.3 µg/m³ (WHO, 2010) and the tolerable concentration in air related to an excess lifetime risk of 1:1,000,000 was selected.

*2.6 Accuracy and Conservatism*

Screening-level algorithms have been developed that predict indoor air concentrations as a result of groundwater and/or soil contamination. Screening aims at identifying contaminated soils that should be further investigated for the need of remediation and/or the presence of an intolerable health risk as a result of indoor air contamination. To be useful in this respect, screening-level algorithms should be sufficiently conservative so that they produce very few false-negative predictions but they should not be overly conservative because on the latter case they might have insufficient discriminatory power (Provoost et al., 2009).

Provoost et al. (2009) define accuracy as the algorithms ability to predict air concentrations that are in close proximity to the observed air concentrations. The accuracy and conservatism of screening-level algorithms is objectified by calculating the Maximum relative Error (ME), Root Mean Squared Error (RMSE) and Coefficient of Residual Mass (CRM), as described by Loague and Green (1991), for the paired predicted and observed air concentrations. These criteria were applied in Provoost et al. (2008b, 2009), and also in this study, for inter-algorithm comparison and provided a ranking of the algorithms as to their accuracy.

The three statistical criteria provide insight in the suitability of the algorithm for regulatory purposes. The $O$ in each of the formulas means the observed concentration, $P$ the predicted concentration and $n$ the number of cases.

1) Maximum relative Error (ME):

$$ME = \frac{\max_{i=1}^{n}\left[abs\left(O_i - P_i\right)\right]}{\bar{O}}$$

(1)

The ME provides the maximum difference between the observed and predicted concentrations. If a lower ME is calculated in comparison with another algorithm, than the maximum difference between $O$ and $P$ is smaller and hence the higher the accuracy of the algorithm.

2) Root Mean Squared Error (RMSE):

$$RMSE = 100 \frac{\sqrt{\dfrac{\sum_{i=1}^{n}\left(O_i - P_i\right)^2}{n}}}{\bar{O}}$$

(2)

The RMSE provides a measure of the average difference between $O$ and $P$ for each calculation. If a lower RMSE is calculated in comparison with another algorithm, than the average difference between $O$ and $P$ is smaller and hence the higher the accuracy of the algorithm.

3) Coefficient of Residual Mass (CRM):

$$CRM = -\frac{\left[\sum_{i=1}^{n} O_i - \sum_{i=1}^{n} P_i\right]}{\sum_{i=1}^{n} O_i}$$

(3)

The CRM provides a measure whether the algorithm over- or under-predicts in comparison to observations. If the CRM has a positive (+) value the predicted concentration by the algorithm are frequently higher when compared to the observed concentration, and vice versa. The lower the CRM, the closer the observation and predictions are in close proximity, and the more accurate the particular algorithm.

Furthermore, for each of the algorithms and contaminants the paired deterministic and 95 percentile probabilistic predicted soil air and indoor air concentration will be compared. Comparison is made by calculating the CRM with the formular:

$$CRM = -\frac{\left[\sum_{i=1}^{n} De_i - \sum_{i=1}^{n} Pr_i\right]}{\sum_{i=1}^{n} De_i}$$

(4)

$De_i$ means the deterministic concentration whereas $Pr_i$ the probabilistic predicted 95 percentile concentration. The CRM provides in this perspective a measure of which algorithms deterministically predict concentrations that are under or above the probabilistic 95 percentile predicted soil or indoor concentrations and therefore indicates the conservatism of each algorithm's default parameter set. A positive CRM value indicates a lower deterministic predicted concentration in comparison to the probabilistic 95 percentile concentration, and vice versa for positive CRM values. The outcome of the analysis will be presented in figures. The 95 percentile value is frequently used in probabilistic risk assessments as the measure for a sufficient protection of the exposed population (ITRC, 2008; CCME, 1996; Ferguson, 1999; Swartjes, 1999; Provoost et al., 2013).

*2.7 Uncertainty and Variability*

The dominant parameters, resulting from the sensitivity analysis, were grouped in two different ways to investigate the overall importance of uncertainty or variability. Firstly parameters are allocated to the group physical chemical, soil or building properties. Secondly to either uncertainty or variability and a justification is provided for each of the dominant parameters in a table. Hereto, literature was consulted to contrast the allocation of input parameters to uncertainty or variability and findings will be presented in the same table.

## 3. Results

*3.1 Probabilistic and Deterministic Analysis*

A box-and-whisker plot is provided for each combination of the three contaminants (benzene, ethylbenzene, trichloroethylene) and algorithms (six). The box-and-whiskers plot display the minimum, 25 percentile, median, 75 percentile and maximum predicted soil or indoor air concentration, as well as the predicted deterministic concentration (●) and, in the case of indoor air concentrations, the tolerable concentration in air (◇) for the pertaining contaminant. Also for each of the contaminants the observed (measured) soil air and indoor air concentrations are displayed to contrast with the predictions. The box-and-whiskers plots provide an insight in the spread (range of the values from the highest to the lowest value), and the midspread (range of middle 50% of the values) or also called inter-quartile range. The location of the median line relative to the 25 and 75 quartiles indicates the amount of skewness or asymmetry in the data.

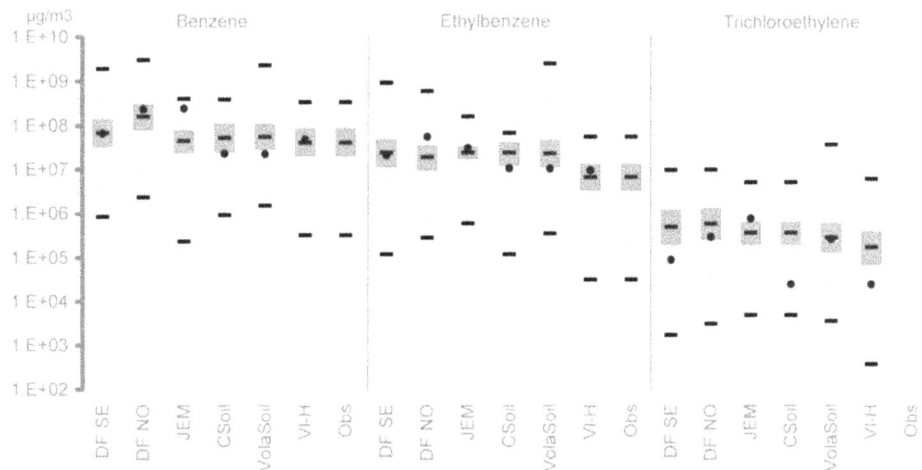

Figure 1a. Box-and-whiskers plot for predicted and observed soil air concentrations by algorithm and contaminant

DF SE: Dilution Factor algorithm from Sweden, DF NO: Dilution Factor algorithm from Norway, JEM: Johnson and Ettinger model, Vl-H: Vlier-Humaan, Obs: observed concentrations, Box plot: — minimum, median or maximum concentration, box is 25 or 75 percentile concentration, ● deterministic concentration

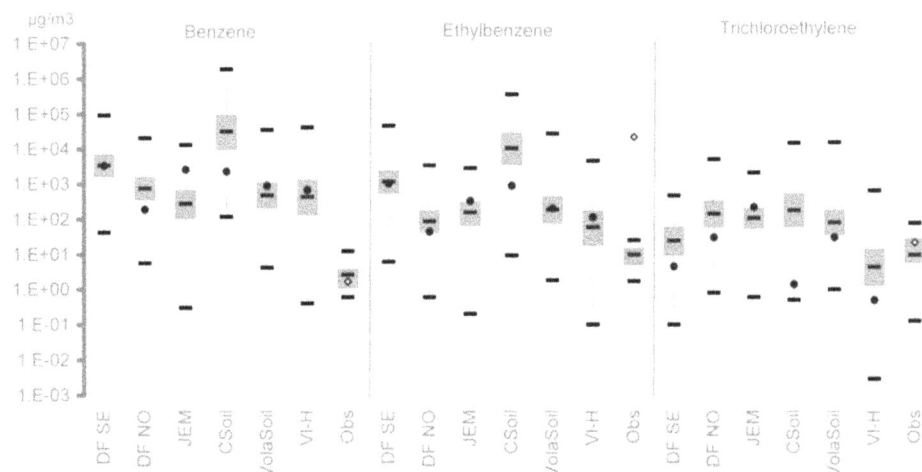

Figure 1b. Box-and-whiskers plot for predicted and observed indoor air concentrations by algorithm and contaminant

DF SE: Dilution Factor algorithm from Sweden, DF NO: Dilution Factor algorithm from Norway, JEM: Johnson and Ettinger model, Vl-H: Vlier-Humaan, Obs: observed concentrations, Box plot: — minimum, median or maximum concentration, box is 25 or 75 percentile concentration, ● deterministic concentration, ◇ tolerable concentration in air

Figure 1a reveals, for all algorithms and aromatic hydrocarbons (benzene and ethylbenzene from the site Astral), an overall higher or equal midspread for the probabilistic predicted soil air concentration when compared to the midspread of the observed soil air concentrations. For the chlorinated hydrocarbon trichloroethylene (from the site CDOT) no comparison could be made with observed soil air concentrations as they were not published. Some of the box-and-whiskers plots show a negative skewness towards the higher concentrations. With some exception the deterministically predicted concentrations are in the probabilistic midspread.

Figure 1b indicates that all algorithms have a higher or equal probabilistic midspread for the predicted indoor air concentration when compared to the mid-spread of the observed indoor air concentrations, with the exception of the Vl-H algorithm for the contaminant trichloroethylene. In general the box-and-whiskers plots for predicted indoor air concentrations do not show a particular positive or negative skewness.

*3.2 Accuracy and Conservatism*

The statistical parameters ME, RMSE and CRM provide a way for inter-algorithm comparison and a ranking of their accuracy. Figure 2a-d present the outcome of the three statistical criteria for soil air and indoor air predictions.

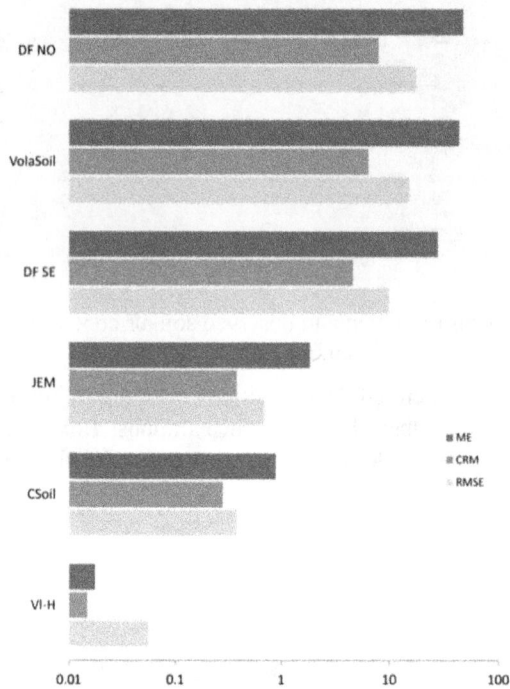

Figure 2a. Accuracy of algorithms for soil air concentrations

DF SE: Dilution Factor algorithm from Sweden, DF NO: Dilution Factor algorithm from Norway, JEM: Johnson and Ettinger model, Vl-H: Vlier-Humaan, ME: maximum relative error, RMSE: root mean squared error, CRM: coefficient of residual mass

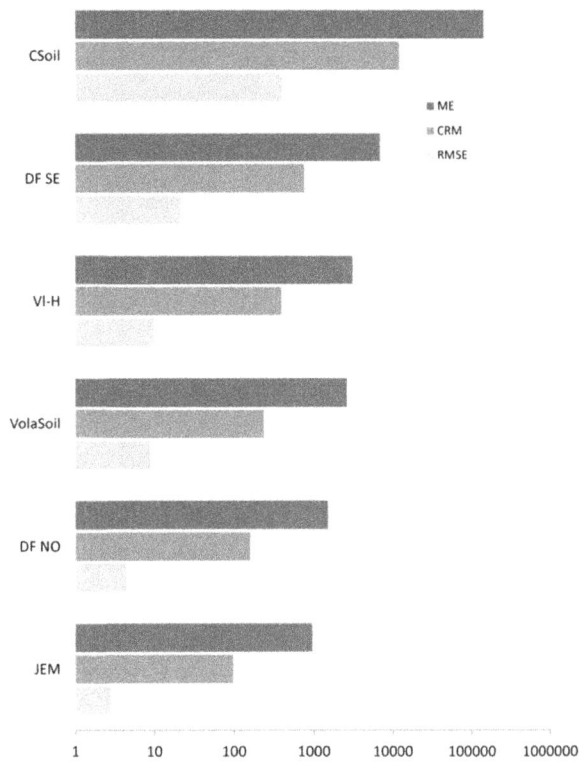

Figure 2b. Accuracy of algorithms for indoor air concentrations

DF SE: Dilution Factor algorithm from Sweden, DF NO: Dilution Factor algorithm from Norway, JEM: Johnson and Ettinger model, Vl-H: Vlier-Humaan, ME: maximum relative error, RMSE: root mean squared error, CRM: coefficient of residual mass

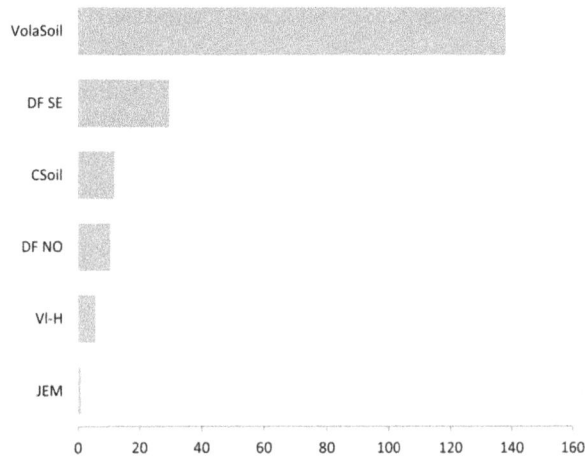

Figure 2c. Conservatism (CRM) for soil air by algorithms and contaminant

DF SE: Dilution Factor algorithm from Sweden, DF NO: Dilution Factor algorithm from Norway, JEM: Johnson and Ettinger model, Vl-H: Vlier-Humaan, CRM: coefficient of residual mass

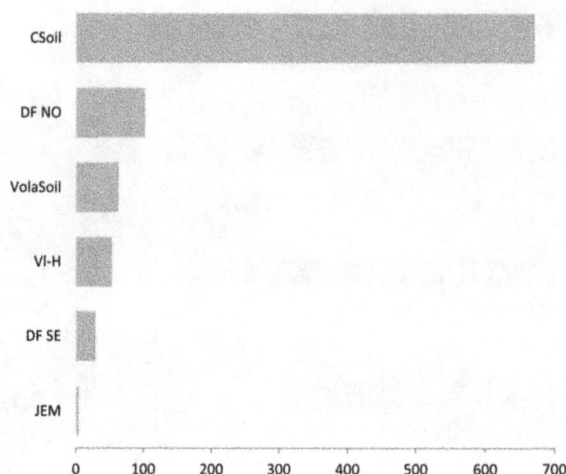

Figure 2d. Conservatism (CRM) for indoor air by algorithms and contaminant

DF SE: Dilution Factor algorithm from Sweden, DF NO: Dilution Factor algorithm from Norway, JEM: Johnson and Ettinger model, Vl-H: Vlier-Humaan, CRM: coefficient of residual mass

Figure 2 provides the accuracy for each of the algorithms for the soil air (2a) and indoor air (2b). Algorithms are ranked from low (top) to high (bottom) accuracy. The lower the ME and RMSE value, the smaller the difference between observed and predicted values and therefore the more accurate the algorithm. The algorithms that most frequently over-predict (less accurate) the observed soil air concentrations are the DF NO, VolaSoil and the DF SE algorithms, while the JEM, CSoil and Vl-H have a higher accuracy. For predicting the indoor air concentrations JEM, DF NO, VolaSoil and Vl-H seem to have a higher accuracy than DF SE and CSoil.

Figures 2c and d show the CRM for each of the algorithms when comparing the median observed concentration with the deterministic predicted soil air (2c) and indoor air (2d) concentration. A positive CRM indicates a higher deterministically predicted concentration when compared to the median observed concentration. A low CRM value indicates that the deterministic predicted and median observed concentrations are in closer proximity from each other. A higher positive CRM value therefore indicates a more conservative algorithm.

The deterministic predicted soil air concentrations are overall higher than the median observed concentrations with the exception of VolaSoil and CSoil for benzene. For aromatic hydrocarbons the least conservative algorithms are Vl-H, VolaSoil and CSoil, followed by the DF SE, whereas JEM and the DF NO are the most conservative algorithms. For the predictions of the aromatic hydrocarbon concentrations in the indoor air the DF NO, Vl-H and VolaSoil are the least conservative algorithms, followed by the JEM, CSoil and DF SE. For trichloroethylene the DF SE, CSoil and Vl-H are not conservative (negative CRM), as they predict deterministic concentrations that are below the median observed concentration. VolaSoil and the DF NO are more conservative, followed by JEM, which is the most conservative. A difference is observed in the ranking between aromatic and chlorinated hydrocarbons.

The comparison shows that the JEM and Vl-H are the more conservative algorithms while maintaining a level accuracy (slightly over-predict). The DF NO and SE, the VolaSoil, and CSoil algorithm have a lower accuracy (mostly over-predict when compared to observations) and conservatism.

Table 2. Parameters contributing most to the variation of soil air (%)

| | DF SE | | | DF NO | | | JEM | | | CSoil | | | VolaSoil | | | VI-H | | |
|---|---|---|---|---|---|---|---|---|---|---|---|---|---|---|---|---|---|---|
| | B | EB | TCE | B | EB | TCE | B | EB | TCE | B | EB | TCE | B | EB | TCE | B | EB | TCE |
| **Physical chemical properties** | 30.5 | 31.1 | 63.2 | 21.7 | 17.4 | 51.1 | | 75.8 | 8.2 | 33.2 | 31.4 | 70.1 | 6.8 | 7 | 41 | 23.3 | 12.2 | 72.2 |
| Organic carbon-water partitioning coefficient | 29.6 | 18.1 | 63.2 | | | | | | | 28.3 | 18.6 | 52.9 | | | | | | |
| Octanol-water partition coefficient | | | | 19.6 | 4 | 51.1 | | | | | | | | | | 17.9 | 1.9 | 45.1 |
| Henry's coefficient | 0.9 | 13 | | 2.1 | 13.4 | | 51.4 | 8.2 | | | | | | | | | | |
| Solubility | | | | | | | 24.4 | | | 4.6 | 7.5 | 16.8 | 5.9 | 5.3 | 41 | 4.9 | 8.1 | 26.8 |
| Vapour pressure | | | | | | | | | | 0.3 | 5.3 | 0.4 | 0.9 | 1.7 | | 0.5 | 2.2 | 0.3 |
| **Soil properties** | 69.4 | 68.8 | 36.6 | 78.1 | 82.5 | 48.6 | 99.8 | 23.8 | 90.3 | 66.3 | 68.2 | 29.4 | 93.2 | 93 | 59 | 76.4 | 87.5 | 27.1 |
| Water filled porosity | | | | | | | | | | | | | | | | | | 0.2 |
| Air filled porosity | | | | | | | | | | | | | | | | | | 0.2 |
| Initial concentration (soil) | 41.9 | 38.9 | 25 | 52 | 47.6 | 32.3 | 60.6 | 15.5 | 61.9 | 40.7 | 39.8 | 20 | 93.1 | 92.6 | 58.2 | 47.1 | 50 | 11.6 |
| Organic carbon fraction | 27.5 | 29.9 | 11.6 | 26.1 | 34.9 | 16.3 | 39.2 | 4.4 | 28.4 | 25.6 | 28.4 | 9.4 | 0.1 | 0.4 | 0.8 | 29.3 | 37.5 | 15.1 |
| Soil temperature | | | | | | | | 3.9 | | | | | | | | | | |
| **Total** | 99.9 | 99.9 | 99.8 | 99.8 | 99.9 | 99.7 | 99.8 | 99.6 | 98.5 | 99.5 | 99.6 | 99.5 | 100 | 100 | 100 | 99.7 | 99.7 | 99.3 |

DF SE: Dilution Factor algorithm from Sweden, DF NO: Dilution Factor algorithm from Norway, JEM: Johnson and Ettinger model, Vl-H: Vlier-Humaan, B: benzene; EB: ethylbenzene; TCE: trichloroethylene

Table 3. Parameters contributing most to the variation of indoor air (%)

| | DF SE | | | DF NO | | | JEM | | | CSoil | | | VolaSoil | | | VI-H | | |
|---|---|---|---|---|---|---|---|---|---|---|---|---|---|---|---|---|---|---|
| | B | EB | TCE | B | EB | TCE | B | EB | TCE | B | EB | TCE | B | EB | TCE | B | EB | TCE |
| **Physical chemical properties** | 30.5 | 31.1 | 63.2 | 17.9 | 14.4 | 37.8 | | 13.5 | 6.5 | 9.4 | 5.3 | 56.5 | 4.4 | 5.5 | 28.8 | 8.2 | 4.3 | 39.6 |
| Organic carbon-water partitioning coefficient | 29.6 | 18.1 | 63.2 | | | | | | | 8.3 | 3.7 | 43.3 | | | | | | |
| Octanol-water partition coefficient | | | | 16.3 | 3.1 | 37.8 | | | | | | | | | | 6.2 | 0.9 | 23.8 |
| Henry's coefficient | 0.9 | 13 | | 1.6 | 11.3 | | | 8.3 | 6.5 | | | | | | | | | |
| Solubility | | | | | | | | 5.2 | | 1.1 | 1.6 | 12.9 | 4.4 | 4.2 | 28.8 | 2 | 3.4 | 15.5 |
| Vapour pressure | | | | | | | | | | | 0.3 | | | 1.3 | | | | 0.3 |
| **Soil properties** | 69.4 | 68.8 | 36.6 | 66.1 | 71 | 61.9 | 77.8 | 53.3 | 89.8 | 89.1 | 92.8 | 43.1 | 76.8 | 77.2 | 70.6 | 28.7 | 30.7 | 16.7 |
| Total porosity | | | | | | | 1 | | 6.4 | | | | | | | | | |
| Water filled porosity | | | | | 0.3 | 11.8 | 1 | 3.3 | 11.3 | 33.5 | 38.6 | 9.1 | | | 13.1 | | | 0.9 |
| Air filled porosity | | | | | 0.3 | 11.9 | | | | 34.4 | 39.5 | 9.3 | | | 14 | | | 1 |
| Air permeability | | | | | | | 30 | 43.9 | | | | | 11.5 | 11.1 | | | | |
| Initial concentration (soil) | 41.9 | 38.9 | 25 | 44.7 | 40.5 | 25 | 29.3 | 4 | 49.7 | 11.7 | 8.4 | 16.5 | 65.3 | 66.1 | 42.5 | 16.8 | 18.3 | 6.9 |
| Fraction organic carbon | 27.5 | 29.9 | 11.6 | 21.4 | 29.9 | 12.3 | 17.5 | 1.1 | 22.4 | 7.6 | 4.5 | 7.3 | | | | 11.9 | 12.4 | 7.8 |
| Depth to contaminant (groundwater/soil layer) | | | | | | | | | 0.9 | 1.9 | 1.8 | 0.9 | | | 1 | | | 0.1 |
| **Building properties** | | | | 15.5 | 14.6 | | 21.2 | 32.4 | 2.8 | 1.1 | 1.1 | 0.35 | 17.8 | 16 | 0.55 | 62 | 63.5 | 43.4 |
| Intrusion rate of pore air | | | | 12.4 | 11.3 | | | | | | | | | | | | | |
| Indoor air exchange rate | | | | 3.1 | 3.3 | | 1.8 | 2.6 | 1 | 1.1 | 1.1 | 0.35 | 3.6 | 2.5 | 0.3 | 1.8 | 1.9 | 0.1 |
| Soil-building pressure differential | | | | | | | 18.8 | 28.7 | 0.5 | | | | 8.6 | 9 | 0.25 | | | |
| Floor-wall seam crack width | | | | | | | 0.6 | 1.1 | 1.3 | | | | 5.6 | 4.5 | | | | |
| Fraction pores in concrete | | | | | | | | | | | | | | | | 9.7 | 8.9 | 6.5 |
| Fraction air in concrete | | | | | | | | | | | | | | | | 50.5 | 52.7 | 36.8 |
| **Total** | 99.9 | 99.9 | 99.8 | 99.5 | 100 | 99.7 | 99 | 99.2 | 99.1 | 99.6 | 99.2 | 99.95 | 99 | 98.7 | 99.95 | 98.9 | 98.5 | 99.7 |

DF SE: Dilution Factor algorithm from Sweden, DF NO: Dilution Factor algorithm from Norway, JEM: Johnson and Ettinger model, Vl-H: Vlier-Humaan, B: benzene; EB: ethylbenzene; TCE: trichloroethylene

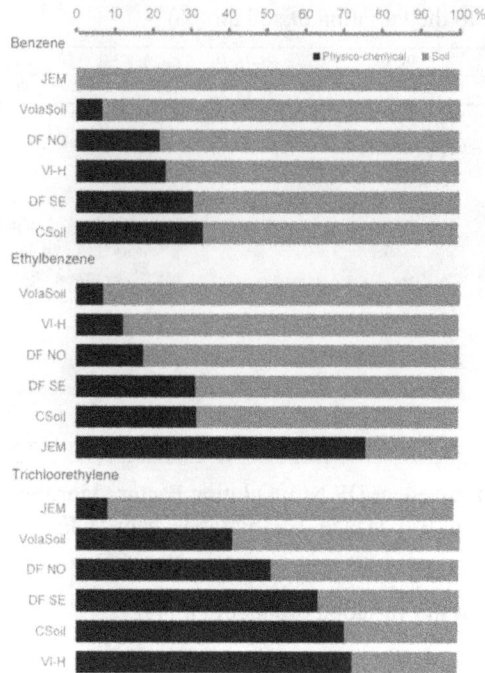

Figure 3a. Stack bars of the percentage that physico-chemical and soil parameter values contribute to the variation in soil air concentrations by algorithm and contaminant

DF SE: Dilution Factor algorithm from Sweden, DF NO: Dilution Factor algorithm from Norway, JEM: Johnson and Ettinger model, Vl-H: Vlier-Humaan

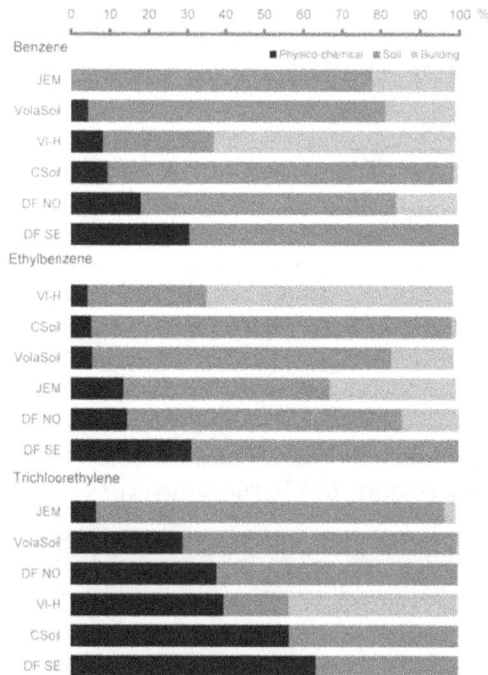

Figure 3b. Stack bars of the percentage that physico-chemical, soil and building parameter values contribute to the variation in indoor air concentrations by algorithm and contaminant

DF SE: Dilution Factor algorithm from Sweden, DF NO: Dilution Factor algorithm from Norway, JEM: Johnson and Ettinger model, Vl-H: Vlier-Humaan

*3.3 Sensitivity Analysis*

3.3.1 Grouping by Parameter Type

The sensitivity analysis allows for the ranking of dominant parameters to the variation in predicted air concentrations. Parameters were first grouped by physico-chemical, soil or building parameters (Table 2) resulting in an overall contribution of the group to the total variation (Figures 3a, b).

Figure 3a shows that the soil air concentration for benzene and ethylbenzene are driven by the soil parameters and for trichloroethylene increasingly by the physico-chemical parameters, with the exception of JEM. Table 2 provides details on what individual parameters contribute most. The dominant parameters that contributing to the variation in soil air concentrations are for physical-chemical properties, depending on the algorithms and contaminant: the organic carbon-water partitioning coefficient (18-63%), octanol-water partition coefficient (2-51%), Henry's coefficient (0.9-51%), solubility (5-41%) and vapour pressure (0.3-5%). For soil properties the dominant parameters were: initial concentration (12-93%) and organic carbon fraction (0.1-39%). Table 2 shows that the dominant parameters (± 10) drive for > 99% the variation in the soil air concentration.

Figure 3b shows that the most dominant parameters contributing to the variation in indoor air concentration are for benzene and ethylbenzene soil and building parameters and for trichloroethylene soil and physico-chemical parameters. Table 3 reveals that for the physical-chemical properties the most dominant parameters are, depends on the algorithms and contaminant considered: the organic carbon-water partitioning coefficient (3.7-63%), octanol-water partition coefficient (1-38%), Henry's coefficient (1-13%) and solubility (1.1-29%). For the soil properties the water (0.3-38%) and air filled porosity (0.3-38%) (correlated), air permeability (11-44%), initial concentration (4-66%) and fraction organic carbon (1-30%) are dominant parameters. For the building properties the intrusion rate of pore air (±12%), soil-building pressure differential (0.3-29%) and fraction air in concrete (37-53%) drive the variation. Table 3 shows that the dominant parameters (±19) account for > 98% the variation in the indoor air concentration.

3.3.2 Grouping by Uncertainty and Variability

Dominant parameters were in addition grouped by parameters that are either uncertain or variable as shown in Figure 4. Table 4 provides a justification for the allocating of parameters to one of the groups. The allocation of parameters was, besides expert judging, based on literature. Table 4 shows that algorithm parameters are not consistently attributed to variability or uncertainty. The table provides a justification for each of the dominant parameters with their allocation.

The contribution to the variation in predicted soil air concentration (Figure 4a) is for all algorithms and contaminants dominated by variability (> 60%). Overall, the contribution to the variation in soil air concentrations is driven by variability, however differences are observed between algorithms and contaminants. The main parameter that contributes to variability is the organic carbon fraction (0.1-39%), depending on the algorithm and contaminant, whereas for variability the contributing parameters are organic carbon-water partitioning coefficient (18-63%), octanol-water partition coefficient (2-51%), Henry's coefficient (1-51%), solubility (5-41%), and initial concentration (soil) (12-93%).

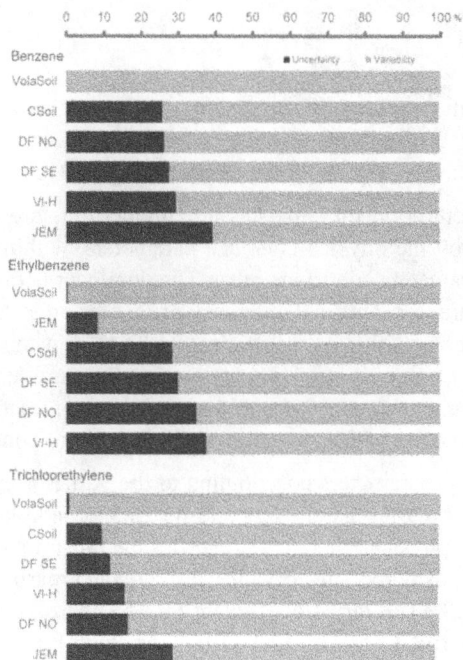

Figure 4a. Stack bars of the percentage that uncertain and variable parameter values contribute to the variation in soil air concentrations by algorithm and contaminant

DF SE: Dilution Factor algorithm from Sweden, DF NO: Dilution Factor algorithm from Norway, JEM: Johnson and Ettinger model, Vl-H: Vlier-Humaan

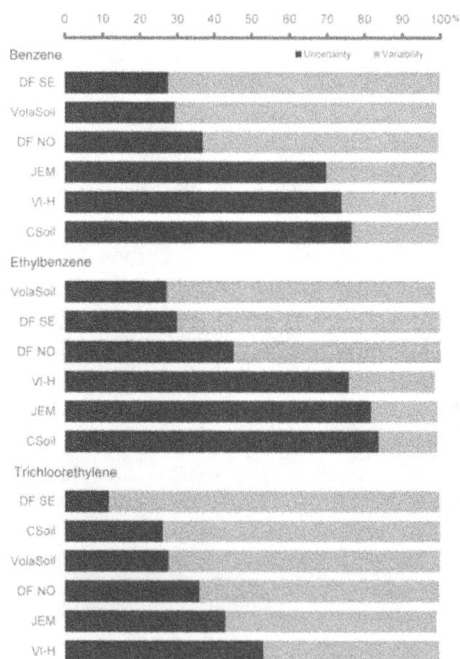

Figure 4b. Stack bars of the percentage that uncertain and variable parameter values contribute to the variation in indoor air concentrations by algorithm and contaminant

DF SE: Dilution Factor algorithm from Sweden, DF NO: Dilution Factor algorithm from Norway, JEM: Johnson and Ettinger model, Vl-H: Vlier-Humaan

Table 4. Allocation of the dominant parameters that contribute most to the variation

| Parameter | Type | Allocated to | Justification | Weaver &Tillman, 2005 | Tillman & Weaver, 2007 | Webster & Mackay, 2003 | Oberg & Bergback, 2005 |
|---|---|---|---|---|---|---|---|
| Soil temperature | Soil | Variable | B | Uncertain | | Variability | Spatial variability |
| Octanol-water partition coefficient | Chemical | Variable | C | Constant | | Variability | |
| Henry's coefficient | Chemical | Variable | C | Constant | | Variability | |
| Solubility | Chemical | Variable | C | Constant | | Variability | |
| Vapour pressure | Chemical | Variable | C | Constant | | Variability | |
| Organic carbon-water partitioning coefficient | Chemical | Variable | C | Uncertain | | Variability | |
| Initial concentration | Soil | Variable | B | Uncertain | | Spatial variability | Uncertain |
| Depth to contaminant | Soil | Variable | B | Uncertain | Variability | Spatial variability | Uncertain |
| Water filled porosity | Soil | Uncertain | A | Uncertain | | Variability | Spatial variability |
| Air filled porosity | Soil | Uncertain | A | Uncertain | | Variability | Spatial variability |
| Air permeability | Soil | Uncertain | A | Uncertain | | Variability | Spatial variability |
| Total porosity | Soil | Uncertain | A | Uncertain | Variability | Variability | Spatial variability |
| Organic carbon fraction soil | Soil | Uncertain | A | Uncertain | | Variability | Spatial variability |
| Intrusion rate of pore air | Building | Uncertain | D | Potentially uncertain | Variability | | |
| Indoor air exchange rate | Building | Uncertain | D | Uncertain | Variability | | |
| Soil-building pressure differential | Building | Uncertain | D | Uncertain | | | |
| Floor-wall seam crack width | Building | Uncertain | D | Uncertain | Variability | | |
| Fraction pores in concrete | Building | Uncertain | D | Potentially uncertain | | | |
| Fraction air in concrete | Building | Uncertain | D | Potentially uncertain | | | |

[a] The parameter was derived from on site measured soil properties and values were taken from literature to complement the measurement. Deriving a parameter from other soil properties introduces uncertainty; [b] The parameter in a region varies spatially and temporally, even over short distances. The soil is not homogenous and thus the parameter can be described with a distribution of values (variable). The parameter was measured on site and additional values were taken from literature to complement the measurement; [c] Chemical parameter values are measured and derived from literature. Values were collected for different reference temperatures and are therefore variable; [d] The building parameters were not measured on site and a range of values was taken from literature. The parameter also varies between houses and are thus considered to be uncertain.

A clear trend in the contribution of parameters to the variation of indoor air concentrations between algorithms or contaminants could not be established. However, for the aromatic hydrocarbons the prediction of the indoor air concentration by the algorithms JEM, Vl-H and CSoil are predominantly driven by uncertainty (> 70%), while for the other algorithms (the DF SE, NO and VolaSoil) variability tends to dominate. For the chlorinated hydrocarbon trichloroethylene variability generally dominates the contribution of the variation to the indoor air concentration, over uncertainty (Figure 4b). The main contributors to the uncertainty are, depending on the algorithm and contaminant, water filled porosity (0.3-39%), air filled porosity (0.3-40%), organic carbon fraction (1-30%), soil-building pressure differential (0.3-29%), and fraction air in concrete (37-53%). For variability these are: octanol-water partition coefficient (1-38%), Henry's coefficient (1-13%), solubility (1-29%), organic carbon-water partitioning coefficient (4-63%) and initial concentration (soil) (4-66%).

## 4. Conclusions

### 4.1 Accuracy

The soil air concentration is most accurately predicted by Vl-H, followed by CSoil and JEM, whereas higher values for the 3 criteria are observed for the DF SE, VolaSoil and DF NO. Table 5 summarises what parameters contribute to the variation in predicted soil air concentration when contributions over 20% are considered (see Table 2). No distinction is made between the three contaminants.

Table 5. Summary of parameters contributing > 20% to the variation in soil air

| Parameter | Group | Type | DF SE | DF NO | JEM | CSoil | VolaSoil | Vl-H |
|---|---|---|---|---|---|---|---|---|
| Octanol-water partition coefficient | PC | Var | | X | | | | X |
| Henry's coefficient | PC | Var | | | X | | | |
| Solubility | PC | Var | | | X | | X | X |
| Organic carbon-water partitioning coefficient | PC | Var | X | | | X | | |
| Initial concentration | S | Var | X | X | X | X | X | X |
| Organic carbon fraction soil | S | Unc | X | X | X | X | | X |

Var: variability, Unc: uncertainty, PC: physico-chemical property, S: soil property, DF SE: dilution factor algorithm from Sweden, DF NO: dilution factor algorithm from Norway, JEM: Johnson and Ettinger model, CSoil: CSoil algorithm, VolaSoil: VolaSoil algorithm, Vl-H: Vlier-Humaan algorithm

Variation in the predicted soil air concentrations are for all algorithms driven by the soil parameters initial soil concentration (variable) and organic carbon fraction soil (uncertain). These parameters are routinely measured during site investigations and vary between sample points. The contribution from variable physico-chemical parameters depends on the algorithms, and these parameters are not routinely measured on site as they are contaminant specific.

All selected sites, aromatic hydrocarbons and statistical criteria considered the algorithm with the highest accuracy for predicting a soil air concentration close to the observed concentration is Vl-H, CSoil and JEM. The JEM and CSoil algorithms have low values for the ME, CRM and RMSE, where Vl-H has the lowest values.

Table 6. Summary of parameters contributing to the variation in indoor air

| Parameter | Type | Group | DF SE | DF NO | JEM | CSoil | VolaSoil | Vl-H |
|---|---|---|---|---|---|---|---|---|
| Octanol-water partition coefficient | Var | PC | | X | | | | X |
| Henry's coefficient | Var | PC | | | | | X | |
| Solubility | Var | PC | | | | | | |
| Organic carbon-water partitioning coefficient | Var | PC | X | | | X | | |
| Initial concentration | Var | S | X | X | X | | X | |
| Water filled porosity | Unc | S | | | | X | | |
| Air filled porosity | Unc | S | | | | X | | |
| Air permeability | Unc | S | | | X | | | |
| Organic carbon fraction soil | Unc | S | X | X | X | | X | |
| Soil-building pressure differential | Unc | S | | | X | | | |
| Fraction air in concrete | Unc | B | | | | | | X |

Var: variability, Unc: uncertainty, PC: physico-chemical property, S: soil property, B: building property, DF SE: dilution factor algorithm from Sweden, DF NO: dilution factor algorithm from Norway, JEM: Johnson and Ettinger model, CSoil: CSoil algorithm, VolaSoil: VolaSoil algorithm, Vl-H: Vlier-Humaan algorithm

Table 6 summarises what type of parameters contribute to the variation in predicted indoor air concentration when considering contribution over 20% (see Table 3). No distinction is made between the three contaminants. Variation in the predicted indoor air concentrations are for DF SE, DF NO, JEM and VolaSoil driven by the soil parameters initial soil concentration (variable) and the organic carbon fraction in the soil (uncertain). Both parameters are routinely measured on site during site investigations. Predicted concentrations for CSoil are driven by the organic carbon-water partition coefficient (variable physico-chemical parameter), water and air filled porosity (uncertain soil parameters). These parameters are not routinely measured on site. Vl-H's predicted indoor air concentrations are mainly driven by the physico-chemical parameter octanol-water partition coefficient (variable) and building parameter fraction air in concrete (uncertain). Neither is measured on site.

All selected sites, contaminants and statistical criteria considered the algorithm with the highest accuracy for predicting an indoor air concentration close to the observed concentration is the JEM, DF NO, VolaSoil and Vl-H. All four algorithms have similar values for the ME, CRM and RMSE. The DF SE and CSoil have higher values for the three criteria, and are therefore considered less accurate.

*4.2 Conservatism*

To determine the conservatism of each of the algorithms the CMR was calculated by comparing the deterministic predicted air concentration to the median observed concentration. The algorithm with the most conservative deterministic parameter set for predicting the soil air concentration for the two aromatic hydrocarbons are the JEM and DF NO, followed by the DF SE, CSoil and VolaSoil, and finally Vl-H. For the aromatic hydrocarbons the most conservative algorithms for predicting the indoor air concentration are the DF SE, followed by JEM and CSoil. For the chlorinated hydrocarbon JEM was by far the most conservative followed by VolaSoil and the DF NO. The algorithms Vl-H, CSoil and DF SE predicted median concentrations below the median observed concentration, and are considered less conservative.

## 5. Discussion

The findings from this study suggest that the screening-level algorithms that have a relative high degree of conservatism (less false negative predictions) are the JEM, DF NO, VolaSoil and Vl-H. Of these four algorithms the JEM and VolaSoil have a higher accuracy (discriminative power) as well. For these two algorithms different parameters, that are variable and uncertain, contribute to the variation in indoor air concentration and differences were observed between the contaminants. The results are in line with the major findings of other related studies.

van Wijnen and Lijzen (2006) performed a probabilistic analysis on the VolaSoil algorithm for chlorinated contaminants. The study reports that predictions of the indoor air concentrations were driven by the depth of the groundwater table and the groundwater concentration. The variation between predicted and observed air concentrations was similar for tetrachloroethylene and trichloroethylene. Especially for sites with high tetrachloroethylene concentrations in the groundwater VolaSoil over-predicted the indoor air concentration, the opposite was observed for sites with low (near detection limit) concentrations. The overall conclusions were that the algorithm reasonably well predicted the tetrachloroethylene and trichloroethylene concentrations. Results from this study indicate that for VolaSoil the initial concentration in the soil is the main parameters that contribute to the variation in the indoor air concentration, and for trichloroethylene in addition the physico-chemical parameter solubility. Results from the present study did not indicate the depth to the source as an important parameter, but the contribution of other parameters are in line with de van Wijnen and Lijzen study.

Bakker et al. (2008) evaluated the new VolaSoil algorithm that now includes the JEM algorithm for benzene and tetrachloroethylene. The results for the two algorithms differ less than a factor three. JEM in comparison to VolaSoil predicted slightly lower concentrations for a sandy soil and almost the same for a loamy soil. The contribution to the variation in indoor air concentrations is for VolaSoil the soil-building pressure difference, depth to the source, total porosity, air permeability, indoor air exchange rate and solubility. For JEM contributing parameters were depth to the source, vapour pressure, solubility, groundwater concentration, soil-building pressure difference, surface floor area and indoor air exchange rate. Results from this study confirm that JEM and VolaSoil predict indoor air concentrations which are in close proximity. For the VolaSoil and JEM algorithm similar parameters drive the predictions, with the exception of VolaSoil for the total porosity and depth to the source, and for JEM the depth to the source, vapour pressure and surface floor area. The contaminants of concern in the present study include aromatic hydrocarbons and other sites which could partially clarify the differences observed. Furthermore the study from Bakker predicted concentrations from a groundwater contamination while the present study considers soil pollution.

The algorithm Vl-H was subjected to an uncertainty analysis by Seuntjens et al. (2001) and the variation of the indoor air concentration was predominantly driven by the water filled porosity, indoor air exchange rate, fraction basement air in indoor air, air filled porosity, organic carbon-water partitioning coefficient, octanol-water partition coefficient and height of the crawl space. The general conclusion was that for benzene and trichloroethylene the deterministically predicted soil screening values were not considered as sufficiently conservative. The findings in the present study show different parameters that mainly drive the predictions of the indoor air concentration, though there is some overlap. The conclusion of the low conservatism of Vl-H is in line with this study.

The probabilistic analysis provides a more in-depth view of the importance of variability and uncertainty, and the contribution of each varies depending on the contaminant or site considered. The probabilistic approach provides a better prioritisation for further actions needed, such as reducing the uncertainty by additional measurements, and a better insight into the matter which parameters dominate risk estimates, when compared to the deterministic approach (Johnson & MacDonald, 2010). "On the other hand it may be noted that Monte Carlo assessment is limited to the uncertainty and variability that can be quantified and expressed as probabilities, and a reasonable basis on which to ascribe a probability distribution function to parameters may not be available. The interpretation of results by decision makers might not always be straightforward as there is no rule-of-thumb to decide on the acceptance of the variation in predicted air concentrations in relation to uncertainty and variability of input parameters" (van der Sluijs et al., 2004; EPA, 1997; Morgan & Henrion, 1990; Saltelli, Tarantola, Campolongo, & Ratto, 2004; Burmaster & Anderson, 1994). Another limitation of the Monte Carlo approach is that the contribution of uncertainty and variability, however distinct phenomena with different interpretation and implications, are mixed in one output distribution (Ragas et al., 2009) unless a 2D simulation is performed in which the contribution from uncertainty and variability are dealt with separately in the calculations.

## 6. Recommendations

The data in this study are derived from two sites and therefore extrapolating to the application of the algorithms to other sites need to be done with caution.

Adapting the default parameter set for some of the less conservative algorithms should be considered to decrease the possibility for false negative deterministic predictions. The different default parameter sets from each algorithm showed for some deterministically predicted concentrations that not all of them were sufficiently conservative. For the chlorinated hydrocarbon trichloroethylene, the default parameter set of Vl-H, CSoil and DF SE might be adapted to arrive at a higher deterministically predicted indoor air concentration if more conservatism is required.

The sensitivity analysis revealed that depending on the algorithms and contaminants different parameters drive the variation in the air concentration. Against a similar background Fisher et al. (2002) advises to use more than one algorithm to account for the uncertainty and variability and explains differences.

Research in probabilistic risk assessment for additional sites with either predominantly a soil or a groundwater contamination is needed, especially for chlorinated hydrocarbons. A two-dimensional sampling-based technique (nested simulation) may be used to further separate the contribution from uncertainty and variability to the variation in predictions. This is not a trivial matter and a nested simulation is much more complex as to interpretation than a one-dimensional simulation (Krupnick et al., 2006). Nevertheless, the nested simulation will assist in the identification of uncertainties that can be reduced by performing additional research or identify the need to extensive analysis.

## References

Bakker, J., Lijzen, J. P. A., & van Wijnen, H. J. (2008). Site-specific human risk assessment of soil contamination with volatile compounds, RIVM Report 711701049. Retrieved from http://www.rivm.nl/bibliotheek/rapporten/711701049.pdf

Brand, E., Otte, P. F, & Lijzen, J. P. A. (2007). CSOIL 2000: An exposure model for human risk assessment of soil contamination, RIVM report 711701054. Retrieved from www.rivm.nl/bibliotheek/rapporten/711701054.html

Bronders, J., Cornelis, C., Olivier, I., Patyn. J., Provoost, J., Wilczek, D., & Smolders, R. (2000). Descriptive soil research site Astral near Brussels, phase two, 2000/IMS/R/161

Bryan, R. (2000). In-depth review of Colorado (CDOT) facility) data, RBCA Corrective Action Environmental Indicator Forum

Burmaster, D. E., & Anderson, P. D. (1994). Principles of good practice for the use of Monte Carlo techniques in human health and ecological risk assessments. *Risk Analysis, 14*(4). http://dx.doi.org/10.1111/j.1539-6924.1994.tb00265.x

Carlon, C., D'Alessandro, M., & Swartjes, F. A. (2007). Derivation methods of soil screening values in Europe. A review of national procedures towards harmonisation. *European Commission Joint Research Centre, Ispra*

*Italy,       EUR       22805       EN–2007,*       320,       Retrieved       from
http://ies.jrc.ec.europa.eu/uploads/fileadmin/Documentation/Reports/RWER/EUR_2006-2007/EUR22805-
EN.pdf

CCME. (1996). A Protocol for the derivation of environmental and human health soil quality guidelines, Appendix
G, Development of a relationship to describe migration of contaminated vapours into buildings, Canadian
Council of Ministers of the Environment, The National Contaminated Sites Remediation Program. Retrieved
from http://www.ccme.ca/assets/pdf/sg_protocol_1332_e.pdf

Crystal    Ball.    (2000).    Users    manual    2000,    Decisioneering    Inc.    Retrieved    from
http://www.oracle.com/us/products/applications/crystalball/overview/index.html

Cullen, A. C., & Frey, H. C. (1999). *Probabilistic Techniques in Exposure Assessment, A handbook for dealing
with variability and uncertainty in models and inputs* (p. 335). Plenum Press, New York.
http://dx.doi.org/10.1002/sim.958

DeVaull, G., Ettinger, R., & Gustafson, J. (2002). Chemical vapor intrusion from soil or groundwater to indoor air:
Significance of unsaturated zone biodegradation of aromatic hydrocarbons. *Soil and Sediment Contamination,
11*(4), 625-641. http://dx.doi.org/10.1080/20025891107195

DOH. (2006). Guidance for evaluating soil vapor intrusion in the state of New York, New York State Department
of Health - DOH, Center for Environmental Health, Bureau of Environmental Exposure Investigation, New
York.                               Retrieved                               from
http://www.health.ny.gov/environmental/investigations/soil_gas/svi_guidance/docs/svi_main.pdf

EPA. (1997). Guiding principles for Monte Carlo analysis, EPA - Risk Assessment Forum, EPA/630/R- 97/001.
Retrieved from www.epa.gov/raf/publications/pdfs/montecar.pdf

EPA. (2001). Risk Assessment guidance for superfund: Volume III - Part A, Process for conducting Probabilistic
Risk Assessment, Office of Emergency and Remedial Response, U.S. Environmental Protection Agency,
Washington,    DC    20460,    report    number    EPA    540-R-02-002.    Retrieved    from
www.epa.gov/superfund/RAGS3A/index.htm

Ferguson, C. C. (1999). Assessing risks from contaminated sites: Policy and practice in 16 European countries.
*Land Contamination and Reclamation, 7*(2), 33-54.

Filipsson, M. (2011). Uncertainty, variability and environmental risk analysis. Linnaeus University Dissertations,
number       35/2011,       ISBN:       978-91-86491-63-5.       Retrieved       from
http://www.diva-portal.org/smash/get/diva2:405602/FULLTEXT01.pdf

Finley, B., & Paustenbach, D. (1994). The benefits of probabilistic exposure assessment: Three case studies
involving    contaminated    air,    water    and    soil.    *Risk    Analysis,    14*(1),    53-73.
http://dx.doi.org/10.1111/j.1539-6924.1994.tb00028.x

Fisher, B. E. A., Ireland, M. P., Boylandb, D. T., & Critten, S. P. (2002). Why use one model? An approach for
encompassing model uncertainty and improving best practice. *Environmental Modeling and Assessment, 7*,
291-299. http://dx.doi.org/10.1023/A:1020921318284

Fugler, D., & Adomait, M. (1997). Indoor infiltration of volatile organic contaminants: measured soil gas entry
rates and other research results from Canadian houses. *Journal of Soil Contamination, 6*(1), 9-13.
http://dx.doi.org/10.1080/15320389709383542

Hammersley, J. S., & Handscomb, D. C. (1964). *Monte Carlo Methods.* New York: John Wiley and Sons Inc.
http://dx.doi.org/10.1002/bimj.19660080314

Huijbreghts, M. A., Gilijamse, W., Ragas, M. J., & Reijnders, L. (2003). Evaluating uncertainty environmental
life-cycle assessment. A case study comparing two insulation options for a Dutch one-family dwelling.
*Environmental Science and Technology, 37*, 2600-2608. http://dx.doi.org/10.1021/es020971+

ITRC. (2008). Use of risk assessment in management of contaminated sites - overview document. *Interstate
Technology & Regulatory Council.* Risk Assessment Resources Team. Retrieved from
www.itrcweb.org/Guidance/GetDocument?documentID=77

Johnston, J. E., & MacDonald, G. J. (2010). Probabilistic approach to estimating indoor air concentrations of
chlorinated volatile organic compounds from contaminated groundwater: A case study in San Antonio, Texas.
*Environmental Science and Technology*, online first. http://dx.doi.org/10.1021/es102099h

Johnson, P. C. (2005). Identification of application-specific critical inputs for the 1991 Johnson and Ettinger vapor intrusion Algorithm. *Ground Water Monitoring and Remediation, 25*(1), 63-78. http://dx.doi.org/10.1111/j.1745-6592.2005.0002.x

Johnson, P. C., Ettinger, R. A., Kurtz, J., Bryan, R., & Kester, J. E. (2002). Migration of soil gas vapors to indoor air: determining vapour attenuation factors using a screening-level model and field data from the CDOT-MTL Denver, Colorado site, American Petroleum Institute, No. 16. Retrieved from http://www.api.org/ehs/groundwater/upload/16_bull.pdf

Johnson, P. C., Kemblowski, M. W., & Johnson, R. L. (1999). Assessing the significance of subsurface contaminant vapor migration to enclosed spaces: site-specific alternatives to generic estimates. *Journal of Soil Contamination, 8*(3), 389-421. Retrieved from http://openagricola.nal.usda.gov/Record/IND22015155

Kaplan, M. B., Brandt-Rauf, P., Axley, J. W., Shen, T. T., & Sewell, G. H. (1993). Residential releases of number 2 fuel oil: A contributor to indoor air pollution. *American Journal of Public Health, 8*(1), 84-88. http://dx.doi.org/10.2105/ajph.83.1.84

Krupnick, A., Morgenstern, R., Batz, M., Nelson, P., Burtraw, D., Shih, J. S., & McWilliams, M. (2006). Not a sure thing: making regulatory choices under uncertainty, U.S. *Environmental Protection Agency*. Retrieved from http://rff.org/rff/Documents/RFF-Rpt-RegulatoryChoices.pdf

Kuusisto, S. M., & Tuhkanen, T. A. (2001). Probabilistic risk assessment of a contaminated site, International conference on practical applications in environmental geotechnology ecogeo 2000. *Geological Survey of Finland, special paper 32*, 99-105. Retrieved from http://arkisto.gtk.fi/sp/sp32/sp_032.pdf

Little, P. C. (2013). Vapor Intrusion: The Political Ecology of an Emerging Environmental Health Concern. *Human Organisation, 72*(2), 145-155. Retrieved from http://library.calstate.edu/sanfrancisco/articles/record?id=FETCH-proquest_dll_30079778911

Loague, K. M., & Green, R. E. (1991). Statistical and graphical methods for evaluating solute transport models: Overview and application. *Journal of Contaminated Hydrology, 7*, 51-73. http://dx.doi.org/10.1016/0169-7722(91)90038-3

McAlary, T., Provoost, J., Dawson, H., & Swartjes, F. (editor) (2011). Book on Dealing with Contaminated Sites - From Theory towards Practical Application, chapter 10 – Vapour intrusion', January 2011, pages 409-453, Springer Publishers, Netherlands, ISBN 978-90-481-9756-9. http://dx.doi.org/10.1007/978-90-481-9757-6_10

McHugh, T. E., Connor, J. A., & Ahmad, F. (2004). An empirical analysis of the groundwater-to-indoor-air exposure pathway: the role of background concentrations in indoor air. *Environmental Forensics, 5*, 33-44. http://dx.doi.org/10.1080/15275920490424024

McKone, T. E., & Bogen, K. T. (1990). Uncertainty in health risk assessment – An integrated application to tetrachloroethylene, Abst. Pap. Am. Vhem. 200: 32-Chas Part 1. http://dx.doi.org/10.1016/0273-2300(92)90087-p

Morgan, M. G., & Henrion, M. (1990). Uncertainty, A guide to dealing with uncertainty in quantitative risk and policy analysis, Cambridge University Press. http://dx.doi.org/10.1002/(sici)1099-0771(199606)9:2<147::aid-bdm199>3.0.co;2-8

Nassar, I. N., Ukrainczyk, L., & Horton, R. (1999). Transport and fate of volatile organic chemicals in unsaturated, nonisothermal, salty porous media: 2. Experimental and numerical studies for benzene. *Journal of Hazardous Materials, B69*, 169-185. http://dx.doi.org/10.1016/s0304-3894(99)00100-4

Oberg, T., & Bergback, B. (2005). A review of probabilistic risk assessment of contaminated land. *Journal of Soils and Sediments, 5*(4), 213-224. http://dx.doi.org/10.1065/jss2005.08.143

Otte, P. F., Lijzen, J. P. A., Otte, J. G., Swartjes, F. A., & Versluijs, C. W. (2001). Evaluation and revision of the CSOIL parameter set: Proposed parameter set for human exposure modeling and deriving intervention values for the first series of compounds, RIVM report 711701021 (2001). Retrieved from http://rivm.openrepository.com/rivm/bitstream/10029/9658/1/711701021.pdf

OVAM. (2004). Basisinformatie voor risico-evaluatie, Deel 3-H, Formularium Vlier-Humaan, report. Retrieved from http://www.ovam.be

Provoost, J., Bosman, A., Reijnders, L., Bronders, J., Touchant, K., & Swartjes, F. (2009). Vapour intrusion from the vadose zone – seven algorithms compared. Journal of Soils and Sediments - Protection, Risk Assessment, and Remediation. http://dx.doi.org/10.1007/s11368-009-0127-4

Provoost, J., Cornelius, C., & Seuntjens, P. (2002). Estimation of human indoor air exposure from soil contamination with benzene and ethylbenzene using probabilistic risk assessment, Book Risk Analysis III, In Brebbia CA (Ed.), Wit Press, Southampton, United Kingdom. Retrieved from http://www.witpress.com/978-1-85312-915-5.html

Provoost, J., Ottoy, R., Reijnders, L., Bronders, J., Van Keer, I., Swartjes, F., Wilczek, D., & Poelmans, D. (2011). Henry's equilibrium partitioning between ground water and soil air: Predictions versus observations. *Journal of Environmental Protection, 2*, 873-881. http://dx.doi.org/10.4236/jep.2011.27099

Provoost, J., Reijnders, L., & Bronders, J. (2013). Accuracy and conservatism of vapour intrusion algorithms for contaminated land management. *Journal of Environment and Pollution, 2*(2). http://dx.doi.org/10.5539/ep.v2n2p71

Provoost, J., Reijnders, L., Swartjes, F., Bronders, J., Carlon, C., D'Alessandro, M., & Cornelis, C. (2008a). Parameters causing variation between soil screening values and the effect of harmonization. *Journal of Soils and Sediments, 8*(5), 298-311. http://dx.doi.org/10.1007/s11368-008-0026-0

Provoost, J., Reijnders, L., Swartjes, F., Bronders, J., Seuntjens, P., & Lijzen, J. (2008b). Accuracy of seven vapour intrusion algorithms for VOC in groundwater, Journal of Soils and Sediments – Protection. *Risk Assessment, and Remediation, 9*(1), 62-73. http://dx.doi.org/10.1007/s11368-008-0036-y

Provoost, J., Tillman, F. D., Weaver, J. W., Reijnders, L., Bronders, J., Van Keer, I., & Swartjes, F. (2010). Book on Advances in Environmental Research, volume 5, chapter 2 – Vapour Intrusion into Buildings - A Literature Review, Daniels J. A. (Ed.), Nova Science Publishers Inc., New York, ISBN 978-1-61668-744-1. Retrieved from https://www.novapublishers.com/catalog/product_info.php?products_id=14005

Ragas, A. M. J., Brouwer, F. P. E., Buchner, F. L., Hendriks, H. W. M., & Huijbregts, M. A. J. (2009). Seperation of uncertainty and interindividual variability in human exposure modelling. *Journal of exposure science and environmental epidemiology, 19*, 201-212. http://dx.doi.org/10.1038/jes.2008.13

Ririe, G. T., Sweeyey, R. E., & Daugherty, S. J. (2002). A comparison of hydrocarbon vapour attenuation in the field with predictions from vapor diffusion models. *Soil and Sediment Contamination, 11*(4), 529-554. http://dx.doi.org/10.1080/20025891107159

S-EPA. (1996). Development of generic guideline values: Model and data used for generic guideline values for contaminated soils in Sweden, Swedish Environmental Protection Agency, report 4639 (1996).

Saltelli, A., Tarantola, S., Campolongo, F., & Ratto, M. (2004). Sensitivity analysis in practice: A guide to assessing scientific models, John Wiley & Sons publishers. Retrieved from http://onlinelibrary.wiley.com/doi/10.1002/0470870958.fmatter/pdf

Seuntjens, P., Provoost. J., & Cornelis, C. (2001). Uncertainty analysis of Vlier-Humaan, OVAM contract 001100, report in Dutch.

SFT. (1999). Guidelines for the risk assessment of contaminated sites, Norway, report 99:06. Retrieved from http://www.miljodirektoratet.no/old/klif/publikasjoner/andre/1691/ta1691.pdf

Swartjes, F. A. (1999). Risk-based assessment of soil and groundwater quality in the Netherlands: Standards and remediation urgency. *Risk Analysis, 19*(6). http://dx.doi.org/10.1111/j.1539-6924.1999.tb01142.x

Swartjes, F. A. (2002). Variation in calculated human exposure comparison of calculations with seven European human exposure models, RIVM report 711701030/2002. Retrieved from http://rivm.openrepository.com/rivm/bitstream/10029/9288/1/711701030.pdf

Swartjes, F. A. (2007). Insight into the variation in calculated human exposure to soil contaminants using seven different European models. *Integrated Environmental Assessment and Management, 3*(3), 322-332. http://dx.doi.org/10.1002/ieam.5630030303

Swartjes, F. A. (2009). Evaluation of the variation in calculated human exposure to soil contaminants using seven different European models. *Human and Ecological Risk Assessment, 15*, 138-158. http://dx.doi.org/10.1080/10807030802615733

Tillman, F. D., & Weaver, J. W. (2006). Uncertainty from synergistic effects of multiple parameters in the Johnson and Ettinger (1991) vapour intrusion model. *Atmospheric Environment, 40,* 4098-4112. http://dx.doi.org/10.1016/j.atmosenv.2006.03.011

Tillman, F. D., & Weaver, J. W. (2007). Parameter sets for upper and lower bounds on soil-to-indoor-air contaminant attenuation predicted by the Johnson and Ettinger vapor intrusion model. *Atmospheric Environment, 41,* 5797-5806. http://dx.doi.org/10.1016/j.atmosenv.2007.05.033

Van Belle, G. (2008). *Statistical rules of thumb, Wiley Series in Probability and Statistics Wiley* (pp. 99-100). http://dx.doi.org/10.1002/9780470377963

van der Sluijs, J. P., Janssen, P. H. M., Petersen, A. C., Kloprogge, P., Risbey, J. S., Tuinstra, W., & Ravetz, J. R. (2004). RIVM/MNP Guidance for uncertainty assessment and communication: Tool catalogue for uncertainty assessment, Copernicus Institute and RIVM, Utrecht and Bilthoven, report nr: NWS-E-2004-37. Retrieved from http://www.nusap.net/downloads/toolcatalogue.pdf

van Wijnen, H. J., & Lijzen, J. P. A. (2006). Validation of the VOLASOIL model using air measurements from Dutch contaminated sites - concentrations of four chlorinated compounds, RIVM, Netherlands, RIVM report 711701041/2006. Retrieved from http://www.rivm.nl/bibliotheek/rapporten/711701041.pdf

Waitz, M. F. W., Freijer, J. I., Kreule, P., & Swartjes, F. A. (1996). The VOLASOIL risk assessment model based on CSOIL for soils contaminated with volatile compounds, RIVM report 715810014. Retrieved from http://www.rivm.nl/bibliotheek/rapporten/715810014.pdf

Warmink, J. J., Jansen, J. A. E. B., Booij, M. J., & Knol, M. S. (2010). Identification and classification of uncertainties in the application of environmental models. *Environmental Modelling and Software, 25,* 1518-1527. http://dx.doi.org/10.1016/j.envsoft.2010.04.011

Weaver, J. W., & Tillman, F. D. (2005). Uncertainty and the Johnson-Ettinger Model for vapor intrusion calculations, EPA/600/R-05/110, U.S. Environmental Protection Agency, Office of Research and Development, Washington, USA. Retrieved from http://nepis.epa.gov/Exe/ZyPURL.cgi?Dockey=P1000BEM.TXT

Webster, E., & Mackay, D. (2003). Defining uncertainty and variability in environmental fate models, CEMC Report No. 200301, Canadian Environmental Modelling Centre, Trent University, Ontario, Canada. Retrieved from http://www.trentu.ca/academic/aminss/envmodel/CEMC200301.pdf

WHO. (1996). Environmental Health Criteria 186, Ethylbenzene, ISBN 9241571861. Retrieved from http://apps.who.int/bookorders/anglais/detart1.jsp?codlan=1&codcol=16&codcch=186

WHO. (2010). WHO Guidelines for indoor air quality, ISBN 9789289002134. Retrieved from http://www.who.int/indoorair/publications/9789289002134/en/

# Flower Bud Abscission Reduced in *Hibiscus Sabdariffa* by Radiation from GSM Mast

Ayoola O. Oluwajobi[1,2], Olamide A. Falusi[1] & Nuha A. Zubbair[2]

[1] Department of Biological Sciences, School of Natural and Applied Sciences, Federal University of Technology, Minna, Nigeria

[2] Department of Biology, Institute of Applied Sciences, Kwara State Polytechnic, Ilorin, Nigeria

Correspondence: Ayoola O. Oluwajobi, Department of Biological Sciences, School of Natural and Applied Sciences, Federal University of Technology, Minna, Nigeria.

**Abstract**

Radiation from GSM mast is regarded as a harmless but has recently generated a lot of controversies on its bio safety. Roselle naturally has about 50 % flower bud abscission rate, however, this research work is aimed at investigating the influence of radiation from 900 MHz GSM mast on the rate of flower bud production and abscission in *Hibiscus sabdariffa*. The plants exposed to the mast radiation generally produced significantly lower number of flower buds per plant (ranging from 80.00±9.62 to 106.80±8.17) than the control plants (107.80±15.21). Exceptions were found in locations A 100 m, 400 m and location B 100 m where the values were slightly higher (129.20±34.97, 138.40±19.50 and 118.80±15.80 buds / plant respectively). Abscission in control plants was 55.6 % per plant while in the exposed plants, a range of 40.33 to 48.21 % were obtained. Weak correlations exist between EMF intensities and abscission rate. The significant (p ≤ 0.05) reductions obtained in flower bud abscission in *Hibiscus sabdariffa* due to radiation from the GSM mast creates ecological stability for the plant whereby it enhances fruit formation and therefore increases the yield of the plant. If properly harnessed the radiation may play useful roles in the generation of variability and also in plant improvement.

**Keywords:** GSM, mast, radiation, flower bud, abscission, *Hibiscus sabdariffa*, variability

## 1. Introduction

Roselle (*Hibiscus sabdariffa* L.) belongs to the family Malvaceae of the Angiospermae and it is one of the most important and popular food, medicinal and industrial plants whose calyx and leaves are widely eaten. Its members are rich in anthocyanin and organic acids (Hong & Wrost-lad, 1990; Gomez-Leyva et al., 2009). It is also widely used as additives in the manufacture of several products such as liquor, jellies and jams (Akindahunsi & Olaleye, 2003) and the plant is drought tolerant (Torres-Moranet et al., 2011). *Hibiscus sabdariffa* is an herbaceous annual plant which is generally cultivated for its fruit (calyces) and leaves. It is a tropical region plant but now widely grown in many regions of the world (Duke, 1983; Morton, 1987). It is also a popular vegetable crop grown widely in Nigeria with several varieties (Falusi, 2004; 2008). Cisse et al. (2009) reported that the plant play important roles in income generation and subsistence among rural farmers in developing countries through the sale of the calyces and leaves.

Qi, Chin, Malekian, Berhane and Gager (2005); Hussein, Shahein, El-Hakim and Awad (2010) reported that *Hibiscus sabdariffa* has considerable industrial, pharmaceutical and economic values in several countries around the world. It is commonly used to make jelly, jam, juice, wine, syrup, gelatin, pudding, cake, ice cream and flavours. Many medicinal applications of the plant have been developed around the world in the treatment of hypertension, pyrexia, liver damage and leukaemia because of its high content of polyphenols (Tseng et al., 2000; Odigie, Ettarh, and Adigun, 2003). Also, it is a source of vegetable oil (from the seeds) that is low in cholesterol and rich in other phytosterols and tocopherols. The global characteristics of Roselle and seed oil allow important industrial applications for this oil. These characters represent added values for the culture of this plant (Mohamed, Fernadez, Pineda & Aguilar, 2007). The fleshy calyces of the flowers (sepals) have pleasant acid taste and are used as a beverage crop in many countries in form of mild laxatives.

Roselle is popularly grown for its fruits; especially the calyces, however, Falusi *et al.* (2014) reported a relatively low fruit production in six accessions of Roselle grown in Nigeria. Except in only one accession where a mean fruit set of 58.16 / plant was obtained, other accessions produced less than 50 fruits / plant, mostly a range from 33.50 to 58.16 fruits / plant.

Many factors are important for fruiting of crops which ranges from genetic (intrinsic) to the various environmental conditions. Abortion of flowers, flower buds and fruits according to Wien, Tripp and Hernandez-Armenta (1989) is an important factor that influences yield in many crops including pepper. They also identified several environmental stresses associated with fruit abortion. These include heat, drought and low light conditions however; with the recent introduction of mobile telephony, the increasing proliferation of telecommunications antennae and the attendant increase in complaints associated with the system has called for serious public scrutiny, despite the benefits of the system to humanity.

Although the transmitting power of the telecommunications system is relatively very low, a weak electromagnetic radiation with no heating effect, its pulsating nature of emissions which synchronises with the physiological systems is of concern. Hyland (2000, 2003), a physicist and executive member of International Institute of Biophysics in Germany, has earlier reported that the pulsed microwave radiation used in the GSM systems of telecommunications is characterised by a number of particularly well defined frequencies, a feature that can greatly enhance its impact on the biochemistry of the body and facilitate its discernment against the heat radiation that is emitted by the body which actually depends on its physiological temperature. There have been reports of the impact of the emissions from GSM antennae on both man, animals and plants (Santini, Santini, Le Ruz, Danze & Seigne, 2003; Panogopoulos & Margaritis, 2006; Haggerty, 2010).

In view of the above, this study examines the impact of the radiation from GSM mast on flower bud abscission in *H. sabdariffa*.

## 2. Materials and Methods

### 2.1 Experimental Sites

Three locations were chosen each with at least a 900 MHz GSM mast and they were free from obstructions such as buildings and trees. The locations were named A (06° 31′ 36.9″E and 09° 39′ 17.8″N), B (06° 27′ 35.2″E and 09° 32′ 15.9″N) and C (06° 32′ 16.7″E and 09° 37′ 13.6″N) while another location (06° 22′ 29.5″E and 09° 42′ 31.6″N), without any GSM signals was chosen as the control site.

### 2.2 Techniques

The soil used in raising the plants was collected from a garden, mixed with poultry droppings and left for four weeks to properly decompose and homogenise. The soil was turned weekly.

Five viable seeds were sowed in each of the eleven litres sized plastic bucket containing 9000 cm$^3$ of the soil and the buckets were placed at 100 m, 200 m, 300 m and 400 m from the mast. The buckets were labelled according to location and distance from the mast. For instance, the three buckets placed in location A at 100 m, 200 m, 300 m and 400 m away from the mast were labelled $1A_1$, $1A_2$, $1A_3$; $2A_1$, $2A_2$, $2A_3$; $3A_1$, $3A_2$, $3A_3$ and $4A_1$, $4A_2$, $4A_3$ respectively. The plants in locations B, C and the control were all labelled following the same patterns. However, in the control location, the three spots chosen were five hundred meters apart. Cages made of wood and chicken mesh were placed to delineate and also to protect the experiments at each site from ranging animals such as goats, sheep and cattle. The experimental design used was Randomised Block Design. The number of flower buds produced in 30 days were estimated by visual counting while the number of buds aborted within the same period were estimated using the formula below:

$$\% \text{ Bud abortion} = \frac{\text{Number of buds aborted}}{\text{Total number of buds produced}} \times 100 \tag{1}$$

The intensity of the radiofrequency radiations from the GSM mast was measured in each location and at the specific distances (100 m, 200 m, 300 m and 400 m) considered for the experiment. The instrument used was the Acoustimeter (RF meter), Model AM-10 manufactured by EMFields, UK. The measurement sensitivity is 0.02 – 6.0 V/m (1 – 100,000 µW/m$^2$). An average power density and peak hold (measured as the average of 1024 samples in 0.35 secs) were recorded for each study spot throughout the period of the experiment.

## 3. Results

The control plants produced a mean flower bud of 107.80±15.21 per plant in 30 days (Table 1) while plants from other locations and distances to the mast produced significantly lower number of buds (ranging from 80.00±9.62

to 106.80±8.17) except in locations A 100 m, 400 m and location B 100 m where the values were slightly higher (129.20±34.97, 138.40±19.50 and 118.80±15.80 buds / plant respectively).

Table 1. Number of flower bud production in thirty days in *H. sabdariffa* in relation to distance from GSM mast

| Locations | Control | Distance (m) | | | |
| | | 100 | 200 | 300 | 400 |
|---|---|---|---|---|---|
| A | 107.80±15.21$^{bc}$ | 129.20±34.97$^{cd}$ | 80.00±9.62$^{a}$ | 87.20±8.41$^{ab}$ | 138.40±19.50$^{d}$ |
| B | 107.80±15.21$^{b}$ | 118.80±15.80$^{b}$ | 88.40±6.54$^{a}$ | 80.40±12.99$^{a}$ | 106.80±8.17$^{b}$ |
| C | 107.80±15.21$^{d}$ | 101.00±9.92$^{cd}$ | 93.60±3.05$^{bc}$ | 78.60±8.38$^{a}$ | 83.80±7.79$^{ab}$ |

Means in the same row with different superscripts are significantly different (p ≤ 0.05).

The percentage bud abortion per plant within 30 days of bud emergence in *H. sabdariffa* showed that there was more bud abortion in the control plants (50.66 %) than those obtained from the plants exposed to ray emissions from the 900 MHz GSM antennae as shown in Table 2. Plants from location A gave a range of 40.33 % (100 m) to 48.04 % (200 m). In location B, it was 43.18 % (100 m) to 48 % (300 m) while in location C; it was 45.48 % (100 m) to 48.21 % (400 m). All samples exposed to radiations from GSM mast in all locations gave significantly lower abscission rate than the control (p ≤ 0.05). Table 3 shows very weak positive correlations between percentage bud abortions and EMF intensities of the ray emissions at various distances from the mast while figure 1 shows the photograph of aborted flower buds.

Table 2. Flower bud abscission (%) in thirty days in *H. sabdariffa* in relation to distance from 900 MHz GSM mast

| Locations | Control | Distance (m) | | | |
| | | 100 | 200 | 300 | 400 |
|---|---|---|---|---|---|
| A | 50.66±6.42$^{c}$ | 44.33±3.91$^{a}$ | 48.04±3.32$^{b}$ | 47.45±5.39$^{b}$ | 45.43±3.38$^{a}$ |
| B | 50.66±6.42$^{c}$ | 43.18±3.39$^{a}$ | 47.74±6.41$^{b}$ | 48.00±4.91$^{b}$ | 44.22±5.24$^{a}$ |
| C | 50.66±6.42$^{c}$ | 45.48±8.25$^{a}$ | 47.53±3.74$^{b}$ | 47.73±1.78$^{b}$ | 48.21±7.08$^{b}$ |

Means in the same row with different superscripts are significantly different (p ≤ 0.05)

Table 3. Correlation coefficient (r) between flower bud abortion in *H. sabdariffa* and EMF intensities in study locations in relation to distance from GSM Mast

| Distance (m) | | Location A | | Location B | | Location C | |
| | | Mean ± SD | r | Mean ± Sd | r | Mean ± Sd | r |
|---|---|---|---|---|---|---|---|
| 100 | EMF (V/m) | 0.80±0.19 | -0.662 | 0.41±0.11 | -0.457 | 0.85±0.13 | -0.121 |
| | % Bud abortion | 44.33±3.91 | | 43.18±3.39 | | 45.48±8.25 | |
| 200 | EMF (V/m) | 0.70±0.05 | 0.119 | 0.47±0.24 | -0.016 | 0.92±0.31 | 0.037 |
| | % Bud abortion | 48.04±3.32 | | 47.74±6.41 | | 47.53±3.74 | |
| 300 | EMF (V/m) | 1.12±0.22 | -0.053 | 0.61±0.26 | 0.467 | 1.08±0.8 | -0.454 |
| | % Bud abortion | 47.45±5.39 | | 48.00±4.91 | | 47.73±1.78 | |
| 400 | EMF (V/m) | 0.90±0.29 | 0.085 | 0.72±0.35 | 0.269 | 1.05±0.47 | 0.339 |
| | % Bud abortion | 45.43±3.38 | | 44.22±5.24 | | 48.21±7.08 | |

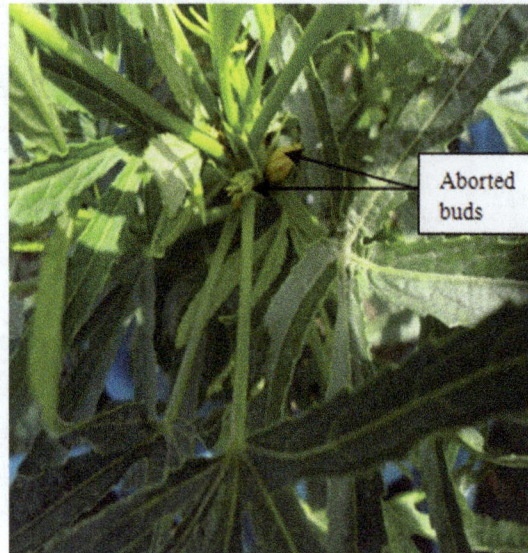

Figure 1. Aborted flower bud in *H. sabdariffa*

## 4. Discussion

Flower buds are the sexual reproductive structures of the plants that eventually gives rise to the seeds and fruits. Mohamed *et al.* (2011) reported that gamma irradiation of *H. sabdariffa* showed significant difference on the number of fruits per plant among the varieties at $p < 0.01$. Marcelis, Heuvelink and Baan Hofman-Eijer (2004) reported that shading, changes in temperature, plant density, position of earlier formed fruits and leaf pruning, all affected flower and fruit abortion in sweet pepper (*Capsicum annuum* L.). Flower bud production was significantly lower in plants exposed to radiation from the GSM mast than the number obtained for the unexposed plants. This reduction will normally affect the total number of fruits produced by the plant, hence its yield. However, flower bud abscission in 30 days showed that the abortion of buds in the control plant (50.66 %) was significantly higher than those obtained in all the plants exposed to GSM mast radiation (40.33 to 48.21 %) at $p \leq 0.05$. Aborted buds cannot develop into fruits; therefore the reduction of bud abortion tend to counteract the impact of the reduced bud production in the plant, thus an advantage to *H. sabdariffa* as it will boost fruit formation and by implication yield of the plant.

## 5. Conclusion

This study showed that the radiation from 900 MHz GSM mast has impact on *H. sabdariffa* by reducing the flower bud production and abscission. The reduction in flower bud formation is a negative impact while the reduction in its abscission is a positive impact. These attributes of the radiation thus created an ecological stability in the plant, and if properly harnessed may play useful roles in the generation of variability and also in plant improvement programmes.

### Acknowledgements

The assistance of Messrs Jerry Simon and Felix Maigida of the Federal University of Technology, Minna, Nigeria towards the success of the experiment is acknowledged.

### References

Akindahunsi, A. A., & Olaleye, M. T. (2003). Toxicological investigation of aqueous-methanolic extract of the calyces of *Hibiscus sabdariffa* L. *J Ethnopharmacology, 89*, 61-164. http://dx.doi.org/10.1016/S0378-8741(03)00276-9

Cissé, M., Dornier, M., Sakho, M., N'Diaye, A., Reynes, M., & Sock, O. (2009). Lebissap (*Hibiscus sabdariffa* L.) composition et principales utilisations. *Fruits, 64*(3), 179-193. http://dx.doi.org/10.1051/fruits/2009013

Duke, J. A. (1983). *Handbook of Energy Crops*. Retrieved February 21, 2014, from http://www.hort.purdue.edu/newcrop/ duke energy

Falusi, O. A. (2004). Collecting Roselle (*Hibiscus sabdariffa* L.) Germplasm in Nigeria. *J Arid Agric, 14*, 81-83.

Falusi, O. A. (2008). Inheritance of Characters in Kenaf (*Hibiscus canabinus*). *Afri. J. Biotechnol, 7*(7), 904-906.

http://dx.doi.org/10.5897/AJBO7.694

Falusi, O. A., Dangana, M. C., Daudu, O. A. Y., Oluwajobi, A. O., Abejide, D. R., & Abubakar, A. (2014). Evaluation of some Rossele (*Hibiscus sabdariffa* L.) germplasm in Nigeria. *International Journal of Biotechnology and Food Science, 2*(1), 117-121.

Gomez-Leyva, J. F., Acosta, L. A. M., Muraira, I. G. L., Espino, H. S., Ramirez-Cervantes, F., & Andrade-Gonzalez, M. (2008). Multiple shoot regeneration of Roselle (*Hibiscus sabdariffa* L.) from shoot apex culture system. *International Journal of Botany, 4*(3), 326-330. http://dx.doi.org/10.3923/ijb.2008.326.330

Haggerty, K. (2010). Adverse Influence of Radio Frequency Background on Trembling Aspen Seedlings: Preliminary Observations. *International Journal of Forestry Research,* 1-7. http://dx.doi.org/10.1155/2010/836278

Hong, V., & Wrost-lad, O. (1990). Use of HPLC separation/photodiode array detection for characterisation of anthocyanin. *Journal of Agriculture and Food Chemistry, 38,* 708-715. http://dx.doi.org/10.1021/jf00093a026

Hussein, R. M., Shahein, Y., El-Hakim, & Awad, H. M. (2010). Biochemical and Molecular Characterisation of three coloured types of roselle (*Hibiscus sabdariffa*). *Journal of American Science, 6*(11), 726-733.

Hyland, G. J. (2000). The Physics and Biology of Mobile Telephony. *The Lancet, 356,* 1833-1836. http://dx.doi.org/10.1016/S0140-6736(00)03243-8

Hyland, G. J. (2003). *How exposure to GSM and TETRA Base Stations Radiations can adversely affect Humans.* Retrieved August 6, 2014, from http://www.mastsanity.org/documents/ cluster2003.doc

Marcelis, L. F., Heuvelink, M., & Baan Hofman-Eijer, E. (2004). Flower and fruit abortion in sweet pepper in relation to source and sink strength. *Journal of Experimental Botany, 55*(406), 2261-2268. http://dx.doi.org/10.1093/jxb/erh245

Mohamed, R., Fernadez, J., Pineda, M., & Aguilar, M. (2007). Roselle (*Hibiscus sabdariffa*) seed oil a rich source of γ-tocopherol. *Journal of Food Science, 72,* 207-211. http://dx.doi.org/10.1111/j.1750-3841.2007.00285.x

Morton, J. (1987). Roselle. In *Fruits of Warm Climates.* Retrieved August 7, 2013, from http://www.hort.purdue.edu/newcrop/ morton/roselle.html

Odigie, I. P., Ettarh, R. R., & Adigun, S. (2003). Chronic administration of aqueous extract of *Hibiscus sabdariffa* attenuates hypertension and reverses cardiac hypertrophy in 2K-1C hypertensive rats. *Journal of Ethnopharmacol*ogy, *86,* 181-185. http://dx.doi.org/10.1016/S0378-8741(03)00078-3

Panagopoulos, D. J., & Margaritis, L. (2006). Effects of different kinds of electromagnetic fields on the offspring production of Insect. *2nd International Workshop on Biological effects of Electromagnetic field,* pp. 348-452.

Qi, Y., Chin, K. L., Malekian, F., Berhane, M., & Gager, J. (2005). Biological Characteristics, Nutritional and Medicinal Values of Roselle, *Hibiscus sabdariffa.* Circular: *Urban Forestry Natural Resources and Environment* No 604. Retrieved May 14, 2014, from http://www.suagcenter.com

Santini, R., Santini, P., Le Ruz, P., Danze, J. M., & Seigne, M. (2003). Survey study of people living in the vicinity of Cellular Base Stations. *Electromagnetic Biology and Medicine, 22*(1), 41-49. http://dx.doi.org/10.1081/JBC-120020353

Torres-Moran, M. I., Escoto-Delgadillo, M., Ron-Parra, J., Parra-Tovar, G., Mena Munguia, S., Rodriguez-Garcia, A. ... Castellanos-Hernandez, Y. (2011). Relationships among twelve genotypes of Roselle (*Hibiscus sabdariffa* L.) cultivated in western Mexico. *Industrial Crops and Products, 34,* 1079-1083. http://dx.doi.org/10.1016/j.indcrop.2011.03.020

Tseng, T., Kao, T., Chu, C., Chou, F., Lin, W., & Wang, C. (2000). Induction of apoptosis by *Hibiscus* Protocatechuic Acid in Human Leukemia Cells via Reduction of Retinoblastoma (RB) Phosphorylation and Bcl-2 Expression. *Biochemical Pharmacology, 60,* 307-315. http://dx.doi.org/10.1016/S0006-2952(00)00322-1

Wien, H. C., Tripp, K. E., & Hernandez-Armenta, R. (1989). Abscission of reproductive structures in pepper: Causes, mechanisms and control. In S. K. Green, (Ed.), *Tomato and pepper production in the tropics* (pp. 150-165). Taipei, Taiwan, R.O.C: Asian Research and Development Center.

# 3

# Organochlorine Pesticides in Sediment-Dwelling Animals from Mangrove Areas of the Calabar River, SE Nigeria

Orok E. Oyo-Ita[1], Bassey O. Ekpo[1], Peter A. Adie[2] & John O. Offem[1]

[1] Environmental & Petroleum geochemistry Research Group (EPGRG), Department of Pure and Applied Chemistry, University of Calabar, Nigeria

[2] Department of Chemistry, Benue State University, Makurdi, Nigeria

Correspondence: Orok E. Oyo-Ita, Environmental & Petroleum geochemistry Research Group (EPGRG), Department of Pure and Applied Chemistry, University of Calabar, Nigeria.

## Abstract

Sediment-dwelling biota such as mollusks (clam) and crabs collected from mangrove areas of the Calabar River are important routes of exposure to organochlorine pesticides (OCPs) contamination. Residual levels of OCPs including HCHs, DDTs, heptachlor, heptachlor epoxide, aldrin, endrin ketone, entrin aldehyde, dieldrin, endosulfan, endosulfan sulphate, methoxychlor, were determined in these organisms. The results revealed the OCP loads to be predominated by DDTs and HCHs (much of which was derived from illegal usage of GAMMALIN 20 for fishing) with the overall means of 49.6 and 35.1 ng/g wet weight (ww), respectively, at 100% frequencies of occurrence. Concentrations of other OCP components were generally low and were not detected in all biota samples. This probably reflects low utilization of these OCPs in the region and/or low bioaccumulation potential in the biota species. In general, the OCP concentrations were higher in freshwater mollusks and crabs than in brackish water, indicating that freshwater biota were more easily influenced by OCPs than their brackish water counterparts. One way analysis of variance (ANOVA) indicated no significant relationship between lipid content (LC) or body size of organisms and contaminant load, probably because of the non-equilibrium situation: smaller animals accumulated more OCPs than their larger counterparts, suggesting different uptake and elimination rates for these compounds. Biota-sediment accumulation factors (BSAFs) for DDTs and HCHs varied among the organisms and were in the ranges 1.02 – 9.78 and 0.74 – 8.72, respectively, indicating probably that HCHs were less bio-available in the river than DDTs. They were generally lower for highly polluted site (UMA; freshwater area) and higher for areas of low anthropogenic pressure (MR- brackish water area). Risk assessment matched against various standards clearly showed that the biota were highly contaminated with HCHs and DDTs, and may pose serious health threats to local inhabitants of the catchments. Furthermore, other selected OCPs such as heptachlor and dieldrin may in addition pose life-time cancer risk, especially to residents of the riverine/coastal communities who often consume more of these organisms than those living inland.

**Keywords:** mangrove sediment, biota, GAMMALIN 20, fate and risk assessment

## 1. Introduction

Organochlorine pesticides (OCPs) including hexachlorocyclohexanes (HCHs: α-HCH – which account for about 60% of technical mixture; β-HCH - having the highest toxicity and accumulation potential for aquatic organisms; γ-HCH – constituting about 100% lindane which is a major component of GAMALIN-20; and δ-HCH - comprising about 20% of the technical mixture), together with dichlorodiphenyltrichloroethane (DDT) and its metabolites, dichlorodiphenyldichloroethane (DDD) and dichlorodiphenyldichloroethylene (DDE) are not only ubiquitous in the environment, but are also capable of posing serious health threat to resident macro-fauna and humans through bioaccumulation and bio-magnification via food chain (Yang, Lui, Su, Ou, Liu, Cheng, & Hofmanni, 2006; Guo, Zeng, Wu, Meng, Mai, & Luo, 2007; Guan., Wang, Ni, & Zeng, 2009; Liu., Zhang, Tao, & He, 2010). This is due to their resistance to environmental alteration and potential toxicity. Thus, their use has been banned in many countries of the world as they are considered priority pollutants and included in the 2001 Stockholm Convention of the United Nation (Grimalt. van-Drooge, Ribes, Vilanova., Fernandez, & Appleby,

2004) to which Nigeria was a signatory. Despite this global ban, these chemicals are still commonly used in developing countries such as Nigeria. Therefore, contrary to the developed countries scenario, the body tissue loads in contemporary population in developing countries are expected to be high.

In Nigeria for the past few decades, a very good percentage of imported OCP products were applied for malaria vector control. A minor proportion was used for agriculture. This is due to the dearth of large-scale commercial farming and the prevalence of subsistence farming. The circumstances have today changed and a reverse trend is gradually emerging as the Cross River State government recently embarked on large-scale cultivation of pineapple, cocoa and other fruits within Akamkpa locality (upper Cross River state) for subsequent processing and export. This has triggered large scale importation and application of new sources of DDTs as well as mosquito treated nets in the region.

An important route of exposure to OCPs from various sources is via food consumption. Different studies have shown that exposure to OCPs has become a serious public health concern (Liu et al., 2010). For example, many authors (eg. Yang et al., 2006; Guo et al., 2007; MacIntosh, Spengler, Ozkaynak, Tsai, & Ryan, 1996) have observed that certain proportion of the adult population may be exposed to OCPs in seafood products at levels above the permissible levels by Council of the National Academy of Sciences (NAS). Environmental Working Groups have also found that children exposure to OCPs may cause immediate adverse outcomes (Wiles, Davies & Campbell, 1998).

Risk assessments of human health have been undertaken worldwide by many researchers to examine the potential risk associated with exposure to toxic compounds in different environmental compartments and seafood products including crab and mollusk (e.g. Yang et al., 2006; de Souza, Machado Torres, Meire, Neves, Couri, & Serejo, 2008). The importance of this biota to human health globally makes it imperative that information concerning the state of OCPs contamination in the biota be urgently obtained for the region. The information is needed to highlight the impact of OCPs on consumers, particularly on the inhabitants of the riverine/coastal communities commonly observed to consume more of this biota than their counterparts living hinterland. The health consequences associated with historical usage and application of new sources of DDTs and HCHs to the environment via aquatic organisms consumption have spurred many researchers worldwide to undertake quality assessment of this biota. The present study is the first quality assessment undertaken on crabs and mollusks from the Calabar River.

From the global standpoint, OCPs levels are relatively well established but on a local scene, there is still little or no data for many sites, particularly in the case for Sub-Saharan tropical African mangrove region. As part of the coastal environment, mangroves may be exposed to organic contamination from various sources including industrial wastes and agricultural runoff, etc. Because of its distinctive characteristics, mangrove ecosystem appears to be a favorable environment where uptake, accumulation and conservation of organic contaminant can take place (Ke, Wong, & Tam, 2005). It is also a nursery, development and feed area for many aquatic organisms (Mille, Guiliano, Asia, Malleret, & Jalaluddin, 2006; de Souza et al., 2008). To the best of our knowledge, little or no data concerning levels of OCPs in crab and mollusk from Sub-Saharan African mangrove areas are available. The present study will therefore contribute to the extension of data available for mangrove zones worldwide.

According to Yang et al., (2006), the controlling factors that determine the accumulation of organic contaminants in aquatic biota include sex, age, body size, lipid content and health status of organism. In addition to these physiological factors, nature of chemical compounds and their bioavailability have also been demonstrated to play a major role in bioaccumulation (Muncaster, Omokaro, Holzolehner, & Udensi, 1990). Due to the observed relationship between bioaccumulation and biological abnormalities in aquatic biota (Yang et al., 2006; de Souza et al., 2008), it is necessary that information on the bioavailability of OCPs be obtained for the region. Aquatic biota, on exposure to sediment impacted by organic contaminants, indiscriminately ingests this sediment while foraging, and may transfer these contaminants to higher trophic levels through the food chain.

Mollusks and crabs have been demonstrated as bio-indicators of the extent of contamination of many river systems (Zuloaga, Prieto, Usobiaga, Sarkar, Chatterice, Bhattacharya, Adam, & Satpathy, 2009; Liu et al., 2010). The species of these biota investigated in the present study are abundant groups of sediment-dwelling macro-zoobenthos within the region as exemplified by their high consumption rate. They are also readily distinguishable from other species and constitute a major ingredient in almost every meal prepared in Nigeria. Many researchers have demonstrated the usefulness of different species of crabs (e.g. *Scylla serratab* and *Ovalipes punctatusb*) and mollusks (e.g. *Solen grandis* and *Scapharca subcrenata*) as bio-indicators of organic and metallic pollution in estuaries and rivers (Guo et al., 2007; de Souza et al., 2010). However, this study reports for the first time in the study area the use of the species, *Musculin securis* and *Egeria radiate* (mollusk-clam) as well as *Callinectes amicola* and *Micropipus depurator*(crabs) as bio-indicators of OCPs contamination.

The main objectives of the present study were to: (1) determine the extent of OCPs contamination, (2) determine factors that influence the accumulation of OCPs in sediment-dwelling animals, (3) determine biota-sediment accumulation factors (BSAFs) in order to assess the bio-availabilities of these compounds in the river and (4) carry out risk assessment related to the consumption of the biota in order to identify potential health risks.

## 2. Samples and Methods

### 2.1 Study Area and Samples

Detail about the study area is as described by CRBDA (1982), Asuquo (1989), Löwenberg & Künzel (1992), Ekpo & Ibok (1998), Oyo-Ita, Offem, Ekpo & Adie (2013). Briefly, the lower reaches of the Imo River (where biota and associated sediment samples was collected-MR site; Figure (1) are influenced by semi-diurnal ocean tides. The inhabitants of the riverine/coastal communities of the area engage in fishing activities mostly to meet domestic/local needs (a good number of them are involved in illegal use of GAMMALIN 20 - a chemical substance known to contain high levels of lindane - to catch/kill fish). On the other hand, some of them, particularly dwellers of the upper Cross River region (close to where biota and associated sediment samples were collected-UMA site; Figure(1) are subsistent farmers, cultivating vegetables, pineapple, banana, plantain, cocoa, etc. Widespread application of pesticides in agriculture was not a common practice by these peasant farmers, until the advent of large-scale commercial agriculture recently introduced by the Cross River State government involving utilization of old and new DDT sources. As a result, runoff of agricultural waste could significantly contribute to the pollutants load in the river, especially near area where both biota and associated sediment samples were collected-UMA).

The current relatively high cost of biota may be associated with their gradual extinction occasioned by enhanced anthropogenic activity.

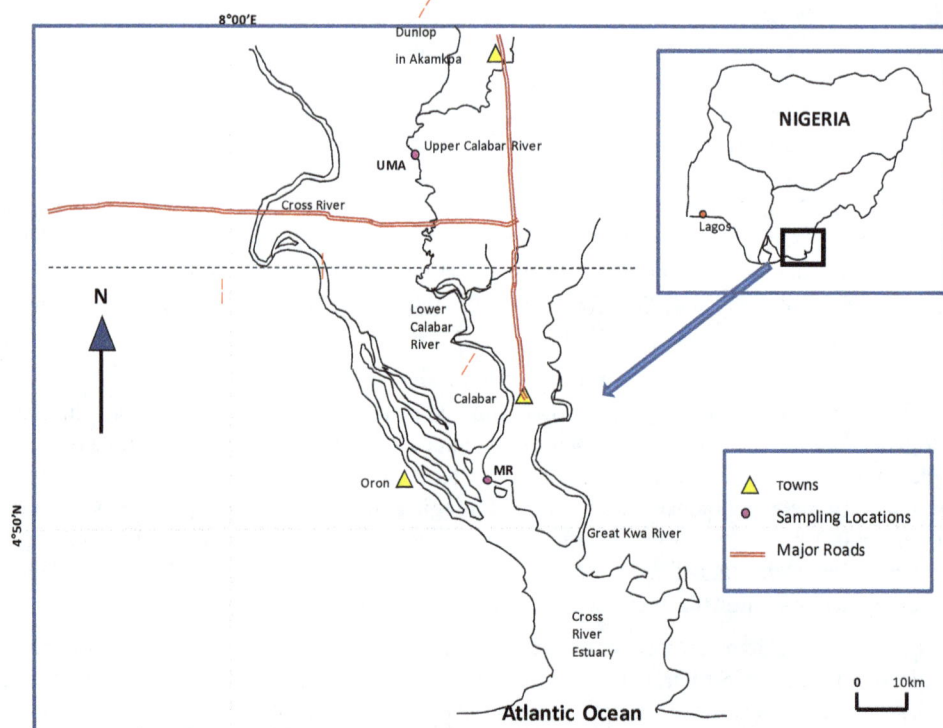

Figure 1. Map of the study area indicating sample locations

Information on current fishery products (including crabs and mollusks) caught from Calabar River is not available. However, in 2008, the USAID/West Africa reported the total 579,537 metric tons of fishery products caught for Nigeria as highest in West Africa (USAID, 2008). In order to reduce over-dependence on imported seafood products, the Nigerian government recently embarked on extensive commercial fishery for both domestic consumption and export.

Sediment-dwelling animal samples (crabs and mollusks-clam) were captured on April, 2011 towards the mouth of the river (MR; brackish water) and upper river mangrove area (UMA; fresh water) using a woody pike and/or by hand, with the associated sediments simultaneously (using stainless steel trowel) (Figure 1; Yang et al., 2006). These two sites were chosen on the basis of assumed differential in the degree of OCPs contamination. For instance, higher influx of OCPs from agricultural run-off is expected to occur at UMA relative to MR site. Each biota contained at least 20 individuals of different sizes. Two sizes (small $\leq 2$ and large $\geq 2$; determined by measuring the weight and length of the animals) were chosen. At the UMA site (characterized by input from agricultural waste, commercial fishery activity and garbage discharge), crab and mollusk species collected were *Micropipus depurator and Musculin secures*, respectively, whereas the species *Egeria Radiate* (Mollusk-clam) and *Collinectes amicola* (crabs) were collected at the MR site (characterized by offshore oil spill influx, commercial fishery activity and vehicle emission) for analysis. The biota samples were separated according to their type and left overnight in filtered seawater. This allowed them to empty their guts without losing hydrophobic contaminants. The following morning, the animals were removed, freeze-dried and stored at -20 °C before further processing.

## 2.2 Materials and Quality Assurance

Pesticides mixture standards, and *o,p'*-DDD, *o,p'*-DDE as well as *o,p'*-DDT (Acustandards) purchased from Supelco, Bellefont, PA, USA were used for identification and quantification. Surrogate standard, 2,4,5,6-tetrachloro-*m*-xylene obtained from Ultra Scientific (New Kingstown, RI, USA. SX-3, Bio-Beads was used in gel permeation chromatograph. All organic solvents were re-distilled in glass assemblage before use. The following analytical grade solvents were used for the analysis: dichloromethane (DCM), methanol (MeOH), n-hexane, acetone and toluene. Silica (80-100 mesh) and alumina (100-200 mesh) were extracted with MeOH:DCM (1:1) for 72 h in a soxhlet system before use. Sodium sulfate was heated at 450 °C and stored in sealed containers. Distilled water was obtained with a distillation assemblage (Guo et al., 2007).

## 2.3 TOC Determination, Extraction and OCPs Analysis

Prior to lipid analysis, freeze-dried associated sediment samples were repeatedly de-carbonated in hydrochloric acid (37%) until bubbling stopped and then rinsed continually in de-ionised water until neutral pH. TOC determination was performed by flash combustion at 1024 °C and then followed by thermal conductivity detection in triplicate in a CHNS Elemental Analyzer, Carlo Erbar 1108.

About 5 g of each dried sample (biota and associated sediment) was spiked with 2,4,5,6-tetrachloro-*m*-xylene and extracted in a soxhlet assemblage for 24 h with 200 ml acetone: n-hexane (1:1, v/v). The extract was evaporated *in vauo* to approximately 5 ml at 30 °C. An aliquot (about 10%) of the concentrate was taken out for gravimetric determination of lipid content and the remaining subjected to gel permeation chromatography using a glass column packed with 40 g SX-3 Bio-Beads. The well-packed column, loaded with an extract was eluted with 50% dichloromethane in n-hexane for lipid removal. The fraction containing OCPs was collected and concentrated. Subsequent cleanup was performed with a multi-layer alumina/silica column packed, from bottom to top, with a neutral alumina (6 cm, 3% deactivated) and silica gel (12 cm, 3% deactivated), and anhydrous sodium sulfate (1 cm). The defatted sample was eluted with 70 ml n-hexane: dichloromethane (7:3 v/v). The eluent OCPs were concentrated and quantitatively transferred into a 2 ml vial. This was further concentrated to a final volume of 100 µl under a gentle nitrogen stream. An internal standard was added to the fraction prior to instrumental analysis.

## 2.4 Instrumental Analysis

Instrumental analysis was carried out with a Hewlett-Packard (HP, Avondale, PA, USA) 5890 gas chromatograph and 5973 mass spectrometer operated in the selective ion monitoring mode. A 30 m x 0.25 mm (id) x 0.25 µm (film thickness) DB-5 fused silica capillary column was used for chromatographic separation. The column temperature was programmed from 80 °C (held for 1 min. to 200 °C at the rate of 12 °C /min, followed by an increase at a rate of 10/min to 220 °C, and the temperature was further ramped at a rate of 15 °C /min to 290 °C (held for 10 min). Ultrahigh purity helium (99.99%) was employed as carrier gas at a constant pressure of 10 psi. Splitless injection of 1 µl sample was performed with a 10 min solvent delay time. Injector and detector temperatures were set at 280 °C and 300 °C, respectively.

Mass spectra were acquired in the electron impact mode with an impact voltage of 70eV. Data acquisition and processing were performed with a DOS based HP ChemStation System. The 21 OCPs in the samples were identified on the basis of their retention times and ion fragment profile compared against authentic standards, while quantification was done using multi-point internal calibration method. The correlation coefficient (r) for all calibration curves were greater than 90.969. The limits of detection of the 21 OCPs were defined as signal-noise ratio > 3, ranging from 0.01 to 0.05 ng/g. For each batch of 5-12 samples, a procedural blank, a spiked blank, a pair of spiked matrix sample/duplicate, and a sample replicate were processed. The mean recovery of the surrogates added to each sample was 59.8±12.5%. Recoveries of 21 OCPs in six spiked blanks ranged from 68.2±17.2% for aldrin to 102.7±17.4% for methoxychlor. No target compounds were found in the procedural blanks, and the final concentrations were not corrected with surrogate recoveries.

*2.5 Biota Sediment Accumulation Factor (BSAF) and Health Risk Assessment Index*

In the present study, biota sediment accumulation factors were calculated according to the accompanying equation:

$$BSAF = \frac{C_b / f_l}{C_s / f_{oc}}$$

Where $C_b$ is the biota contaminant concentration (ng/g ww), $f_l$ is biota lipid concentration (fraction by weight), $C_s$ is sediment contaminant concentration (ng/g dw) and $f_{oc}$ is the organic carbon fraction of the sediment (fraction by weight) (Wang, Cheng, Zhang, Piao, Hu, & Tao, 2004).

On the other hand, health risk assessment of human exposure to OCPs via crab and mollusk consumption was calculated using an index referred to as Estimated Daily Intake (EDI), according to the accompanying equation:

$$EDI = \frac{EDC \times CC}{BW} \qquad \text{(Liu et al., 2010)}$$

Where EDC is food daily consumption, CC is mean contaminant concentration and BW is body weight for which 60 kg is typical (IPCS, 2006).

*2.6 Statistical Analysis*

Data for the OCPs were examined statistically in order to determine any significant environmental variation. All statistical analyses were performed using the statistical package, STATISTICA 8.0. These include descriptive statistic (e.g. a measure of dispersion) and analysis of variance. The objectives of the statistical analyses were to define the environmental characteristics of the sediments in relation to the biota, examine the relationship between samples, and assess the extent of the occurrence of OCPs.

Analysis of variance (ANOVA) is a statistical techniques used to investigate the different sources of variability associated with a series of results. It enables the effects of each source to be assessed separately and compares with other(s) using F-test.

To estimate the extent of crab and mollusk consumption in the region, 250 questionnaires were sent to families living in Calabar city and its environs as well as to inhabitants of the riverine/coastal communities. Information on the number of people (including children), consumption frequencies and quantities of crabs and mollusks consumed were included in the questionnaires. The overall consumption information for the region was obtained after statistical analysis of generated data.

## 3. Results and Discussion

The results presented in this study satisfied the acceptance criteria suggested by Burns, Mankiewiez, Bence, Page, & Parker (1997), implying the analytical procedures met the required quality control standard. Characteristic features of the environment, biota sample type and size, and gravimetric data including extractable organic matter (EOM) and lipid content (LC) are presented in Table 1. The mean LC for the mollusks and crabs were in the range 1.72±0.24 – 1.78±0.16% and 1.62±0.41 – 1.78±0.17%, respectively at the MR site, whereas at the UMA site, 1.44±0.32 – 1.56±0.21% and 1.92±0.42 – 2.02±0.67% ranges were respectively recorded for mollusks and crabs. The following OCP compounds were detected in the biota and associated sediment samples: α-HCH, β-HCH, γ-HCH, δ-HCH, heptachlor, heptachlor epoxide, aldrin, endosulfan, endosulfan sulphate, *o,p*-DDE, *p,p*-DDE, *o,p*-DDD, *p,p*-DDD, *o,p*-DDT, *p,p*-DDT, dieldrin, endrin, endrin ketone, endrin aldehyde, and methoxychlor.

## 3.1 OCP Concentrations in Biota and Associated Sediments

The mean concentrations and ranges of OCPs in biota are listed in Table 2, while the isomeric ratios of related DDTs and HCHs as well as total HCH (t-HCH) and total DDT (t-DDT) for the biota are presented in Tables 3 and 4, respectively. Generally, higher t-HCH levels were found in associated sediment samples obtained from both MR and UMA sites than t-DDT. HCH concentrations varied among the biota, and among isomers. α-HCH mean concentrations varied from 2.3 ng/g wet weight (ww) in large size mollusk (LMK) at the MR site to a maximum of 6.4 ng/g ww in small crabs (SCRM) at the UMA site (Table 2). Generally at both sampling sites, small mollusks and crabs accumulated more α-HCH than their larger counterparts. This pattern was also found in other HCH isomers as well as in the t-HCH - sum of all HCH isomers; Figure 2 showing that levels of HCHs in biota may be affected by body size. This observation may be due to the fact that smaller individuals have faster rates of hydrophobic contaminants uptake and slower rates of elimination than their larger counterparts (Yang et al., 2006). In general, despite the fact that γ-HCH degrades ultimately to α-HCH in the environment (Chrysikou, Gemenetzis, Kouras, Manoli, Terzi, & Samara, 2008), the γ-HCH isomer dominated the biota tissues compared with other isomers, indicating continued illegal usage of lindane in the form of GAMMALIN-20 for fishing in the region. Histopathological study carried out by Ezemoye & Ogbomida, (2010) on African catfish showed effects such as distress physical activity, convulsion, eratic swimming, lose of equilibrium and increased breathing after sub-lethal administration of GAMMALIN-20. Therefore, the levels of t-HCH found in the biota are capable of posing serious health threat to the resident organisms. The next higher mean concentrations were exhibited by β-HCH. These high levels of β-HCH could be explained by its relative stability to metabolic processes and resistance to microbial degradation in the environment compared with other HCH isomers (Guo et al., 2007; Guan et al., 2008).

Table 1. Characteristic features of the environment, body size, and lipid contents of biota

| Site feature | Species | Sample code | Weight (g) | LC (%) | Length (cm) | EOM (mg/kg) |
|---|---|---|---|---|---|---|
| Mouth of river (MR); offshore oil spill influx, Vehicle emission, commercial fishery, etc. | Egeria radiate (mollusks-clam) | SMK | 3.2± 0.2 | 1.78± 0.16 | 1.65± 0.2 | 800 |
| | | LMK | 4.4±0.21 | 1.72± 0.24 | 2.17± 0.25 | |
| | Callinectes amicola (crab) | SCRB | 3.1± 0.42 | 1.87± 0.17 | 1.80± 0.31 | |
| | | LCRB | 12.5± 0.5 | 1.62± 0.41 | 3.91± 0.39 | |
| Upper mangrove area (UMA): agricultural waste, commercial fishery,garbage discharge, vehicle emission | Musculin securis (mollusks-clam) | SMM | 3.9±0.25 | 1.58± 0.21 | 1.85± 0.41 | 2350 |
| | | LMM | 5.2± 0.22 | 1.44± 0.32 | 1.97± 0.35 | |
| | Micropipus depurator (crab) | SCRM | 3.1± 0.44 | 2.02±0.67 | 2.0± 0.68 | |
| | | LCRM | 14.4± 0.41 | 1.95± 0.42 | 5.77± 0.52 | |

N/B: SMK= small mollusks from MR site, LMK= large mollusks fron MR site, SCRB= small crab from MR site, LCRB= large crab from MR site, SMM= small molluskr from UMA site, LMM= large mollusks from UMA site, SCRM= small crabs from OMA site, LCRM= large crab from UMA site.

Table 2. Mean and range concentrations (ng/g ww) of OCPs in biota from the Calabar River

| Compound | SMK | LMK | SCRB | LCRB | SMM | LMM | SCRM | LCRM |
|----------|-----|-----|------|------|-----|-----|------|------|
| α-HCH | 5.1(1.2-8.9) | 2.3(0.8-3.9) | 6.3(1.7-9.3) | 3.8(1.3-5.6) | 4.7(1.2-6.7) | 4.1(1.3-5.9) | 6.4(2.1-8.9) | 5.8(1.9-7.2) |
| β-HCH | 11.3(3.5-14.7) | 6.1(1.6-9.1) | 9.2(3.2-13.8) | 7.8(2.6-11.9) | 12.3(4.1-15.2) | 8.1(2.9-12.2) | 9.4(3.1-13.4) | 7.9(2.3-10.2) |
| γ-HCH | 13.3(5.3-17.3) | 10.2(4.7-15.9) | 21.3(10.2-31.2) | 11.6(6.3-17.9) | 31.2(19.5-41.6) | 12.4(3.9-16.9) | 13.1(3.3-17.1) | 9.5(3.6-12.7) |
| δ-HCH | 7.3(2.9-10.6) | 5.5(2.5-10.2) | 6.9(2.2-11.3) | 5.8(1.9-10.8) | 7.2(2.1-11.1) | 5,1(1.1-12.3) | 6.5(1.8-11.8) | 4,6(1.1-9.7) |
| Heptachlor | 0.3(0.1-0.5) | 0.2(0.1-0.4) | 0.5(0.1-0.8) | 0.5(0.1-0.7) | 0.2(0.1-0.5) | 0.1(Nd-0.3) | 0.1(Nd-0.2) | 0.3(0.1-0.6) |
| Aldrin | 0.7(0.2-1.2) | 0.7(0.1-0.9) | 1.1(0.4-2.1) | 0.6(1.9-8.6) | 0.7(0.3-1.1) | 0.6(0.1-0.9) | 0.6(0.1-0.8) | 0.9(0.2-1.7) |
| Heptachlor epoxide | 0.3(0.1-0.6) | 0.3(0.1-0.8) | 0.4(0.1-0.9) | 0.2(Nd-0.4) | 0.4(0.1-0.7) | 0.2(Nd-0.5) | 0.2(Nd-0.6) | 0.3(0.1-0.7) |
| α-Endosulfan | 2.5(0.5-6.5) | 2.3(0.3-6.1) | 0.8(0.2-2.4) | 3.0(0.6-5.7) | 3.5(0.5-6.7) | 2.6(1.2-6.2) | 2.2(0.9-5.8) | 3.1(1.1-7.2) |
| β-Endosulfan | 0.1(Nd-0.2) | Nd | 0.3(Nd-0.5) | 0.1(Nd-0.2) | 0.1(Nd-0.3) | 0.4(0.1-0.6) | 0.2(Nd-0.3) | 0.1(Nd-0.2) |
| o,p-DDD | 7.3(2.6-10.3) | 6.8(1.7-9.8) | 11.1(3.5-14.5) | 8.5(1.6-10.3) | 8.1(2.4-11.8) | 6.5(2.7-9.5) | 7.1(1.8-10.7) | 5.5(1.8-8.7) |
| p,p'-DDD | 6.1(1.9-9.8) | 5.4(0.9-10.1) | 7.1(2.2-9.8) | 6.8(1.6-0.8) | 10.1(3.2-14.7) | 9.5(2.4-12.8) | 9.1(3.1-11.5) | 6.8(1.9-8.9) |
| o,p-DDE | 8.5(2.1-11.8) | 7.7(1.8-10.6) | 9.8(3.3-12.2) | 8.7(2.9-13.2) | 9.9(3.1-13.6) | 9.5(2.1-12.5) | 8.2(1.9-12.8) | 6.2(1.3-9.2) |
| p,p'-DDE | 6.5(1.7-10.3) | 5.8(1.2-9.6) | 7.2(1.8-11.1) | 7.0(0.9-10.6) | 9.5(1.9-12.7) | 9.0(2.3-13.0) | 7.5(1.3-9.9) | 7.1(2.0-9.8) |
| o,p-DDT | 4.3(1.1-6.8) | 0.9(0.3-1.8) | 3.5(0.8-5.9) | 3.0(0.8-6.8) | 7.4(2.8-11.2) | 5.1(1.2-8.9) | 7.7(2.1-10.8) | 6.5(1.6-9.3) |
| p,p'-DDT | 5.5(1.8-8.7) | 3.8(0.8-6.8) | 6.7(1.4-9.3) | 5.1(2.0-7.8) | 4.8(1.1-7.1) | 3.3(0.8-7.9) | 6.5(1.5-9.1) | 6.1(1.8-8.8) |
| DDT | 9.8(3.1-12).5) | 4.8(1.77.2) | 10.2(4.1-14.5) | 8.1(2.6-11.3) | 13.2(5.6-16.7) | 8.5(3.3-12.7) | 14.2(6.718.2) | 12.5(4.815.5) |
| DDE | 15.0(7.8-18.7) | 13.5(6.8-17.6) | 17.0(8.2-20.7) | 15.7(7.6-18.4) | 19.4(9.8-22.3) | 18.5(9.5.206) | 15.7(7.719.6) | 13.3(6.716.2) |
| DDD | 13.4(6.5-17.4) | 12.2(5.9-16.8) | 18.1(8.5-21.3) | 5.3(6.8-17.8) | 8.2(3.8-14.2) | 16.1(7.3-19.7) | 16.2(8.3-19.9) | 13.3(6.8-16.8) |
| Endrin | Nd | Nd | Nd | Nd | Nd | Nd | Nd | Nd |
| Endosulfan sulphate | 0.8(0.3-2.3) | 0.3(0.1-0.7) | 0.9(0.4-3.1) | 0.5(0.2-0.9) | 0.4(0.1-0.8) | 1.2(0.6-2.2) | 0.7(0.3-1.3) | 0.4(0.1-0.8) |
| Endrin ketone | Nd | 0.1(Nd-0.2) | 0.1(Nd-0.3) | 0.2(Nd-0.3) | 0.1(Nd-0.2) | 0.1(Nd-0.2) | 0.1(Nd-0.3) | 0.1(Nd-0.2) |
| Endrin aldehyde | 0.2(Nd-0.4) | 0.2(Nd-0.3) | 1.4(0.5-2.1) | 1.1(0.5-2.0) | 0.8(0.3-1.7) | 0.2(Nd-9,5) | 0.3(0.1-0.7) | 0.3(Nd-0.5) |
| Methoxychlor | 2.8(0.8-4.4) | 3.5(1.5-5.9) | 2.5(1.1-5.2) | 1.6(0.8-2.6) | 1.2(0.5-2.5) | 1.9(0.9-3.1) | 0.9(0.3-1.8) | 0.5(0.2-1.1) |
| Dieldrin | 2,8(0.8-3.1) | 2.2(0.6-2.9) | 0.9(0.2-1.5) | 0.5(0.1-0.9) | 2.1(1.5-2.9) | 2.1(1.3-3.1) | 1.7(0.9-2.5) | 1.3(0.5-2.1) |

N/B: Nd= Not detectable; figures in parenthesis represent concentration ranges.

Table 3. Isomeric ratios of related DDTs and HCHs for the biota

| Sample Code | DDT/(DDD+DDE) | o,p-DDT/p,p'-DDT | DDE/DDD | α-HCH/γ-HCH |
|-------------|---------------|------------------|---------|-------------|
| SMK | 0.3 | 0.8 | 1.1 | 0.4 |
| LMK | 0.5 | 0.2 | 1.1 | 0.2 |
| SCRB | 0.3 | 0.5 | 0.9 | 0.3 |
| LCRB | 0.3 | 0.6 | 1.0 | 0.3 |
| SMM | 0.4 | 1.5 | 1.1 | 0.2 |
| LMM | 0.2 | 1.5 | 1.2 | 0.3 |
| SCRM | 0.4 | 1.2 | 1.0 | 0.4 |
| LCRM | 0.2 | 1.1 | 1.1 | 0.6 |

Table 4. Biota and associated sedimentary t-DDT and t-HCH concentrations, total organic carbon (TOC) contents and BSAFs

| Site | Species | Sample code | BSAF$_{DDT}$ | t-DDT (ng/g) | t-HCH (ng/g) | BSAF$_{HCH}$ | TOC (%) |
|------|---------|-------------|--------------|--------------|--------------|--------------|---------|
| Mouth of the river (MR) | *Egeria radiate* (mollusks-clam) | SMK | 6.47 | 35.7 | 26.8 | 5.18 | NA |
|  |  | LMK | 7.11 | 26.8 | 33.7 | 6.32 |  |
|  | *Callinectes amicola* (crab) | SCRB | 9.78 | 44.7 | 41.2 | 8.70 |  |
|  |  | LCRB | 7.71 | 39.1 | 35.7 | 9.12 |  |
| Upper mangrove area (UMA) | *Musculin securis* (mollusks-clam) | SMM | 1.49 | 46.4 | 35.2 | 1.13 | NA |
|  |  | LMM | 1.65 | 45.2 | 48.8 | 1.71 |  |
|  | *Micropipus depurator* (Crab) | SCRM | 1.13 | 43.3 | 28.3 | 0.86 |  |
|  |  | LCRM | 1.02 | 40.6 | 34.3 | 0.74 |  |
| Associated | NA | ASM | NA | 6.5 | 102.7 | NA | 2.10 |
| Sediment | *NA* | ASU | NA | 32.6 | 1041.7 | NA | 5.21 |

N/B: NA = Not applicable.

Besides the effect of body size on HCHs accumulation in sediment-dwelling organisms, the lipid content (LC) effect was also examined. LC has been shown to be an important factor that determines the accumulation of hydrophobic compounds as they are a primary site for storage of these compounds (Muncaster, Hebert, & Lazar, 1990). According to these authors, the relative size of lipid pool affects movement of these compounds into biota tissues, implying that higher LC causes faster uptake and slower elimination of these contaminants. This gives rise to minimal accumulation of contaminants in low LC tissues compared to high accumulation in those tissues with high LCs (Hanson, Jenson, Appelquist, & Morch 1978). However, this scenario was not observed in the present study. One-way analysis of variance (ANOVA) of HCHs concentrations in the biota at a confidence level of 95% was performed in order to test the significance of the relationship between LC or body sizes of organisms and the contaminant body burden. The calculated F values showed 1.63 (df=7; p=0.05) for body size and 1.85 (df= 7; p=0.05) for LC. These F values were less than the critical F value of 2.01, implying that there were no significant relationships between these two paired variables. The implication here is that it appears there was a non-equilibrium state in which LCs in the biota tissues have not decreased or increased proportionally with contaminant load (Kelly & Campbell, 1994; Yang et al., 2006). The mean t-HCH concentrations ranged from a minimum value of 22.1 ng/g ww in LMK at the MR site to a maximum value of 55.4 ng/g ww in SMM at the UMA site (Figure 2, 3a; mean 35.50±6.99 ng/g ww). The low standard deviation value suggests that there was no significant difference in the concentrations of t-HCH among the biota samples.

Comparing data with those for other regions, it can be seen that while the mean t-HCH levels for both mollusks and crabs in the present study were respectively about 10 and 1.7 orders of magnitude higher than those found for Yangtze estuary, China, Yang et al. (2006) and Liu et al. (2010) reported mean t-HCH concentrations for mollusks in Liaoning Province that are comparable to those found at the mouth of the river (MR site) in our study area. On the other hand, while the mean t-HCH levels recorded for crabs obtained from different rivers of southern China were about 50 times lower than those found in our study area, the levels recorded in crabs from Santos Bay, Sao Paulo, Brazil (with similar environmental characteristics) are comparable (Magalhaces, Taniguchi, Coscaes, & Montone, 2012) to those from the study area.

On the basis of the ultimate degradation of γ-HCH to α-HCH, the ratio of the concentrations of α-HCH to γ-HCH was therefore used to determine the age of contamination of the study sites (Guan et al., 2008). Values of the ratio α-HCH/γ-HCH in the present study varied from 0.02 to 0.95 (mean = 0.19; Table 3), supporting recent and continued use of lindane in the region despite its deleterious effect on the environment. This result agrees with that of Chrysikou et al. (2008) who reported that values of the ratio α-HCH/γ-HCH < 1 are indicative of recent use of lindane.

The mean concentrations and ranges of DDT and its metabolites in the biota samples are summarized in Table 2. DDE (sum of *o,p'*-DDE and *p,p'*-DDE) were the most abundant, followed by DDD (sum of *o,p'*-DDD and *p,p'*-DDD) and then DDT (sum of *o,p'*-DDT and *p,p'*-DDT). For the mollusks, large size individuals (LMK) accumulated less DDD (12.2 ng/g ww) than their small size counterparts (SMK; 13.4 ng/g) at the MR site. Similar DDD accumulation pattern was not only observed for crab samples at the MR site where small size crabs (SCRB) accumulated (18.1 ng/g) more than their large size counterparts (LCRB; 15.3 ng/g), but also for mollusks and crabs at the UMA site (Table 2). The mean concentrations of DDE in LMK and LCRB were lower than those in SMK and SCRB at the MR site, respectively. In the same vein, the mean DDT concentrations were lower in LMK and LCRB (4.8 and 8.1 ng/g ww) than in SMK and SCRB (9.8 and 10.1 ng/g ww) at the MR site. Similarly, the mean DDT was lower in LMM than in SMM at the UMA site.

Generally, LMK accumulated the lowest amounts of total DDT (t-DDT- sum of all DDT isomers and their metabolites) at the MR (30.9 ng/g ww), while the maximum t-DDT accumulation level was found in SCRM at the UMA site (50.8 ng/g ww; mean- 40.23±6.49; Figure 2, 3b). Therefore, despite the fact that both crabs and mollusks are on the same position in the food chain, crabs (irrespective of the species type and geographical location) appear to have accumulated more HCHs and DDTs than mollusks at the two sampling sites. This implies that mollusks are safer to consume than crabs. The mean t-DDT concentrations in mollusks and crabs in the present study are comparable to those found in mollusks and crabs from the Yangtze estuary, and a few southern rivers of China, somewhat suggesting similar extent of historical and new source usage of DDTs in these regions. However, while the mean t-DDT levels in mollusk samples from the Liaonimg Province of China were more than 3 times higher than those found in the present study, t-DDT levels recorded in crabs from Santos Bay, Brazil (Magalhaces et al., 2012) were about 9 times higher than those of the present study.

Although the frequency of occurrence of HCHs and DDTs in both mollusks and crabs was 100%, there was no remarkable difference between levels of t-DDT and t-HCH in both the mollusks and crabs (df=14; F=1.97; p=0.83; 95% confidence interval) and the associated sediment at the two sites (Table 4) except that these species of crabs and mollusks may have greater potential to accumulate DDTs than HCHs or that HCHs are less bio-available in the river than DDTs. Mean concentrations of HCHs and DDTs were respectively about 7 and 5 orders of magnitude higher than the sum of all other miscellaneous OCPs detected in the study area. It was also observed that the mollusk and crab samples collected from the freshwater site (UMA) exhibited higher HCHs and DDTs than those from brackish water site (MR). These results indicate that OCPs accumulation is greater in freshwater mollusks and crabs than in their counterparts in brackish water. This probably reflects the impact of contaminants from agricultural areas which is typically more prevalent in the freshwater zones. These results are consistent with that reported by Liu et al. (2010) who were of the opinion that freshwater fish and mollusk may be more easily influenced than their counterparts in marine water by OCP residues, especially in agricultural areas.

Ratios such as DDT/(DDE+DDD) and DDE/DDD have been used to indicate the extent of DDT degradation and to identify any fresh input in the environment (Lee, Lenabe, & Koh, 2001). In the present study, the ratio DDT/(DDD + DDE) among biota samples ranged from 0.2 to 0.5 with a mean of 0.3, indicating that biodegradation of DDT to its metabolites did not only occur in the sediments prior to ingestion during foraging by the organisms, but also that historical use of DDTs in the region was more prominent in the environment than input from new DDT sources. However, our data revealed that the degree of dehydrochlorination of DDTs was higher than dechlorination, indicating the relative importance of aerobic microbes in the environment. This fact is supported by the DDE/DDD ratios which ranged between 0.9 and 1.2 with a mean of 1.1 for all the biota samples. Qiu, Zhu, Yoa, Hu, & Hu (2005) reported that dicofol (used as pest control for cotton and fruits), characterized with high *o,p'*-DDT/*p,p'*-DDT ratio, could be the new source of DDTs in southern and eastern China. In the present study, biota samples taken from the UMA site had *o,p'*-DDT/*p,p'*-DDT ratios > 1, whereas values < 1 were found for biota samples obtained from the MR site (Table 3). These results may be a reflection of the recent application of dicofol pesticide on the large-scale pineapple farmland and other fruits, particularly in the Akamkpa area (upper Cross River State).

Endosulfans are among the few cyclodiene pesticides used extensively throughout the world as insecticides on crops. In the environment, the cyclic sulfite group of endosulfan gets oxidized to the corresponding sulfate (Endosulfan sulfate), which is more persistent than its parent compound (Kathpal, Singh, Dhankar, & Singh, 1997; Guan et al., 2008; Liu et al., 2010). In Nigeria, endosulfan is currently being used as pest control for cotton and cocoa trees. While α-endosulfan residues were detected in all the biota samples in relatively moderate concentrations, its isomer, β-endosulfan was found in much lower concentrations in all but one sample (Table 2). Generally, there was no regular pattern between the concentrations of endosulfan or its metabolites and body size of biota as observed for both HCH and DDT. Due to the dearth of reports on endosulfan in crabs or mollusks, comparison between mean concentration of Σendosulfan (sum of parent and metabolites) in mollusks in the present study could only be made with that of the Liaoning Province. Our data show values 2 times higher (3.1

ng/g ww) than that reported for the Liaoning Province (1.5 ng/g ww), probably reflecting greater usage of endosulfan on crops in the study area.

Other OCPs in the biota samples were detected in low concentrations, reflecting their relatively low residual levels in the environment and/or relatively low potential for bioaccumulation by the species under consideration. For example, while heptachlor and its metabolites (heptachlor epoxide) were detected in all samples with mean concentrations ranging from 0.1 ng/g ww to a maximum of 0.5 ng/g ww, endrin was not detected in any of the biota samples (Table 2). The non-detection of heptachlor epoxide in the associated sediments and the detection of heptachlor and its epoxide metabolite in comparable amounts suggest the ease of metabolism of heptachlor to heptachlor epoxide in the biota tissues. On the other hand, the non-detection of β-endosulfan and endrin ketone in mollusks at relatively low polluted site (MR) may suggest low capacity of the mollusk species to accumulate these OCPs in relation to the crab species. Although few data are available, for example, those reported by Adie, Ekpo, Oyo-Ita, Offem (2012) on the occurrence of these OCPs in the environment, their frequencies of occurrence in biota from the study area are higher than those found in biota from other regions of the world (e.g. Guo et al., 2007; Liu et al., 2010). Generally, heptachlor represented only 1% of the total OCPs in the biota analyzed.

The mean concentrations of dieldrin in the crabs and mollusks were higher than those of aldrin, indicating that aldrin degradation did not only occur in the biota samples and/or in the environment prior to ingestion, but also suggest that dieldrin is more stable in the ecosystem than aldrin. In all samples, aldrin constituted < 1% of the total OCPs, while the percentage composition of dieldrin varied from 1% to 4%. Endrin metabolites such as endrin aldehyde and ketone were also identified and quantified in almost all biota samples. Unlike in the associated sediment samples where these compounds were undetected at the two sites, most biota samples showed trace levels of the compounds (Table 2), suggesting that metabolism of endrin to its metabolites could possibly have occurred in the body tissue of the biota after initial uptake of endrin via certain pathways. The non-detection of endrin in all the biota samples reflects the effectiveness of the biota enzyme system in metabolizing this compound detected in the associated sediment samples. Endrin metabolites represented about 1 - 2% of the total OCPs measured in all biota samples. The mean concentrations of methoxychlor in both crab and mollusk samples are listed in Table 2 and constituted about 1 - 4% of the total OCPs measured in the biota.

### 3.2 BSAFs and Health Risk Assessment

Biota sediment accumulation factor (BSAF) is a valuable tool used in the present study for predicting bioaccumulation of OCP compounds and represents a measure of bio-availability of these compounds in the river. The total organic carbon (TOC) contents of sediments were 2.10% at the MR site and 5.21% at the UMA site (Table 4). The calculated BSAF values for DDTs were in the range 1.02 –9.78, with the highest value found in small size crabs (SCRB) at the MR site, even though the associated sediment exhibited lower t-DDT concentration and TOC content (Table 4). On the other hand, the lowest BSAF value was observed in LCRM at the UMA site, with higher associated sedimentary DDT concentration and TOC content. Similarly, the BSAF values for HCH ranged between 0.74 and 8.70, with the highest value observed in small size crabs (SCRB) at the MR site, even though the associated sediment exhibited lower t-HCH concentration and TOC content. On the other hand, the lowest BSAF value for HCH was found in large size crabs (LCRM) at the UMA site, with higher associated sedimentary HCH concentration and TOC content. These results are in agreement with those reports by Ferguson & Chandler (1990); Lake, Rubinstein, Lee, Lake, Heltshe, & Pavignano (1990); Yang et al. (2006) who recorded lowest BSAF values in highly polluted sediments with high TOC content and highest values in sediments with low contaminant concentrations and low TOC contents. This scenario reflects the strong affinity DDT and HCH compounds have toward sedimentary organic carbon (OC). However, the observed lower BSAF values recorded for HCHs in relation to DDTs may be a result of their stronger affinity for sedimentary OC. Thus, despite the high loads of HCHs in sediments, this contaminant may therefore be less bio-available to the resident macro-fauna.

Epidemiological survey conducted by many toxicologists (eg. Rogan & Chen, 2005; Beard, 2006) has shown that exposure to OCPs can be associated with various adverse effects on human health and reproduction. For the past 3 decades, studies have shown that the levels of HCHs in aquatic food products have dramatically declined (Guo et al., 2007). However in the present work, HCHs have not only been detected in all biota samples but also their concentrations were relatively high. All samples had residual levels of HCHs higher than the maximum permissible value (10 ng/g) established by the European Union (Binelli & Provini, 2003). These high levels of HCHs suggest that HCHs are still environmentally significant organic contaminants in the SE Niger Delta region of Nigeria. In addition, though HCHs have relatively low persistency in the environment as well as high biodegradability and volatility, continued usage in the region under study may be responsible for the high levels of HCHs measured in the organisms.

As mentioned earlier, the sediment-dwelling animals were also predominated by DDTs. In comparison to the maximum permissible concentration (50 ng/g) established by the European Union on the basis of lipid percentage of aquatic food products, 52.8% of our biota samples were considered overloaded with DDTs. If a relatively stricter limit for fish consumption (14.4 ng/g) recommended by the US Environmental Agency (2006) is used, the residual levels of DDTs in 85.3% of the biota samples exceeded the criterion.

Despite the low levels of other OCP residues in the biota samples, potential health risk associated with the consumption of sediment-dwelling animals in the region cannot be overlooked. The estimated daily intakes (EDIs; which is often taken as a means of assessing human exposure to organic contaminants) of selected OCP compounds (*p, p'*-DDT, dieldrin and heptachlor) were calculated according to Liu et al. (2010). Statistical analysis of generated data provided the overall consumption information for the region and is summarized in Table 5. The amounts ranged between 15.5 and 29.6 g/day, with a mean of 17.7 g/day for crabs, and between 10.5 and 22.4 g/day, with a mean value of 14.5 g/day for the mollusk species. The consumption data recorded in our study were lower than the average value recorded for fishery food from the China Statistical Yearbook (38.90 g/day; NBSC, 2008). On the basis of these generated data, average EDIs for selected OCPs were calculated. The overall mean concentration values for selected OCPs were used for the calculation. Potential threat to public health as a result of exposure to OCPs via biota consumption were assessed by comparison with the US Environmental Protection Agency (USEPA) benchmark concentrations estimated from the cancer slope factors that represent exposure concentrations at which life-time cancer risk is 1 in 1 million (USEPA, 2006). A benchmark concentration shows the daily concentration below which the probability of developing adverse health effect is low (Liu et al., 2010).

The p,p'-DDT and dieldrin in our samples are not only higher than the cancer benchmark concentrations, but also higher than the oral reference dose developed by USEPA, suggesting high probability of developing serious adverse effects through consumption of crabs and mollusks in the region (Dougherty et al., 2000). Besides the p,p'-DDT and dieldrin, the EDI of heptachlor also exceeded the cancer benchmark level. The implication here is that there exist the risks of developing lifetime cancer especially for inhabitants of riverine/coastal communities with potential to consume more sediment-dwelling biota than those living inland. On the other hand, the EDI of each contaminant was slightly higher in crabs than in mollusks (Table 5). This suggests that the mollusk species under study were slightly safer to consume than the species of crabs on the basis of th eir contributions to the EDIs. With limited amounts of data available on persistent organochlorine compounds in sediment-dwelling animals from coastal waters of Sub-Sahara African countries, incorporating consumption and population data from FAO fact sheet (http://post.baidu.com/f?kz=1000212531) in the present study will not only enable comparison of EDIs among the population in African region to be made but also in other regions of the world. Because of the high amounts of mollusks consumed and their high DDT concentrations measured in the present study, the EDI of DDTs via mollusks consumption in the region is higher than those recorded for other regions in the Asian pacific. The comparison may be superficial because mollusks may not entirely represent bivalves, leading to bias results. Nevertheless, the result provides a preliminary baseline for assessing the extent of human exposure to DDTs via aquatic food consumption among different countries in Sub-Sahara Africa. Furthermore, many researchers do not include children population in their health risk assessment, our study was quite encompassing as it did consider children exposure to OCPs. The inclusion of children scenario will contribute significantly to the possibility of developing cancer among the inhabitants of the region.

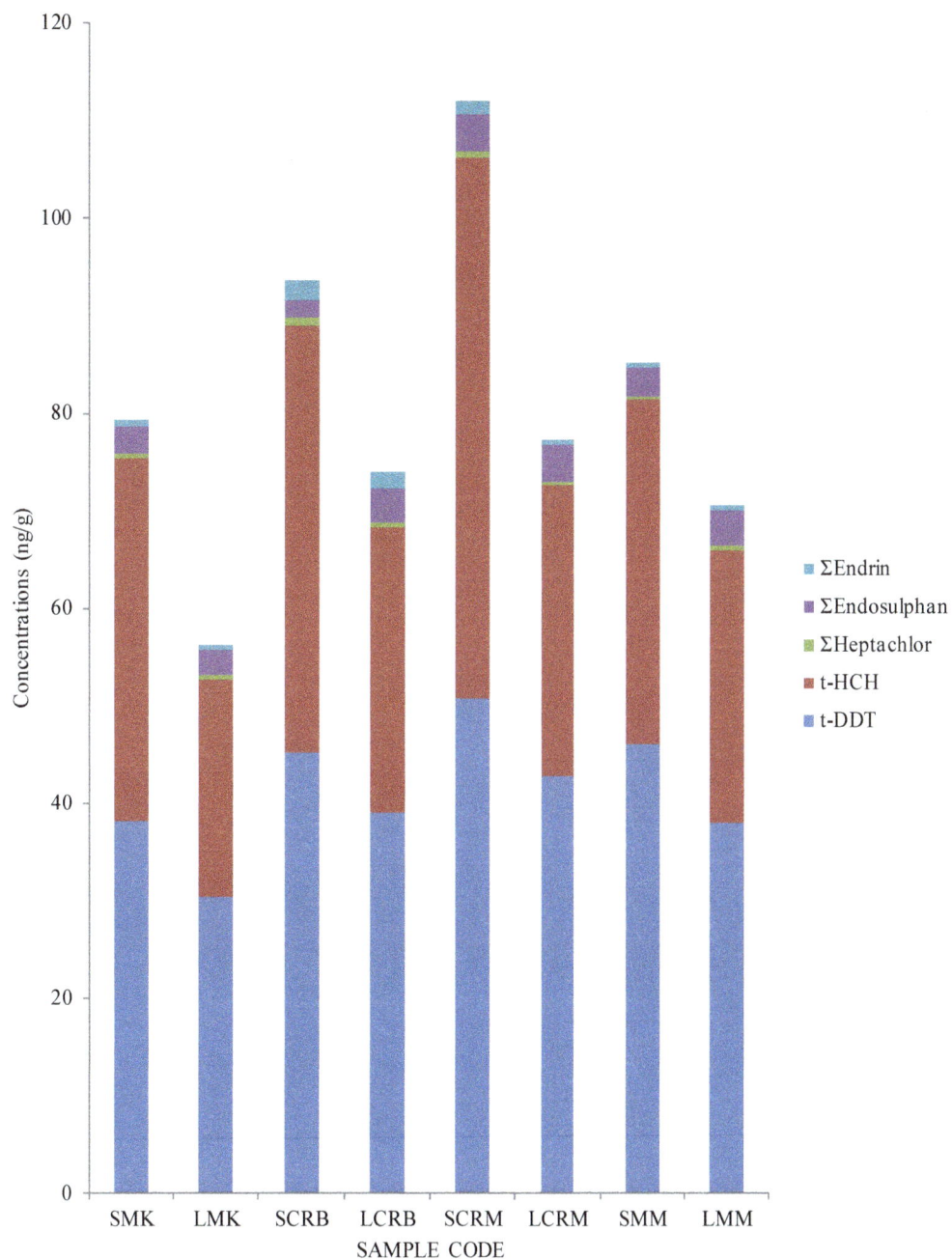

Figure 2. Distribution of OCPs in crabs and mollusks from the Calabar River

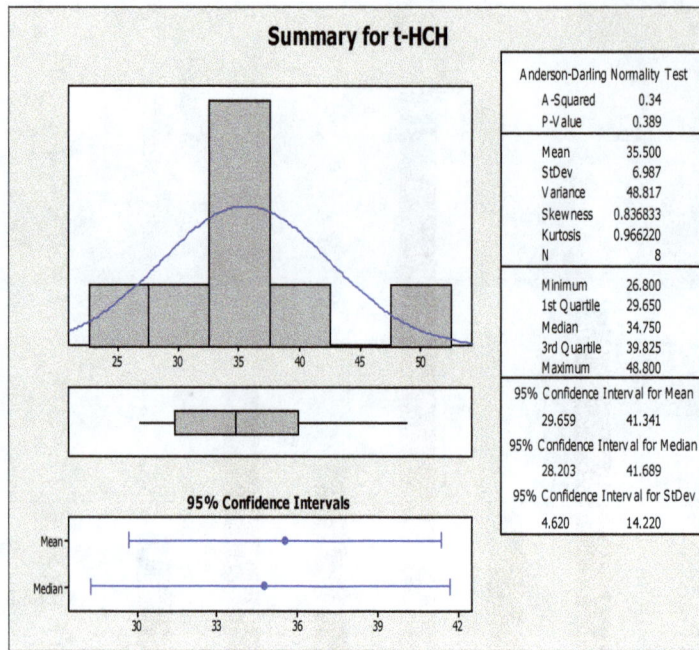

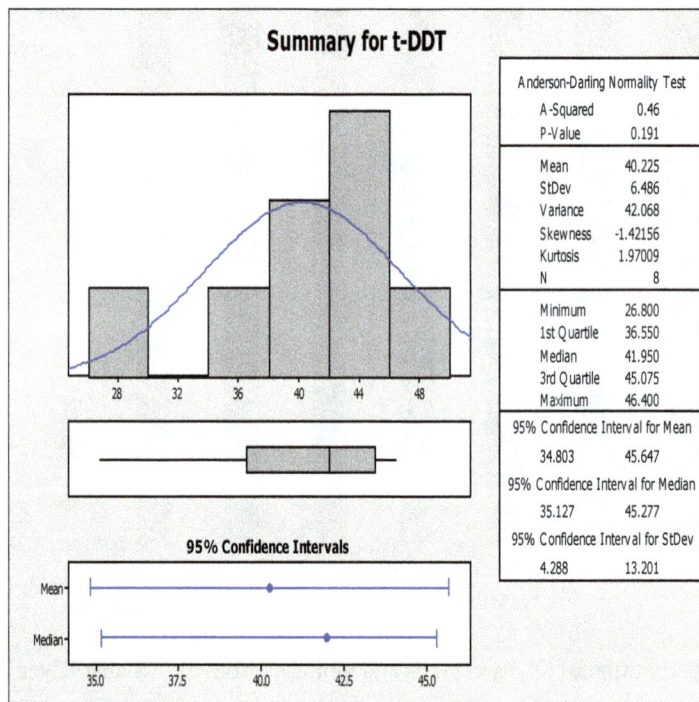

Figure 3a, b. descriptive statistics for (a) t-HCH and (b) t-DDT in biota with normal distribution curve

Table 5. Biota consumption rate and comparison of the estimated daily intake from the present work with cancer benchmark concentration (CBC) and Oral Reference Dose (Oral RfD) developed by the US Environmental Protection Agency

| Contaminant Name | Species | Consumption Rate (g/day) | Average EDI (mg/kg.day) | Oral RfD (mg/kg.day) | CBC (mg/kg.day) |
|---|---|---|---|---|---|
| *p,p'*-DDT | Mollusks | 14.5 | 1.33 | 0.5 | 0.03 |
| | Crabs | 17.7 | 1.46 | --- | --- |
| Dieldrin | Mollusks | 14.5 | 0.13 | 0.5 | 0.00022 |
| | Crabs | 17.7 | 0.15 | --- | --- |
| Heptachlor | Mollusks | 14.5 | 0.42 | 0.05 | 0.0000625 |
| | Crabs | 17.7 | 0.62 | --- | --- |

## 4. Conclusions

The present work has demonstrated the potential of organisms of lower trophic level such as crabs and mollusks to accumulate high levels of OCPs. HCHs and DDTs were the dominant OCPs measured in the biota samples collected from the Calabar River with relatively high levels of t-HCH attributable to the frequent usage of lindane in the form of GAMMALIN 20 for fishing by local fishermen. The distributions of t-HCH and t-DDT in the biota were found to be strongly dependent on geographical positions and their sources. It was also shown that there was no significant difference between lipid contents of biota and concentrations of HCH and DDT and that body size of biota affected these contaminants levels. While $\alpha$-HCH/$\gamma$-HCH ratios showed fresh and continued inputs of HCHs to the river, DDT profile exhibited evidence of extensive degradation to its metabolites (DDD and DDE) with the transformation to DDE being the preferred pathway. BSAFs varied amongst biota and were generally lower in highly polluted site and higher in low impacted areas. Hence, despite the high loads of OCPs in sediments at the UMA site, these contaminants were therefore less bio-available to the sediment-dwelling organisms at this site.

EDI results for selected OCPs revealed life-time cancer risk, especially to inhabitants of the riverine/coastal areas who often consume more sediment-dwelling biota than those living inland. Therefore, for purposes of protecting public health, it is not only urgent for the appropriate government agencies to be mobilized for a remedial action but it is also needful to continually monitor the occurrence and levels of OCPs in the region. Further monitoring efforts should be aimed at identifying other species of crabs and/or mollusks with potential higher OCPs accumulation capacity.

### Acknowledgements

We are thankful to the management and staff of State Key Laboratory of Organic Geochemistry, Guangzhou Institute of Geochemistry, China for accepting to run the GC-Ms analysis of our fractions at no cost and Ola Samuel of Department of Chemistry, Redeemer University of Nigeria, Ogun State for assisting in these runs during his study fellowship in the Institute.

### References

Adie, P. A., Ekpo, B. O., Oyo-Ita, O. E., & Offem, J. O. (2012). Organochkorine pesticides in coastal/mangrove sediment of the Calabar River, SE Nigeria. *African Journal of Environmental Polltion & Health, 9*(2), 52-61.

Asuquo, F. E. (1989). Water quality of the Calabar River, Nigeria. *Tropical Ecology, 30*(1), 31-40.

Beard, J. (2006). DDT and human health. *Science of the Total Environment, 355*, 78-89. http://dx.doi.org/10.1016/j.scitotenv.2005.02.022

Binelli, A., & Provini, A. (2003). POPs in edible clams from Italian and European markets and possible human health risk. *Marine Pollution Bulletin, 46*, 879-886. http://dx.doi.org/10.1016/S0025-326X(03)00043-2

Burns, W. A., Mankiewiez, P. J., Bence, A. F., Page, D. S., & Parker, K. R. (1997). A principal-component and least-square method for allocating Polycyclic aromatic hydrocarbons in sediment to multiple sources. *Environmental Toxicology & Chemistry*, 1119-1131. http://dx.doi.org/10.1002/etc.5620160605

Chrysikou, L., Gemenetzis, P., Kouras, A., Manoli, E., Terzi, E. & Samara, C. (2008). Distribution of persistent organic pollutants, polycyclic Aromatic hydrocarbons and trace elements in soil and vegetation following a large scale landfill fire in northern Greece. *Environment International, 34*, 210-225. http://dx.doi.org/10.1016/j.envint.2007.08.007

Cross River Basin Development Authority (CRDA). (1982). Inventory of Natural sites *conditions*- soils, slopes, hydrology, land use and vegetation throughout the area of operation of the Authority. *Progress report, 4*, 1-45.

De Souza., A. S., Machado Torres, J. P., Meire, R. O., Neves, R. C., Couri, M. S., & Serejo, C. S. (2008). Organochlorine pesticides (OCPs) and polychlorinated biphenyls (PCBs) in sediments and crabs (*Chasmagnathus granulate,* DANA, 1851) from mangroves Guanabara Bay, Rio De Janeiro State, Brazil. *Chemosphere, 73*, 186-192. http://dx.doi.org/10.1016/j.chemosphere.2007.04.093

Dougherty, C. P., Holiz, S. H., Reinert, C., Panyacosit. L., Axrlrad, D. A., & Woodruff, T. J. (2000). Dietary exposures to food contaminants across United States. *Environmental Research, 84*, 170-185. http://dx.doi.org/10.1006/enrs.2000.4027

Ezemonye, L., & Ogbimida, E. (2010). Histopathological effect of GAMALIN 20 on African catfish. *Applied & Environmental Soil Science*, 8-13.

Ferguson, P. L., & Chandler, G. T. (1998). A laboratory and field comparison of sediment polycyclic aromatic hydrocarbons bioaccumulation by the cosmopolitan estuarine polycheate stieblospio benedicti (Webster). *Marine Environmental Research, 45*, 387-401. http://dx.doi.org/10.1016/S0141-1136(98)00101-9

Grimalt, J. O., van Drooge, B. L., Ribes, A., Vilanova, R. M., Fernandez, P., & Appleby, P. (2004). Persistent organochlorine compounds in soils and sediments of European high altitude mountain lakes. *Chemosphere, 54*, 1549-1561. http://dx.doi.org/10.1016/j.chemosphere.2003.09.047

Guan, Y., Wang, J., Ni, H., & Zeng, E. Y. (2009). Organochlorine pesticides and polychlorinated biphenyls in riverine runoff of the Pearl River Delta, China: Assessment of mass loading input source and nvironmental fate. *Environmental Pollution, 157*, 618-624. http://dx.doi.org/10.1016/j.envpol.2008.08.011

Guo, J-Y., Zeng, E. Y., Wu, F.G., Meng, X. Z., Mai, B. X., & Luo, X-J. (2007). Organochlorine pesticides in seafood products from Southern China and health risk assessment. *Environmental Toxicology & Chemistry, 26*(6), 1109-1115. http://dx.doi.org/10.1897/06-446R.1

Hanson, N., Jenson, Y. B., Appelquist, H., & Morch, E. (1978). The uptake and release of petroleum hydrocarbons by the marine mussels Mytilus edulis. *Prog. Water Technology, 10*, 351-359.

International Programme on Chemical Safety (IPCS). (2006). Inventory of IPCS and other WHO pesticide evaluations and summary of toxicological evaluations performed by the joint meeting on pesticide residues (JMPR) through 2005. Retrieved December, 2012, from http://www.who.int/ipcs/publications/jmpr/jmpr_pesticide/en/index.ht ml

Kathpal, T. S., Singh, A., Dhankar, J. S., & Singh, G. (1997). Fate of endosulfan in cotton oil under sub-tropical conditions of northern India. *Pesticide Science, 50*, 21-27. http://dx.doi.org/10.1002/(SICI)1096-9063(199705)50:1<21::AID-PS550>3.0.CO;2-N

Ke, L., Yu, K. S. H., Wong, Y. S., & Tam, N. F. Y. (2005). Spatial and vertical distribution of polyaromatic hydrocarbons in mangrove sediments. *Science of the Total Environment, 340*, 177-187. http://dx.doi.org/10.1016/j.scitotenv.2004.08.015

Kelly, A. G., & Campbell, L. A. (1994). Organochlorine contaminants in liver cod (*Gadus morhua*) and mussles of herring (*Clupea harengus*) from Scottish waters. *Marine Pollution Bulletin, 28*, 103-108. http://dx.doi.org/10.1016/0025-326X(94)90546-0

Lake, J., Rubinstein, N., Lee, I., Lake, C., Heltshe, J., & Pavignano, S. (1990). Equilibrium partitioning and bioaccumulation of sediment associated contaminants in faunal organisms. *Environmental Toxicology & Chemistry, 9*, 1095-1106. http://dx.doi.org/10.1002/etc.5620090816

Lee, K. T., Lenabe, S., & Koh, C. H. (2001). Distribution of organochlorine pesticides in sediments from Kyeonhhi Bay and nearby areas, Korea. *Environmental Pollution, 114*, 207-213. http://dx.doi.org/10.1016/S0269-7491(00)00217-7

Liu, Z., Zhang, H., Tao, M., & He, Z. (2010). Organochlorine pesticides in consumer fish and mollusk of Liaoning province, China: distribution and human exposure implication. *Arch. Environmental Contamination & Toxicology, 59*, 444-453. http://dx.doi.org/10.1007/s00244-010-9504-7

Löwenberg, U. H., & Künzel, T. H. (1992). Investigations on the hydrology of the lower Cross River System, Nigeria. *Anim. Research & Development, 35*, 72-85.

MacIntosh, D. L., Spengler, J. D., Ozkaynak, H., Tsai, L., & Ryan, P. B. (1996). Dietary exposures to selected metals and pesticides. *Environmental Health Perspective, 104*, 202-209. http://dx.doi.org/10.2307/3432790

Magalhaces, C. A., Taniguchi, S., caces, M. J., & Montone, R. C. (2012). PCBs, PBDEs and organochlorine pesticides in crabs Hepatus Pudibundus and Callinectes danae from Santos Bay, State of Sao Paulo, Brazil. *Marine Pollution Bulletin, 64*(3), 662-667. http://dx.doi.org/10.1016/j.marpolbul.2011.12.020

Mille, G., Guiliano, M., Asia, L., Malleret, L., & Jalaluddin, N. (2006). Sources of hydrocarbons in sediments of the Bay of Fort de France (Martinique). *Chemosphere, 64*, 1062-1073. http://dx.doi.org/10.1016/j.chemosphere.2005.12.001

Muncaster, B. W., Hebert, P. D., & Lazar, R. (1990). Biological and physical factors affecting organic contaminants in fresh water mussels. *Arch. Environmental Contaminants & Toxicology, 9*, 23-28.

National Bureau of Statistics of China (NBSC). (2008). China statistical yearbook, 2008. Retrieved February, 2012, from www.stats.gov.cn/tjsj/ndsj/2008/indexch.htm

Oyo-Ita, O. E., Offem, J. O., Ekpo, B. O., & Adie, P. (2013). Anthropogenic PAHs in mangrove sediments of the Calabar River, SE Niger Delta, Nigeria. *Applied Geochemistry, 28*, 212-219. http://dx.doi.org/10.1016/j.apgeochem.2012.09.011

Qiu, X., Zhu, T., Yoa, B., Hu, J., & Hu, S. (2005). Contribution of dicofol to current DDT pollution in China. *Environmental Science & Technology, 39*, 4385-4390. http://dx.doi.org/10.1021/es050342a

Rogan, W. J., & Chen, A. ( 2005). Health risks and benefits of bis(4-chlorophenyl)-1,1,1-Trichloroethane (DDT). *Lancet, 366*, 763-773. http://dx.doi.org/10.1016/S0140-6736(05)67182-6

USAID/West Africa. (2008). *Trade hub annual report.* Retrieved July 30, 2013, from available at http.//www.watradehub.com/resources/resourcesfiles.com

U.S. Environmental Protection Agency (USEPA). (2006). *Guidance for assessing Chemical contaminant data for use in fish advisories, 2*, 231-253.

Risk assessment and fish consumption limits. (2012). Retrieved February, 2012, from http://www.epa.gov/ost/fishadvice/volum2/index.html

Wang, X. J., Chen, J., Zang, Z. H., Piao, X. Y., Hu, J. D., & Tao, S. (2004). Distribution and sources of Polycyclic Aromatic Hydrocarbons in soil Profiles of Tiajin Area, People's Republic of China. *Bulletin of Environmental Toxicology, 73*, 739-748.

Wiles, R., Davies, K., & Campbell, C. (1998). Overexposed organophosphate insecticides in children's food. Washington, DC: *Environmental Working Group.*

Yang, Y., Lui, M., Su, L., Ou, D., Liu, H., Cheng, S., & Hofmanni, T. (2006). HCHs and DDTs in sediment-dwelling animals in the Yangtze Estuary China. *Chemosphere, 62*, 381-389. http://dx.doi.org/10.1016/j.chemosphere.2005.04.102

Zulaga, O., Prieto, A., Usobiaga, A., Sarkar, S. K., Chatterice, M., Bhattacharya, B. D., Adam, M. A., & Satpathy, K. K. (2009). Polycyclic aromatic hydrocarbons in intertidal marine bivalves of Sunderban mangrove wetland, India: an approach to bioindicator species. *Water, Air & Soil Pollution, 201*, 305-331. http://dx.doi.org/10.1007/s11270-008-9946-y

# 4

# Bioaccumulation of Heavy Metals and Hydrocarbons in *Hemichromis Fasciatus* Exposed to Surface Water in Borrow Pits Located Within Onshore Oil Exploration and Production Area

E. N. Vaikosen[1], B. U. Ebeshi[1] & B. Airhihen[2]

[1] Department of Pharmaceutical and Medicinal Chemistry, Faculty of Pharmacy, Niger Delta University, Wilberforce Island, Nigeria

[2] Department of Pharmacology and Toxicology, Faculty of Pharmacy, Niger Delta University, Wilberforce Island, Nigeria

Correspondence: E. N. Vaikosen, Department of Pharmaceutical and Medicinal Chemistry, Faculty of Pharmacy, Niger Delta University, Wilberforce Island, Bayelsa State, Nigeria. E-mail: vaikosen@yahoo.co.uk

**Abstract**

A field bioaccumulation study was carried out. Juvenile *hemichromis fasciatus* in net cages were exposed to contaminated surface water in borrow pits located within oil exploration and production (E & P) installations within the Niger Delta region of Nigeria during wet and dry seasons.

Percentage mortality ranged between 2% and 3.8%. Assuming a first order kinetics and steady state at day 14; toxicokinetic variables were obtained for xenobiotics assessed. The up-take rate constant $k$, ranged from $38.47 \times 10^{-2}$ d$^{-1}$ (pit B, wet season) to $40.82 \times 10^{-2}$ d$^{-1}$ (pit A, dry season) for TPH, while PAHs ranged between $22.40 \times 10^{-2}$ d$^{-1}$ (pit B, wet season) and $24.98 \times 10^{-2}$ d$^{-1}$ (pit A, dry season) for both pits and seasons.

Amongst the water borne metals $k$ for *hemichromis fasciatus* ranged from $4.66 \times 10^{-2}$ d$^{-1}$ to $47.66 \times 10^{-2}$ d$^{-1}$ and $2.70 \times 10^{-2}$ d$^{-1}$ to $33.90 \times 10^{-2}$ d$^{-1}$ in pits A and B respectively during both seasons. The order of uptake was Fe <Pb< Cr < Ba < Cd < Zn < Cu < Ni, while As (arsenic) recorded a zero uptake.

Calculated BCF in both pits and seasons ranged from 2.832 LKg$^{-1}$ to 4.844 LKg$^{-1}$ for TPH and 2.636 LKg$^{-1}$ to 8.0 LKg$^{-1}$ for PAH, while for metals - Ba, Cd, Cr, Cu, Fe , Ni, Pb and Zn, values ranged between 0.211 LKg$^{-1}$ to 71.727 LKg$^{-1}$. *Hemichromis fasciatus* exhibited greater uptake of analytes in the dry season and the amount of heavy metals accumulated were all below the provisional maximum tolerable daily intake (PMTDI).

**Keywords:** bioaccumulation, *hemischromis fasciatus,* hydrocarbons, heavy metals, Toxicokinetics

## 1. Introduction

Studies on the impact of heavy metals and petroleum hydrocarbon in aquatic ecosystems comprising of rivers, streams, lakes, aquatic organisms (fish, shrimp, mollusc etc.) and sediments have become a major global environmental issues. In nature, aquatic organisms are constantly exposed to pollutants such as metals and hydrocarbons due to natural geochemical processes (e.g., weathering of rocks, and leaching) and anthropogenic activities resulting from increase in urbanization, industrialization, agricultural practices, oil exploration and production (E&P) activities as characterized in the Niger Delta region of Nigeria (Ajayi & Osibanjo, 1981; Biney *et al.,*1994; Alloway & Ayres, 1997; Noegrohati, 2006; Godwin *et al.*, 2011;). Often such increases in anthropogenic activities, usually leads to accelerated releases of chemical pollutants into the aquatic environment which poses serious and severe threat to aquatic life and man due to their proved toxicity, persistence, bioaccumulation and biomagnification in the food chain (Tulas *et al.,* 1989; Papagiannis *et al.,* 2004; Martinez-Lopez *et al.*, 2005).

Heavy metals (e.g., Fe, Ni, Cd, Cu, Cr, Ba, etc) and many organics like hydrocarbons and pesticides have been reported in various concentrations in many rivers and creeks in the Niger Delta region of Nigeria (Anyakora *et al.*, 2005; Olowoyo *et al.*, 2010; Godwin *et al.*, 2011; Nwabueze, 2011) and these pollutants have a tendency to accumulate in biota (Landrum *et al.,* 2003; Anyakora & Coker, 2007) and undergo food chain magnification (James *et al.,*1998). It is therefore imperative to monitor the concentration of these pollutants in the environment and to analyze bioaccumulation process in order to assess the possible impact on human health and risk which man faces in such an environment (Kotze *et al.*, 1999).

The use of biological accumulator species in monitoring and assessing the level of contaminants and pollution of our aquatic environment is a major thrust towards knowing the degree to which the various components of our aquatic ecosystem is impacted. Accumulator species such as mollusc and some bentho-pelagic organisms are sedimentary dwellers and have capacity to bioaccumulate relatively large amounts of certain pollutants, even from much diluted solutions without obvious noxious effects. The bioaccumulation of pollutants in organisms is the result of previous uptake from its environment in the past as well as the recent pollution level of the environment in which the organism lives, while the pollutant concentrations in the water only indicate the situation at the time of sampling (Karadede *et al.*, 2004). Chemical pollutants are known to have adverse effects on aquatic environments. A negligible increase in the concentration of chemical pollutants could lead to a drastic effect on the aquatic life. Also, chemicals, which would have been harmless on their own, may become toxic by interacting in the general milieu of contaminated water.

Bioaccumulation starts with the uptake of chemical pollutants across biological membrane and could be investigated via laboratory and field study. Many laboratories bioaccumulation studies have been reported by researchers (Wang & Rainbow, 2006; Noegrohati, 2006; Martin *et al.*, 2007; Kamunde, 2009). However, field bioaccumulative studies, gives a real situation approach, whereby aquatic organisms are exposed to retinue of inorganic and organic compounds that interplays within the natural environment. Oikari, 2006; Crane *et al.*,2007 reported that *in-situ* caging of fish to determine the effects of exposure to contaminants at impacted sites has many advantages over traditional in-lab testing. In a field study using net cages, Goksoyr *et al.,* 1996, reported the accumulation of PAHs in juvenile Flounder (*Platichthysflesus* L.) and Atlantic cod (*Gadusmorhua* L.). The use of caging methodology seems to be a promising way to approach eco-toxicologically relevant problems, such as bioavailability of contaminants, biomarker responses in the field, and dose-response relationships, also under mixed contaminant situations.

The environment is continuously loaded with foreign organic chemicals (xenobiotics) and inorganic compounds released by urban communities and industries (R. van der Oost *et al.,* 2003). The aquatic ecosystem is therefore continuously and seriously threatened by these substances - as it is the ultimate sink for these contaminants by either due to direct discharges or to hydrologic and atmospheric processes (Stegeman & Hahn, 1994).

More so, these effects become more pronounced in non-flowing, receiving water bodies such as lakes, borrow-pits and discharge pits. Some of these man-made pits or lakes are totally submerge during the rainy season especially when rivers around these pits over-flow their banks to cover such pits. Aquatic organisms (like fish) tend to migrate in and out of such receiving pits. In some cases communities have also fished in such pits. Some of the chemical pollutants that could have lasting effects on the natural pits are nutrients, oil and grease, refractory chemical species such as heavy metals, PAHs, etc.

In the Niger Delta region of Nigeria, where onshore operations involving petroleum exploration and production (E & P) activities are carried out within communities such occurrences are prevalence. Physical and chemical characterisations of impacted environmental components in the Niger Delta environment (and indeed in the African continent) are in most cases on air, soil, sediment and surface water, leaving out the biota. To ascertain the chemical characteristics of these environmental components (surface water, sediments, soil, biota and atmosphere) interacting in any ecosystem, the levels of pollution indicators should be evaluated. The chemical characteristics are analysed to indicate the concentration of pollution indicator parameter, selected for evaluation.

There is no literature on field bioaccumulation studies within the Niger Delta region of Nigeria, considering the effect of the E & P operations on the environment. We have adopted the caging method to assess the levels of contamination of the numerous borrow pits dug around this region as a result of various infrastructural development that are on-going - this includes road construction, petroleum exploration and development operations, housing projects, etc. The aim of the study is the use of bio-indicators in assessing the level of contamination in this area and also the use of toxicokinetic variables obtained from both physico-chemical and biological properties of recipient borrow-pits and test organism – *hemichromis fasciatus* respectively, to characterize the effect and risk of onshore exploration and production (E&P) operational activities on communities situated around the Niger Delta region of Nigeria

## 2. Material and Methods

### 2.1 Reagents and Materials

Extraction solvents - dichloromethane (DCM), n-hexane, acetone, petroleum spirit and acetonitrile were all of analytical grade and manufactured by Sigma-Aldrich (St Louis, USA) and Merck (Darmstadt, Germany). Sodium sulphate (anhydrous) and silica gel 60 extra-pure (60 – 120 mesh) for column chromatography were from BDH limited (Poole, England). Concentrated nitric acid used was of analytical grade, manufactured by Merck KGa A of Germany and BDH Limited Poole England. Distilled water used was double distilled (DD). USEPA 16 priority PAHs mixed standard (2000 µg/mL) comprising: naphthalene (Nap), acenaphthylene (Acy), acenaphthene (Ace), fluorene (Flu), phenanthrene (Phe), anthracene (Ant), fluoranthene (Flt), Pyrene (Pyr), benzo[a]anthracene (BaA), chrysene (Chr), benzo[b]fluoranthene (BbF), benzo[k]fluoranthene (BkF), benzo[a]pyrene (BaP), dibenzo[a,h]anthracene (DaA), benzo[g,h,i]perylene (BgP) and indeno[123-cd]pyrene (IP) in dichloromethane: benzene was obtained from SUPELCO, Bellefonte, PA, USA

### 2.2 Instrumentation and Measurements

### 2.2.1 Atomic Absorption Spectrophotometer (AAS)

Heavy metal measurements were carried out using a Varian Atomic Absorption Spectrophotometer (AAS), model SpectraAA 600 with flame system inter-phased to a computer and printer.

### 2.2.2 Gas Chromatography – FID

The hexane reconstituted clean up extract was analysed with a Shimadzu QP – 2010 Gas Chromatography-Flame Ionization Detector (GC-FID). Chromatographic separation was carried out on a 30 m x 0.25 mm id SLB-5MS capillary column with a film thickness of 0.25 µm. The oven temperature was programmed, which was initially held at 70°C for 0.2min, and was increased to 265°C at a rate of 25°C/min, held for 1 min and then raised to 315°C at a rate of 5°C /min and held for 2 min. The flow rate of the carrier gas (helium, 99.99% purity) was kept constant at 1.3 mL/min. Pulsed splitless injection mode at an injection temperature of 250°C was carried out at a pressure of 30psi. Injection volume was 2µl.

All instruments were calibrated before use.

### 2.3 Study Areas and Experimental Procedure

### 2.3.1 Study Area

The study area - Ogba –Egbema Local Government Area of Rivers, lies within latitude $5^0$ 13'N and $5^0$ 22'N and longitude $6^0$ 33'E and $6^0$ 42' North West of the Niger Delta region of Nigeria (Figure 1). The main population is mostly rural communities. There are numbers of oil exploration and production (E & P) facilities and activities located around this area belonging to three major players (multinationals) in the up-stream sector of the oil industry (Figure 2). Some of the installations include oil field, gas plants, flow stations etc. Also the area is one of the prominent and highest oil/gas onshore producing areas of the Niger Delta with over 900 oil wells and about thirteen active oil fields - playing a host to three multinational oil companies. The area is criss-cross with network of pipelines carrying either oil or gas to the flow stations from the different oil wells (Avwiri & Ononugbo, 2010).

Two distinct seasons are recognized namely; the rainy and dry seasons. The vegetation is typically tropical rain-forest with fresh water swamps.

Activities of the communities inhabiting the area consist mainly of crop farming and fishing. One of the predominant fish in the study area is the genus *hemichromis* (or *Hemichromis sp.*).

Field experiment was carried out between March and April 2012 for dry season and August 2012 for wet season.

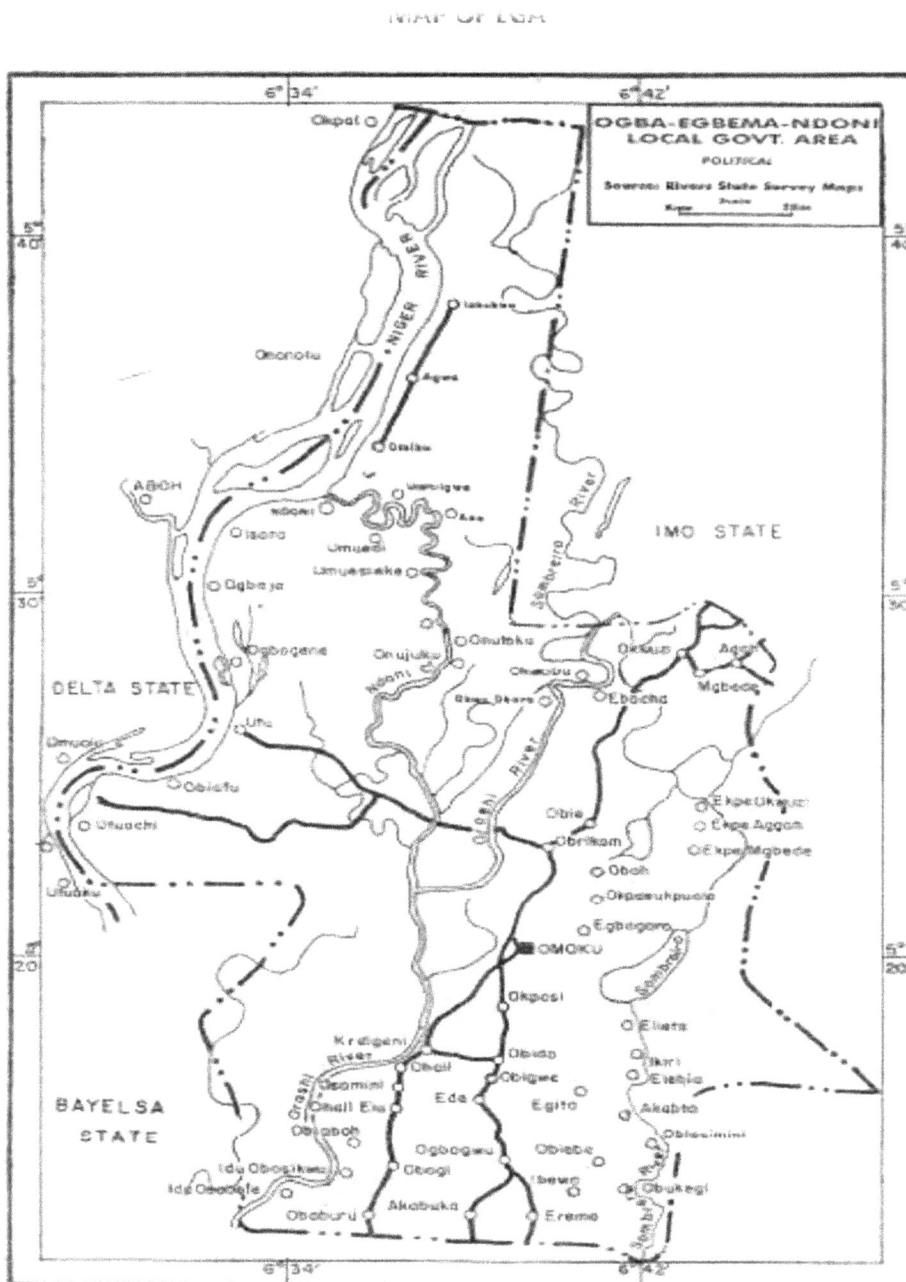

Figure 1. Map of the study area in Egba- Egbama – Ndoni Local Government Area of Rivers State

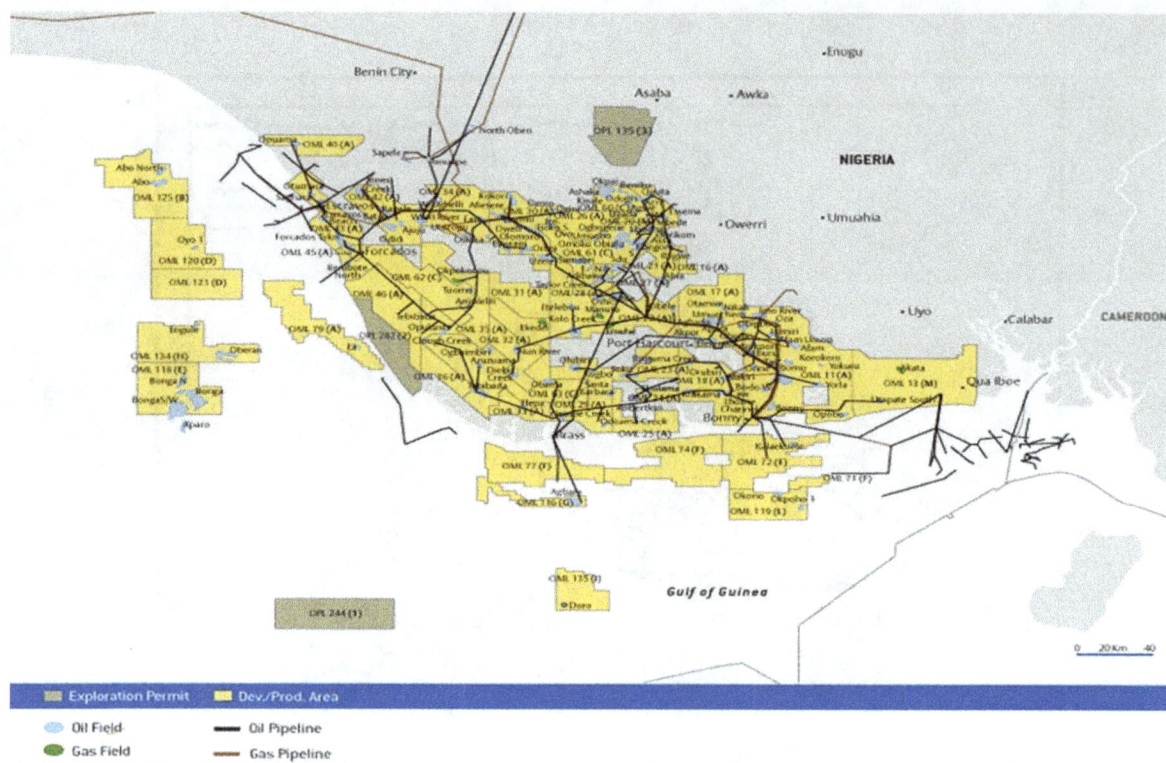

Figure 2. Map showing some onshore active oil fields around the Niger Delta Region of Nigeria

*2.4 Bioaccumulation Test Procedure*

The EPA (2008) and DPR (2002) bioaccumulation test procedures were adopted for *in-situ* (field) test methodology for toxic metals and hydrocarbons for biological monitoring.

Bioaccumulation test was conducted on test organism captured from a pristine environment.

Prior to experimental set up, baseline analysis of toxic metal and hydrocarbon contents in test organisms were determined to ascertain their level of 'cleanliness'.

All organisms were collected from the same pristine environment and were uniform in size (i.e., < 10 g) and age (juvenile organisms - that are post larval, actively feeding, sexually im-matured or spawning). Their lengths were standard and measurement were done from the tip of snout to end of candal peduncle of the fish and none of the test organisms had a length twice more than one another (i.e. the longest fish was not more than twice that of the shortest fish in length). All organisms were well cared for to forestall or avoid unnecessary stress during transportation and acclimatization. Harvested organisms were properly preserved before extraction of analytes (toxic metals and hydrocarbons) and the determination of these analytes were carried out in accordance with specified standard methods.

2.4.1 Test Organism

The test organism used in this study was *hemichromis fasciatus,* a benthopelagic species. It was cultured at The African Regional Aquaculture Centre, Port Harcourt, Nigeria and the specie was duly identified by a Marine Biologist as *hemichromis fasciatus.*

2.4.1.1 Collection of Test Organism and Acclimatization

Test organisms used in this test were captured using a small drag net made of rope (0.5mm thickness). They were of an average weight of 4.98 ± 1.38 g, with an average length of 2.95 ± 0.52cm. The collected organisms were carefully transported to the experimental site with minimal stress on organism and were later transferred into a transparent glass tank with a dimension of 72cm (l) x 48cm (w) x 48cm (h) containing their native water and allowed to acclimatized at a temperature range of $23^0C \pm 2^0C$ in an air condition room for 10 days. The water in the glass tank was continuously aerated during this period and organisms fed with minimal quantity of recommended fish feeds of known lipid and protein content (EPA, 2008) that does not contain any of the toxic metals and hydrocarbon.

## 2.4.2 Test Media

The test media were two water bodies (borrow pits) with dimensions of approximately 72.5 m (l) x 44.2 m (b)×2.5m (d) and 82.7m (l)×45.2m (b)×1.8m (d) for pits A and B respectively. The surface water in the pits were slightly brownish, with thin films of petroleum on the surface of water in some areas, especially in pit A. The odour was characteristic of crude oil.

## 2.4.3 Control Site

The control experiment for this study was set up at Mgbosimiri stream at Akabuka, which is presumably a pristine environment. The major anthropogenic activities at the stream are recreational (swimming), fishing and in some cases sand mining. This stream has not witness any pollution or impact from oil exploitation or exploration exercise. Mgbosimiri stream is a fresh water environment and it is within the study area. Generally, its' vegetation and soil properties are akin to those found around the communities were these facilities are located.

## 2.4.4 Experimental Design

Test was conducted in the field aiming to capture the real aquatic environmental situation – vis-a-viz other variables that could interplay in the study area.

Five (5) nylon net cages with dimensions – 60cm (length) x 45cm (width) x 60 cm (height), with net space size of 3x3mm and rope thickness of 1mm were used for the experiment. They were placed at different locations in the recipient water, at least about 5 meters apart. Each cage was partially lowered into the recipient water body with weights attached to net base at four (4) ends (to restrict being drifted by current movement) from a canoe. Ten (10) acclimatized organisms were then introduced carefully into each cage with the lower part being submerged in recipient water. The cage inlet was then zipped and lowered. The top ends of the cage were tied to four (4) stakes with the aid of a rope (1.5 cm diameter) in a manner that about 5 cm – 10 cm of the top portion of the cage was not submerged.

## 2.4.5 Duration of Study

Test organisms were exposed to recipient water or media in the discharge pits for a duration of 14 days, with an assumption that steady-state equilibrium would have been established between test organism and test media.

## *2.6. Analytical Procedure*

### 2.6.1. Determination of Hydrocarbons Concentration in Test Organism

#### 2.6.1.1 Sample Extraction

EPA methods3540C and 8100 were adopted for sample extraction and determination of hydrocarbon (TPH and PAHs) respectively.

Harvested fish samples from study or sampling pits were air dried at room temperature ($25^0$C) for 10days in a well aerated room, free of hydrocarbon contaminants. A portion of the dried fish was properly grinded and homogenized into an agate mortar using a pestle. 10g dry weight (DW) of the homogenized sample was weighed into a timble and transferred into a soxhlet extractor. Hydrocarbon was extracted using 100 ml of dichloromethane: hexane mixture (3:1) for about 2 hours on a heating mantle. This was allowed to cool. The extract was then concentrated to 5ml in a rotatory evaporator and transferred quantitatively into a 10 ml beaker. The flask was rinsed further with 2 ml of hexane and combined extracts was evaporated completely with the aid of nitrogen gas.

### 2.6.2 Clean up

The residue was re-dissolved with 5 ml of pentane and transferred into chromatography column containing silica gel (60 – 120 mesh). The total petroleum hydrocarbon (TPH) was eluted using pentane. Eluent was then concentrated for Gas Chromatography analysis to give aliphatic and PAHs profiles. All the biogenic hydrocarbons were eliminated during the cleaning exercise.

### 2.6.3 Determination of Heavy Metal in Test Organism

About 1.5g – 2.5g (DW) of homogenized whole fish of the test organism was transferred into a Kjeldah flash (before and after uptake phase). This was digested using 10ml of concentration of $HNO_3$ on a heating mantle. The digest was transferred on cooling into a 50ml volumetric flask quantitatively. Kjeldah flask was rinsed with distilled water and the volumetric flask was made to mark. Solutions were analysed for heavy metals using an

AAS. A blank solution was also prepared containing the 10ml of concentrated nitric, treated as in sample and made to 50 ml mark of volumetric flask.

*2.7 Standard Calibration Graph*

2.7.1 Heavy Metals

A five point calibration graph was performed for each of the heavy metal determined (Pb, Cu, Zn, Cr, As, Fe, Cd, Ni and Ba). The prepared concentrations of heavy metal standards used for the plot varied between 0.005 mg/mL and 10.00 mg/mL depending on the absorbance of the metal. Calibration curves for each metal showed good coefficient of regression ($R^2$) between 0.9987 and 1.0000.

2.7.2 Hydrocarbons

A five point calibration curve was performed using a mixed standard containing twenty (20) poly aromatic hydrocarbons (PAHs). The standards were co-mixed in iso-octane and each concentration contained all the pesticides being analyzed (0.063 ppm, 0.125 ppm, 0.250 ppm, 0.500 ppm and 1.00 ppm). Calibration curves for each PAHs showed good regression coefficient ($R^2$) between 0.9958 and 0.9999. All standards were used as external standards for the identification and quantification of corresponding PAHs present in the test organism.

## 3. Results and Discussion

*3.1 Surface Water Characteristics*

The temperatures of the surface water were $29.7 \pm 2.3^{\circ}C$ and $25.4 \pm 1.0^{\circ}C$ in the dry and wet seasons respectively. These values are typical for the Niger Delta region (Nduka & Orisakwe, 2010; Eziekel *et al.,*2013). The pH, DO and conductivity ranged from 6.48±0.55 to 7.88±0.07, 4.63±0.42mg/L to 8.43 ± 0.31mg/L and 48.33±6.50 to 124.98±2.52 (µs/cm) respectively, while values obtained at the control were (7.25±0.24 and 6.95 ± 0.18, 5.93±0.25 mg/L, 21.28±1.28 and 18.42±0.98 (µs/cm) respectively.

3.1.1 Total Petroleum Hydrocarbon (TPH)

The total petroleum hydrocarbons mean values during the dry and wet seasons were 0.107±0.032 (mg/L) and 0.054±0.037 (mg/L) respectively in borrow pit A, while borrow pit B recorded mean values of 0.087 ± 0.014 (mg/L) and 0.045±0.024 (mg/L) respectively. The control station was 0.0100 ±0.004 (mg/L) and 0.009±0.003 (mg/L) for dry and rainy seasons respectively. These values were all below the DPR/EPA regulated limit of 10mg/L.

3.1.2 Heavy Metals in Water

The mean concentrations of heavy metals in both pits and during dry and wet seasons were Pb 0.002 – 0.014 mg/L; Cu 0.004 – 0.024 mg/L; Zn 0.011 – 0.022 mg/L; Cr 0.003 – 0.014 mg/L, As<0.001 mg/L; Fe 0.168 – 0.243 mg/L; Cd 0.002 – 0.008 mg/L, Ni 0.004 – 0.011 and Ba 0.002 – 0.014 mg/L.

Table 1. Baseline physico-chemical properties of surface water in borrow pits

| PARAMETERS | Borrow Pit A | | Borrow Pit B | | CONTROL | |
|---|---|---|---|---|---|---|
| | Dry Season (Mean ± sd) | Wet Season (Mean ± sd) | Dry Season (Mean ± sd) | Wet Season (Mean ± sd) | Dry Season (Mean ± sd) | Wet Season (Mean ± sd) |
| **Physcio-chemical properties** | | | | | | |
| pH | 6.70 ± 0.07 | 7.88 ± 0.07 | 6.48 ± 0.55 | 6.97 ±0.06 | 7.25±0.04 | 6.95±0.05 |
| Temperature ($^{o}$C) | 29.8 ±1.2 | 25.3±0.3 | 30.5 ± 1.4 | 25.7 ± 0.2 | 26.3±0.6 | 25.4±0.4 |
| Conductivity (µS/cm) | 119.67 ± 1.53 | 97.04 ± 36.40 | 27.33 ±16.5 | 23.78±2.08 | 11.18±0.82 | 127±8.23 |
| DO (mg/L) | 6.53± 0.15 | 6.24 ± 0.42 | 8.13 ± 0.31 | 7.32 ± 0.34 | 4.93±0.25 | 6.4±0.45 |
| TDS (mg/L) | 59.40 ± 0.53 | 48.0 ± 17.32 | 14.30 ± 8.56 | 12.2 ± 0.72 | 6.88±1.23 | 64.8±2.54 |
| COD (mg/L) | 8.68 ± 0.59 | 8.33 ± 0.93 | 8.33 ± 0.45 | 6.83 ± 0.46 | 8.7±1.01 | 6.2±1.21 |
| BOD$_5$ (mg/L) | 5.21 ±0.36 | 4.99 ± 0.55 | 5.00 ±0.27 | 4.09 ± 0.28 | 5.22±0.98 | 3.41±0.29 |
| TSS (mg/L) | 14.13± 2.26 | 12.84± 2.27 | 12.37 ± 2.40 | 16.4 ± 4.26 | 3.14±0.44 | 16.4±1.63 |
| **Petroleum Hydrocarbon** | | | | | | |
| TPH (mg/L) | 0.107±0.074 | 0.054±0.037 | 0.087±0.014 | 0.045±0.024 | <0.001 | < 0.001 |
| PAH (mg/L) | 0.009±0.003 | 0.004±0.002 | 0.011±0.004 | 0.003±0.001 | <0.001 | <0.001 |
| **Heavy Metals** | | | | | | |
| Pb (mg/L) | 0.014±0.001 | 0.009±0.00 | 0.005±0.001 | 0.002±0.001 | <0.001 | <0.001 |
| Cu (mg/L) | 0.024±0.001 | 0.015±0.00 | 0.004±0.001 | 0.009±0.001 | <0.002 | <0.002 |
| Zn (mg/L) | 0.015±0.002 | 0.012±0.001 | 0.013±0.001 | 0.022±0.003 | 0.002±0.001 | <0.001 |
| Cr (mg/L) | 0.014±0.005 | 0.007±0.002 | 0.003±0.00 | 0.002±0.001 | <0.001 | <0.001 |
| As (mg/L) | <0.001 | <0.001 | <0.001 | <0.001 | <0.001 | <0.001 |
| Fe (mg/L) | 0.232±0.027 | 0.168±0.031 | 0.243±0.022 | 0.201±0.012 | 0.031 | 0.029 |
| Cd (mg/L) | 0.008±0.001 | 0.006±0.001 | 0.003±0.001 | 0.002±0.001 | <0.002 | <0.002 |
| Ni (mg/L) | 0.009±0.001 | 0.017±0.003 | 0.004±0.001 | 0.005±0.002 | <0.001 | <0.001 |
| Ba (mg/L) | 0.014±0.002 | 0.012±0.002 | 0.004±0.001 | 0.002±0.001 | <0.001 | <0.001 |

## 3.2 Field Bioaccumulation Study

### 3.2.1 Flesh Tainting and Mortality

Taint in fish is evaluated by its flavour (or taste), odour and physical appearance (organoleptic test).Flavour, odour and colour are physical properties and can be imparted to a fish as a result of its immediate surroundings and this can render the fish unmarketable. Flesh tainting of aquatic organism could be used to evaluate the quality of recipient water body.

In this study *hemichromis fasciatus* was tainted. The extent of flesh tainting was moderate in both pits and seasons.

There was mortality of the test organism in all the discharge pits. The percentage mortality in the pits ranged between 2 and 3.8 %, with zero mortality in the control (Figure 3). The highest percentage mortality was observed in pit A during dry season.

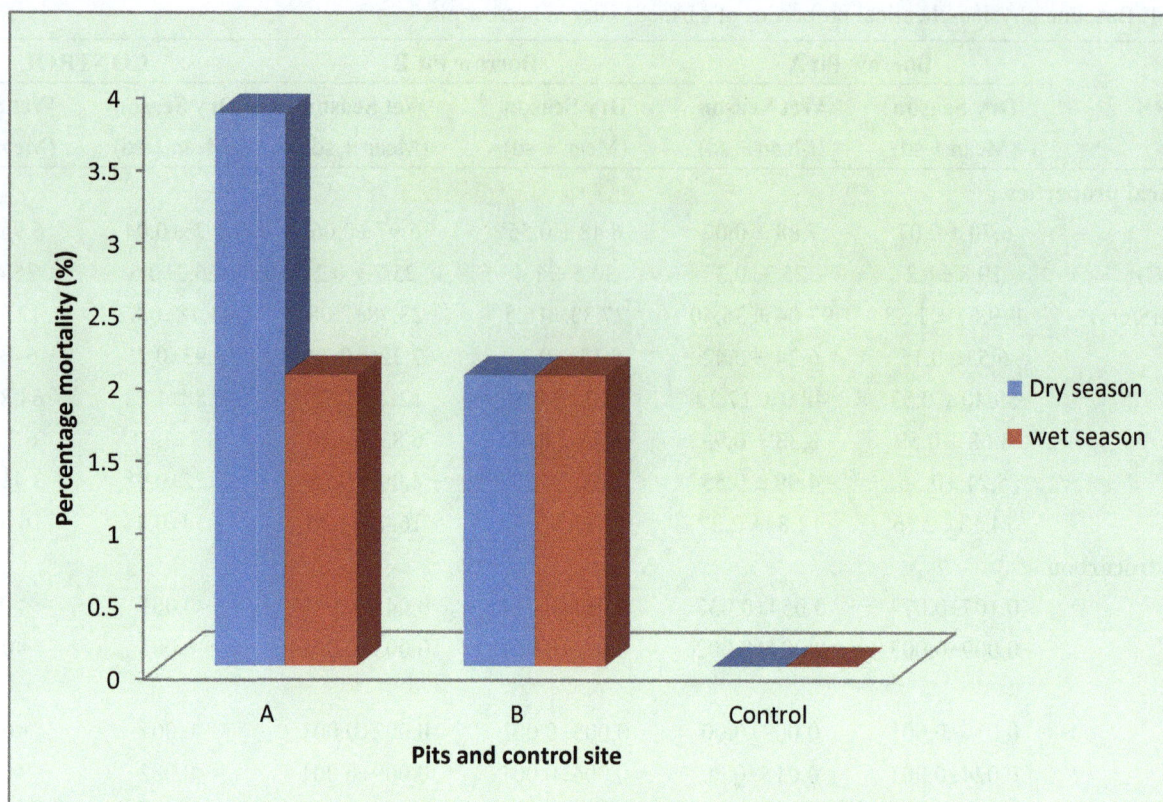

Figure 3. Percentage mortality of test organism in pits and control site

### 3.3 Bioaccumulation of TPH, PAHs and Heavy Metals

The results of bioaccumulation test of *hemichromis fasciatus* for pits A and B and control station for dry and wet seasons are presented in Figure 4. Values obtained are in wet weights (WW).

3.3.1 Hydrocarbon (TPH and PAH)

The TPH content in *hemichromis fasciatus* ranged between $0.218 \pm 0.052$ mg/kg and $0.303 \pm 0.013$ mg/kg in both pits and seasons (Figure 4). The highest bioaccumulation was observed in pit A, during the dry season, while the least was in the wet season in pit B. The mean concentrations for the control location were $0.009 \pm 0.003$ mg/kg and $0.010 \pm 0.004$ mg/kg for wet and dry seasons respectively.

The mean concentration of TPH in the pits for both seasons showed significant increase when compared to values obtained for the control samples. This suggested that there was bioaccumulation of TPH in *hemichromics fasciatus* in pits.

It is pertinent to note that the level of TPH determined in whole fish samples is higher than the recorded for the surface water samples in the pits. *Hemichromis fasciatus* is benthopelagic in nature and can easily contact the sedimentary environment because of its bottom dwelling character. Also, concentrations were relatively higher in the dry season, when compared to the wet season. Davies *et al.*, 2006 reported that many aquatic organisms have the ability to accumulate and bio-magnify contaminants in the environment. Polycyclic aromatic hydrocarbons (PAHs) mean concentration in pit A were $0.033 \pm 0.013$ mg/kg and $0.032 \pm 0.018$ mg/kg during the dry and wet seasons respectively, while pit B recorded an accumulation of $0.029 \pm 0.026$ mg/kg and $0.023 \pm 0.009$ mg/kg for dry season and wet seasons respectively.

Animal studies have shown that TPH have some toxicological effect on the blood, immune system, liver, spleen, kidneys, lungs, reproductive system and even developing foetus (USEPA 2000).

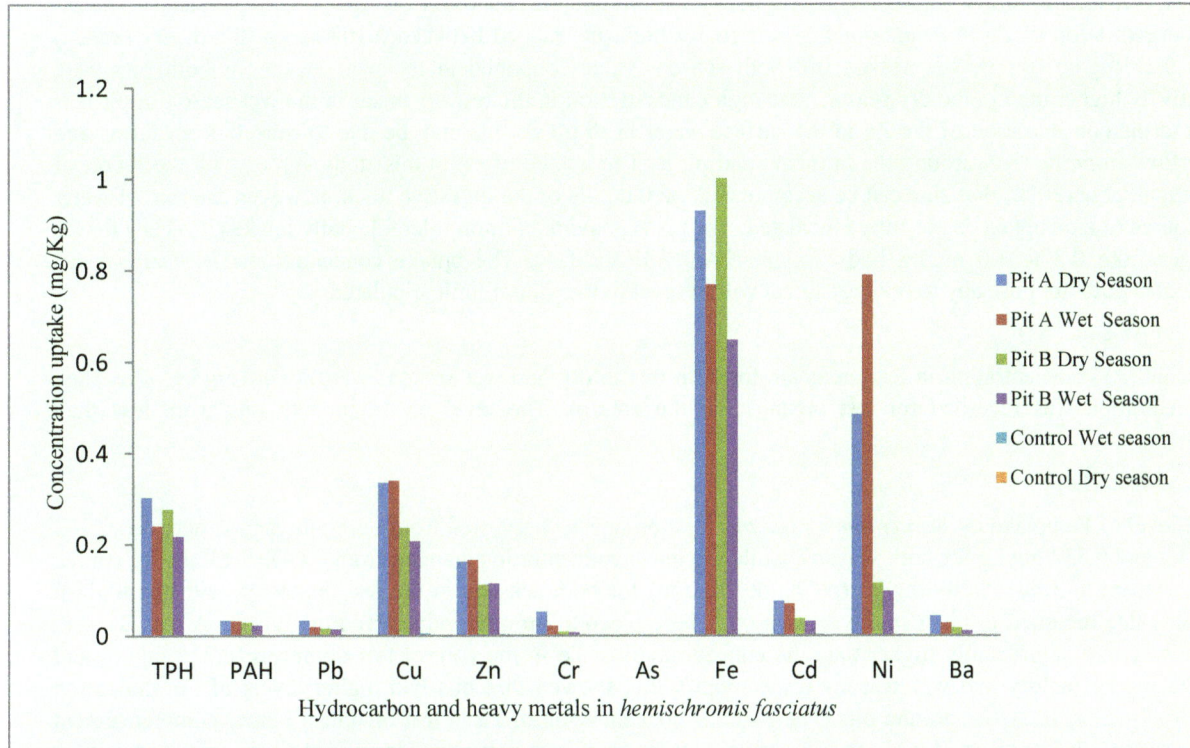

Figure 4. Bioconcentration of hydrocarbons and heavy metals in *hemischromis fasciatus* over 14 days exposure

### 3.3.2 Heavy Metals

### 3.3.2.1 Lead (Pb)

The concentration of Pb in *hemichromis fascistus* for pits A and B ranged from 0.004±0.002 mg/kg to 0.033 ± 0.014 mg/kg during both seasons. The highest value was obtained in the season in pit A during the dry season.

The level of Pb in test organism before the test was <0.001mg/kg. These slight increase in the test organism indicated minimal bioaccumulation. The low Pb bioaccumulation values recorded for *hemichromis fasciatus* tend to agree with the findings of Wong *et al.* (1978) and Eisler (1988b) - that aquatic organisms such as fish can bioconcentrate Pb from water, but does not bioacumulate and tends to decrease with increasing trophic levels in freshwater habitat. The mean dietary exposure of Pb to children (1 – 4 years) and adult are 0.03 – 9.0 µg/kg/day and 0.02 – 3.0 µg/kg/day respectively.

### 3.3.2.2 Copper (Cu)

The concentration of Cu in test organism (*hemichromis fasciatus*) during the dry season were 0.335±0.048 mg/kg (pit A) and 0.237±0.039 mg/kg (pit B), while the wet season recorded 0.339 ±0.047 mg/kg (pit A) and 0.207 ± 0.016 mg/kg (pit B). The highest value was recorded in pit A during the wet season and may have resulted from run-off/wash-off due to E&P operations. The mean concentrations of Cu in "clean" test organism was 0.003±0.001mg/kg (baseline concentration), thus giving a net up-takeof 0.333 mg/kg (dry season) and 0.337 mg/kg (wet season) in pit A and 0.235 mg/kg (dry season) and 0.205 mg/kg (wet season) for pit B, while the control recorded 0.005±0.002 mg/kg and <0.002 mg/kg for dry and wet seasons respectively. These increases in the test environments compared to values obtained in the control depicted bioconcentration of Cu by the test organism. Copper has been reported to bioaccumulate in many different organs in fish and mollusks (Owen 1981).

The provisional maximum tolerable daily intake (PMTDI) of Cu is 0.5 mg/kg body weight (FAO/WHO, 2011). For an average body weight of 60 kg, the maximum daily intake for an adult human is 3.0mg Cu. To exceed this toxicological limit over 8.85kg of *hemichromic fasciatus* has to be consumed per day.

### 3.3.2.3 Zinc (Zn)

The concentration of Zn in the test organism was 0.162 ± 0.028 mg/kg (pit A, dry season), 0.165 ± 0.029 mg/kg (pit A, wet season), 0.111 ±0.005 mg/kg (pit B, dry season) and 0.214 ± 0.011 mg/kg (pit B, wet season). The net bioconcentration of Zn in *hemichromis fasciatus* for both pits ranged between 0.108mg/kg (pit B, dry season) and 0.211mg/kg (pit B, wet season) for both seasons. Values obtained in the wet season for both pits were relatively higher than in the dry season. The high concentration in the test organism in the wet season in pit B is due to the concentration of the Zn in the surface water in th pit B, this may be due to run-off from farms and overflow from the river around the farmland and pit B. The uptake of Zn in this study agrees with the report of Murugan *et al* (2008) that zinc can be accumulated via the gills or the digestive track, however, the role of water as source of zinc uptake is not fully elucidated. The provisional maximum tolerable daily intake (PMTDI) for Zn ranges from 0.3 to 1.0 mg/kg body weight (FAO/WHO, 2011). The uptake concentrations in *hemichromis fasciatus* does not pose any toxicology threat with respect to maximum limit stipulated.

### 3.3.2.4 Arsenic (As)

Arsenic (As) concentration in test organism for both pits in dry and wet seasons were <0.001mg/kg. The same concentration was recorded for test organism at the control. The level of As in both pits were less than 0.001mg/l.

### 3.3.2.5 Iron (Fe)

The level of Fe uptake by *hemichromis fasciatus* (test organism) harvested from both pits ranged between 0.384 mg/kg and 0.737 mg/kg for both seasons, while the mean concentrations ranged from 0.647±0.152mg/kg (pit B, wet season) to 1.001±0.209 mg/kg (pit A, dry season) for both seasons in the test organism, with the highest mean value recorded in pit A in the dry season. These bioconcentration values recorded in pits A and B, were relatively and significantly higher than the concentration of Fe in the control test organism (0.236 mg/kg and 0.242 mg/kg for dry and wet seasons respectively), this showed that pits had higher levels of Fe, indication anthropological activities around pits. The baseline concentration of Fe in test organism before commencement of test was 0.263 mg/kg and when this is compared to levels in test organism, significant decrease was observed. This significant decrease connoted depuration or loss of Fe from *hemichromis fasciatus* (test organism) in the control environment. Iron is an essential trace metal required by all forms of life, especially in the synthesis of haem protein and many enzyme systems. The FAO/WHO average daily intake for male and female are 17 mg/day and 9 -12 mg/day respectively.

### 3.3.2.6 Chromium (Cr)

The bioconcentration of Cr in *hemichromis fasciatus* during both seasons ranged between 0.007±0.001mg/kg and 0.052±0.004 mg/kg in both pits. The highest mean concentration was recorded at pit A in the dry season. Uptake concentrations in the dry season were higher than for the wet season.

### 3.3.2.7 Cadmium (Cd)

The mean concentration uptake of Cd in test organism *(Hemichromis fasciatus)* for dry and wet seasons was 0.076 ± 0.008 mg/kg (pit A, dry season), 0.070 ± 0.005 mg/kg (pit A, wet season), 0.038 ± 0.008 mg/kg (pit B, dry season) and 0.032 ± 0.007 mg/kg (pit B, wet season). Cadmium is bioaccumulated at all trophic levels, especially in the livers and kidneys of fish (Sindayigaya, *et al.* 1994; Sadiq 1992). The provisional tolerable monthly intake (PTMI) for Cd is 0.025mg/kg body weight (ie 1.5 mg/adult of an average body weight of 60 kg) (FAO/WHO 2010). Cadmium levels in test organism were significantly below the PTMI stipulated by the JECFA of FAO/WHO.

### 3.3.2.8 Nickel (Ni)

The concentration of Ni in the test organism (*hemichromis fasciatus*) in the pits A&B ranged between 0.097 ± 0.007mg/kg and 0.789 ± 0.0039 mg/kg for both seasons. The highest uptake was observed in the wet season in pit A, while the least mean concentration was obtained in pit B during the wet season.

There were significant changes in the concentration ranges obtained in this study during uptake, when compared to base level concentration of Ni in *hemichromis fasicatus* (<0.001mg/kg) and levels determined from the control site. This indicated bioaccumulation. Javid*et al.,* (2000) reported the bioaccumulation of Ni in three species of fish - *Catlacatla, Labeorohita* and *Cirrhinamrigala* , with bioaccumulation concentration of 10.12 ±1.27 mg/Kg, 14.41 ±2.24 mg/Kg and 13.46 ±3.39 mg/Kg respectively, on exposure to 0.05 mg/L of Ni for 96 hours. These values were significantly higher when compared to values obtained in this study for *hemichromicfasciatus*.

### 3.3.2.9 Barium (Ba)

The concentration and net uptake levels of Ba in the test organism (*hemichromis fasciatus*) in the dry and wet season were 0.043 mg/kg (pit A, dry season), 0.028 mg/kg (pit A, wet season), 0.017 mg/kg (pit B, dry season,) and 0.011 mg/kg (pit B, wet season).

The concentration of Ba in *hemichromis fasciatus* for the control was <0.001mg/kg for wet and dry seasons, this was relatively and significantly lower to values obtained in the test organisms from both pits, thus indicating bioaccumulation. Previous studies have shown that Ba does not bioaccumulate easily and concentrations in higher species rarely exceed 10 mg/kg (Moore 1991).

*3.4 Bioaccumulation Factor and Rate Constant*

3.4.1 Uptake Rate Constant

The uptake rate constants for each of the heavy metals and hydrocarbons (TPH and PAHs) were calculated, assuming first-order kinetics.

$$C_t = C_{equil}(1 - e^{-kt})$$

Where $C_t$ and $C_{equil}$ are concentrations at time t (d) and steady state respectively, k is the rate constant ($d^{-1}$).

The uptake rate constants for TPH, PAHs and heavy metals in *hemichromis fasciatus* in this study were obtained applying the field kinetic model and also assuming a steady-state after 14 days of exposure.

The up-take rate constants are presented on Table 2 for TPH, PAHs and the heavy metals.

The up-take rate constant *k*, of TPH ranged from $38.47 \times 10^{-2}$ $d^{-1}$ (pit B, wet season) to $40.82 \times 10^{-2}$ $d^{-1}$ (pit A, dry season) for both pits and seasons, while PAHs ranged between $22.40 \times 10^{-2}$ $d^{-1}$ (pit B, wet season) and $24.98 \times 10^{-2}$ $d^{-1}$ (pit A, dry season).

Uptake of TPH and PAHs seemed slightly higher in the dry season than wet season in both pits. The higher concentrations of both organics in the dry season may have resulted to higher uptake rate in the dry season and the order of bioaccumulation rate was TPH>PAHs in both pits and seasons. In aquatic system, uptake and elimination of contaminants ceases when a steady-state or equilibrium is reached between test organism and surrounding media.

Amongst the metals, the uptake rate constants for water borne metals by *hemichromis fasciatus* ranged from $4.66 \times 10^{-2}$ $d^{-1}$ to $47.66 \times 10^{-2}$ $d^{-1}$ and $2.70 \times 10^{-2}$ $d^{-1}$ to $33.90 \times 10^{-2}$ $d^{-1}$ in pits A and B respectively during both seasons. The highest and least uptake rates were recorded in both pits for Ni and Fe respectively for both seasons. The order of rate of uptake constant was Fe <Pb< Cr < Ba < Cd < Zn < Cu < Ni, while As (arsenic) recorded a zero uptake. The detection limit of instrument was <0.001 mg/l.

Table 2. Concentration bioaccumulation (up-take), Bioaccumulation Factor and rate constant of TPH, PAH and heavy metals in *Hemichromics fasciatus* after 14 days exposure to surface water in borrow pits

| | Concentration Up-take (mg/Kg)(ww) | | | | Rate Constant (k)(d⁻¹) | | | | Bioaccumulation Factor (BCF) (Lkg⁻¹) | | | |
|---|---|---|---|---|---|---|---|---|---|---|---|---|
| | Borrow Pit A | | Borrow Pit B | | Borrow Pit A | | Borrow Pit B | | Borrow Pit A | | Borrow Pit B | |
| | Dry season | Wet season | Dry season | Wet season | Dry season | Wet season | Dry season | Wet season | Dry season | Wet season | Dry season | Wet season |
| **Petroleum Hydrocarbon** | | | | | | | | | | | | |
| TPH | 0.303 | 0.240 | 0.277 | 0.218 | $40.82\times10^{-2}$ | $39.15\times10^{-2}$ | $40.18\times10^{-2}$ | $38.47\times10^{-2}$ | 2.832 | 4.444 | 3.184 | 4.844 |
| PAH | 0.033 | 0.032 | 0.029 | 0.023 | $24.98\times10^{-2}$ | $24.76\times10^{-2}$ | $24.06\times10^{-2}$ | $22.40\times10^{-2}$ | 3.667 | 8 | 2.636 | 7.667 |
| **Heavy Metals** | | | | | | | | | | | | |
| Pb | 0.033 | 0.019 | 0.011 | 0.004 | $28.87\times10^{-2}$ | $21.04\times10^{-2}$ | $12.80\times10^{-2}$ | $9.90\times10^{-2}$ | 2.357 | 0.211 | 2.2 | 2 |
| Cu | 0.333 | 0.337 | 0.235 | 0.205 | $33.65\times10^{-2}$ | $33.73\times10^{-2}$ | $31.55\times10^{-2}$ | $30.18\times10^{-2}$ | 13.878 | 22.467 | 16.786 | 22.778 |
| Zn | 0.159 | 0.162 | 0.108 | 0.211 | $31.26\times10^{-2}$ | $31.40\times10^{-2}$ | $33.28\times10^{-2}$ | $28.69\times10^{-2}$ | 10.60 | 13.5 | 8.308 | 9.591 |
| Cr | 0.052 | 0.022 | 0.009 | 0.007 | $28.23\times10^{-2}$ | $22.08\times10^{-2}$ | $15.70\times10^{-2}$ | $13.90\times10^{-2}$ | 3.714 | 3.143 | 3 | 3.5 |
| As | <0.001 | <0.001 | <0.001 | <0.001 | - | - | - | - | - | - | - | - |
| Fe | 0.667 | 0.504 | 0.737 | 0.384 | $6.65\times10^{-2}$ | $4.66\times10^{-2}$ | $7.37\times10^{-2}$ | $2.70\times10^{-2}$ | 2.875 | 3.006 | 3.037 | 1.911 |
| Cd | 0.076 | 0.07 | 0.038 | 0.032 | $30.94\times10^{-2}$ | $30.35\times10^{-2}$ | $25.99\times10^{-2}$ | $24.76\times10^{-2}$ | 9.5 | 11.667 | 12.667 | 16 |
| Ni | 0.484 | 0.789 | 0.115 | 0.097 | $44.17\times10^{-2}$ | $47.66\times10^{-2}$ | $33.90\times10^{-2}$ | $32.68\times10^{-2}$ | 53.778 | 71.727 | 28.750 | 19.4 |
| Ba | 0.043 | 0.028 | 0.017 | 0.011 | $26.87\times10^{-2}$ | $23.81\times10^{-2}$ | $20.24\times10^{-2}$ | $17.13\times10^{-2}$ | 3.071 | 2.333 | 4.25 | 5.5 |

At steady-state, Fe had the lowest uptake rate constants for both seasons and pits; rate constants were $6.65\times10^{-2}$ d⁻¹ (pit A, dry season), $4.66\times10^{-2}$ d⁻¹ (pit A, wet season), $7.37\times10^{-2}$ d⁻¹ (pit B, dry season) and $2.70\times10^{-2}$ d⁻¹ (pit B, wet season). This relatively low rate of uptake constants for Fe by *hemichromis fasciatus* is likely due to its initial high concentration in test organism (0.263 mg/kgFe) and as well as in the surface water of the pits (0.201 – 0.243 mg/kgFe), this situation may have resulted to a gradual attainment of equilibrium between the test organism and the surrounding medium (surface water in pits), thereby leading to slow uptake rate for Fe over a period of 14 days.

3.4.2 Bioaccumulation Factor

The degree to which bioaccumulation occurs is normally expressed as bioaccumulation or bioconcentration factor (BAF/BCF). In an aquatic environment, BAF/BCF quantitatively describes bioaccumulation in fish or aquatic organism and it is defined as the field observed ratio of the concentration of a given chemical in biota to the concentration in corresponding water (Mackay & Fraser, 2000) or as the dimensionless ratio of wet-weight contaminant concentration in fish or aquatic organism ($C_F$), to the corresponding contaminant concentration in the water ($C_W$). It also describes the equilibrium reached between uptake and depuration of contaminant by fish and it is the ratio of respective rate constants for those processes.

$$BCF = C_F \times C_W^{-1}$$

The BCF calculated in this study is based on a steady state system at day 14.

The BCF values (dry and wet seasons) for hydrocarbons and heavy metals are presented in Table 3.

BCF values calculated for *Hemichromics fasciatus* in both pits and seasons ranged from 2.832 LKg⁻¹ to 4.844 LKg⁻¹ for TPH and 2.636LKg⁻¹ to 8.0 LKg⁻¹ for PAH. These values are significantly below the OEHHA recommended BCF defaults of 583 for PAH in fish. Several studies have reported BCFs in fish for benzo[a]pyrene as PAH. Species such as bluegill sunfish have produced BCF values ranging between 224 and 490 (Spacie *et al.*, 1983, McCarthy and Jimenez, 1985) for PAH. Balk *et al.*, 1984, reported BCF ranging from 50 to 80000 in internal organs, with the greatest bioconcentration occurring in the gallbladder and bile on exposing Northern pike fish for 10 or 21 days to benzo[a]pyrene in water.

Amongst the metals, BCF values for Pb, Cd and Cr ranged from 0.211 - 2.357 LKg⁻¹, 9.50 – 12.667LKg⁻¹ and 3.0 – 3.714 LKg⁻¹ respectively in both pits and seasons. From previous study, Taylor (1983), reported that vertebrate fish species that were exposed to Cd have shown a BCF of <20. While Fish BCFs ranging from 5.1 to 300 have been reported for lead (SCAQMD., 1988). The arithmetic mean of these values (155) is recommended as the default BCF for lead (OEHHA, 2000). Also, BCF values ranging from 22 to 200 have been reported for Cd in brook trout fish (US EPA., 1985 and Atchison *et al.*, 1977), while for Cr, BCF values ranging from 1-3.4 have been reported (OEHHA, 2000). The OEHHA recommended default BCF for Cd and Cr are 366 and 2 respectively. Other BCF values in this study ranged as follows; Cu, 13.878 – 22.778LKg$^{-1}$; Zn 8.308 – 13.50 LKg$^{-1}$, Fe 1.911 – 3.037 LKg$^{-1}$, Ni 19.4 – 71.727 LKg$^{-1}$ and Ba 2.333 – 5.5LKg$^{-1}$. In the bioaccumulation studies of some heavy metals in *oreochromis nilotycus* BCF ranges of 38 – 56, 179 – 199 and 21 – 24 were reported by Noegrohati (2006) for Cu, Zn and Cd respectively. These values are comparable with values obtained in this study for *hemichromics fasciatus.*

It has been reported that chemical substances having a BCF or BAF >1000 are characterized by a tendency to accumulate in organisms (Moss and Boethling, 1999), however, in this study, the highest BCF value was 71.727 LKg$^{-1}$ for Ni (pit A and wet season), this was about 14-fold below the stipulated value of 1000.

### 3.5 Seasonal Variation

As shown in Table 1, concentrations of analytes were generally higher in the dry season than the wet season in both pits, except for Ni (pit A) and Zn (pit B), however these changes were considered not very significant or drastic. Relatively, lower concentrations in the wet season may have been due to dilutions from rainfall, especially for the metal analytes that are waterborne in nature. The increases recorded for the wet season may have resulted from ruff-off and washout from some the oil exploration and production facilities and farmlands where fertilizers were applied within this location. Besides, there is the possibility of an overflow from the stream close to the pit A, as often witnessed at periods when rainfall is high. However, these seasonal changes in concentrations of analytes were distinctly expressed in the level of Ni uptake in *hemichromics fasciatus* (Figure 4) during the rainy season. There were no sharp changes in the up-take rate constants between the seasons in each pit (Table 2), as changes were considered insignificant.

Finally, there were strong positive correlations between dry and wet seasons (Figure 5). BCF values obtained for the xenobiotics showed strong seasonal correlation between dry and wet seasons for both pits and between pits. Correlation coefficient ranged between 0.6743– 0.9945 (p <0.005).

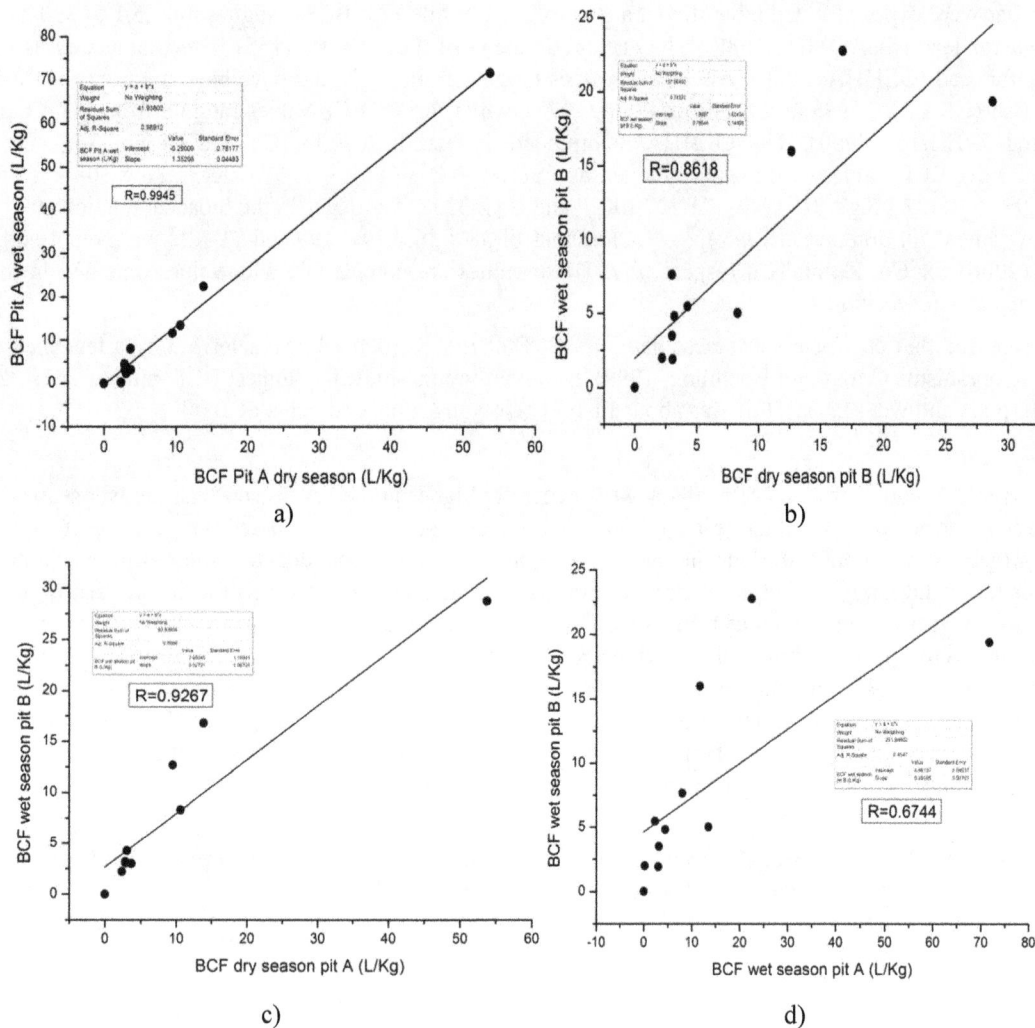

Figure 5. (a) Seasonal correlation between BCF dry and wet seasons for pit A (b) Seasonal correlation between BCF dry and wet seasons for pit B (c) Correlation between pits A and B for dry season (d) Correlation between BCF pits A and B for wet season

## 4. Conclusion

The results presented from this study clearly demonstrated that the borrow pits around the Niger Delta region are contaminated and aquatic organism in such water bodies are also contaminated. *Hemichromics fasciatus* exhibited greater uptake of analytes in the dry season than the wet season and the amount of heavy metals accumulated were all below the provisional maximum tolerable daily intake (PMTDI) as stipulated by the JECFA of FAO/WHO. The toxicokinetic parameters obtained for both pits and season in this study does not pose any toxicological threat to man if fish from such pits are consumed.

### Acknowledgements

The map showing the onshore active fields of the study area was obtained from factsbook.eni.com/node/494, the authors therefore wish to acknowledge Eni Petroleum Nigeria Limited owners of this site. We also thank the management of Technology Partners International (Nig.) Limited for the use of their laboratory facilities.

### References

Ajayi, S. O., & Osibanjo, O. J (1981). Pollution studies in Nigerian Rivers, II. Water Quality of some Nigerian rivers. *Environ. Pollut.,* (Series B)2, 87-95.

Alloway, B. J., & Ayres, D. C. (1997). *Chemical principles of environmental pollution* (2nd ed., pp. 200-240). Chapman and Hall, London: Blackie Academic and professional.

American Society for Testing and Materials. (1988). Standard Practice for conducting Bioconcentration Tests with Fishes and Saltwater Bivalve Molluscs. ASTM E-1022-84.

Anyakora, C., & Coker, H. (2007).Assessment of polynuclear aromatic hydrocarbon content in four species of fish in the Niger Delta by gas chromatography/mass spectrometry. *African Journal of Biotechnology, 6*(6), 737-743.

Anyakora, C., Ogbeche, A., Palmer, P., & Coker, H. (2005). Determination of polynuclear aromatic hydrocarbons in marine samples of Siokolo Fishing Settlement. *Journal of Chromatography A, 1073*(1-2), 323-330 http://dx.doi.org/10.1016/j.chroma.2004.10.014

Atchison, G. J., Murphy, B. R., Bishop, W. E., McIntosh, A. W., & Mayes, R. A. (1977). Trace metal contamination of bluegill (*Lepomis macrochirus*) from two Indiana lakes. *Trans Am Fish Soc, 106*, 637-640. http://dx.doi.org/10.1577/1548-8659(1977)106<637:TMCOBL>2.0.CO;2

Avwiri, G. O., & Ononugbo, C. P. (2010). Assessment of the naturally occurring radioactive material (norm) content of hydrocarbon exploration and production activities in Ogba/Egbema/Ndoni Oil/gas field, Rivers state, Nigeria. Proceedings of the 1st International Technology, Education and Environment Conference (c) African Society for Scientific Research (ASSR).

Balk, L., Meijer, J., DePierre, J. W., & Appelgren, L. E. (1984). The uptake and distribution of [3H]benzo[a]pyrene in the Northern pike (Esox Lucius). Examination by whole-body autoradiography and scintillation counting. *Toxicol. Appl. Pharmacol, 74*, 430-449. http://dx.doi.org/10.1016/0041-008X(84)90296-5

Biney, C., Amazu, A. T., Calamari, D., Mbome, I. L., & Naeve, H. (1994). Review of Heavy cadmium exposure. *Aquat.Toxicol. 91*, 291-301

Crane, M., Burton, G. A., Culp, J. M., Greenberg, M. S., Munkittrick, K. R., Ribeiro, R., Salazar, M. H., & St-Jean, S. D. (2007). Review of aquatic in situ approaches for stressor and effect diagnosis. *Integr. Environ. Assess. Manag. 3*, 234-245. http://dx.doi.org/10.1897/IEAM_2006-027.1

Davies, O. A., Allison, M. E., & Uyi, H. S. (2006). Bioaccumulation of heavy metals in water, sediment and perinwinkle (Tympanotonusfuscatusvar radula) from the Elechi creek, Niger Delta. *African Journal of Biotechnology, 5*(10), 968-973.

EGASPIN (Environmental Guidelines and Standards for Petroleum Industry in Nigeria), Revised Edition, 2002.

EPA. (1980). Ambient water quality criteria for lead. U.S. Environ. Protection Agency Rep. 440/5-80-057.151pp. Avail.from Natl. Tech. Infor. Serv., 5285 Port Royal Road, Springfield, Virginia 22161.

Ezekiel, E. N., Abowei J. F. N., & Charles, E. (2013). Effects of Flooding on Amassoma Flood Plain Sediments Niger Delta, Nigeria. *App. Sci. Report, 4*(1), 173-180.

FAO/WHO. (2010). Joint FAO/WHO Expert Committee on Food Additives (JECFA) Seventy-third meeting Geneva, 8-17 June 2010.

FAO/WHO. (2011). Joint food standards programme codex committee on contaminants in foods Fifth Session. The Hague, the Netherlands, 21 - 25 March 2011.

Ghosh, T. K., & Kshirsagar, D. G. (1973). Selected heavy metals in seven species of fishes from Bombay offshore areas. *Proc. Nat. Acad. Sci. India, 63*(B III), 350-311.

Godwin, J., Vaikosen, N. E., Njoku, C. J., & Sebye, J. (2011). Evaluation of some heavy metals in tilapia nicolitica found in selected rivers in Bayelsa state. *EJEAFChe, 10*(7), 2451-2459.

Goksøyr, A., Beyer, J., Husoy, A. M., Larsen, H. E., Westrheim, K., Wilhelmsen, S., & Klungsoyr, J. (1994). Accumulation and effects of aromatic and chlorinated hydrocarbons in juvenile Atlantic cod (Gadus morhua) caged in a polluted fjord (Sorfjorden, Norway). *Aquat Toxicol, 29*, 21-35. http://dx.doi.org/10.1016/0166-445X(94)90045-0

James, R., Sampath, K., & Selvamani, P. (1998). Effect of EDTA on reduction of Copper toxicity in Oreochromismossambicus (Peters). *Bull. Environ. Contam. Toxicol., 60*, 487-449 http://dx.doi.org/10.1007/s001289900651

Jardine, T. D., MacLatchy, D. L., Fairchild, W. L., Chaput, G., & Brown, S. B. (2005). Development of a short-term in situ caging methodology to assess long-term effects of industrial and municipal discharges on salmon smolts. *Ecotoxicol. Environ. Saf., 62*, 331-340. http://dx.doi.org/10.1016/j.ecoenv.2004.12.006

Kamunde, C. (2009). Early subcellular partitioning of cadmium in gill and liver of rainbow trout

(Oncorhynchusmykiss) following low-to-near-lethal waterborne cadmium exposure. *Aquat. Toxicol., 91,* 291-301. http://dx.doi.org/10.1016/j.aquatox.2008.10.013

Karadede, H., Oymak, S. A., & Unlu, E. (2004). Heavy metals in mullet, Liza abu, and catfish, Silurustriostegus, from the Ataturk Dam Lake (Euphrates), *Turkey. Environ. Int., 30,* 183-188. http://dx.doi.org/10.1016/S0160-4120(03)00169-7

Kotze, P., Dupreez, H. H., & Van Vuren J. H. (1999). Bioaccumulation of Copper and Zinc in Oreochromis Mossambicus and Clarias Garipimus from the Olifants Rivers, Mpumalanga, South Africa. *Water SA,* 25(1), 99-110.

Landrum, P. F., Guilherme, R., Lotufo, G. R., Duane, C., Gossiaux, D. C., Michelle, L., Gedeon, M. L., & Lee, J.-H. (2003). Bioaccumulation and critical body residue of PAHs in the amphipod, Diporeia spp.: additional evidence to support toxicity additivity for PAH mixtures. *Chemosphere, 51,* 481-489. http://dx.doi.org/10.1016/S0045-6535(02)00863-9

Mackay. D., & Fraser, A. (2000) Kenneth Mellanby Review Award. Bioaccumulation of persistent organic chemicals: mechanisms and models. *Environ Pollut, 110,* 375-391. http://dx.doi.org/10.1016/S0269-7491(00)00162-7

Martin, C. A., Luoma, S. N., Cain, D. J., & Buchwalter, D. B. (2007). Cadmium ecophysiology in seven stonefly (Plecoptera) species: delineating sources and estimating susceptibility. *Environ. Sci. Technol. 41,* 7171-7177. http://dx.doi.org/10.1021/es071205b

Martinez-Lopez, E., Maria-Mojica, P., Martinez J. E., Carlvo, J. F., Romero, D., Garcia-Fernandez, A. J. (2005). Cadmium in feathers of adults and blood of nestlings of three raptor species from a non-polluted Mediterranean forest, southeastern Spain. *Bull. Environ. Contam. Toxicol., 74,* 477. http://dx.doi.org/10.1007/s00128-005-0610-6

McCarthy, J. F., & Jimenez, B. D. (1985). Reduction in bioavailability to bluegills of polycyclic aromatic hydrocarbons bound to dissolved humic material. *Environ. Toxicol. & Pharma., 4,* 511-521.

Moore, J. W. (1991). *Inorganic Contaminants of Surface Waters, Research and Monitoring Priorities.* Springer-Verlag, New York. http://dx.doi.org/10.1007/978-1-4612-3004-5

Moss, K., & Boethling. (1999). EPA's New Chemicals Program PBT Chemical Category. 401M St, SW Mailcode 7405, Washington, DC, 20460, US.

Murugan, S. S., Karuppasamy, R., Poongodi, K., & Puvaneswari, S. (2008). Bioaccumulation pattern of zinc in fresh water fish Channapunctatus (Bloch) after chronic exposure. *Turkish Journal of Fisheries and Aquatic Sciences, 8,* 55-59.

Nduka, J. K., & Orisakwe, O. E. (2010). Water quality issues in the Niger Delta of Niger: Polyaromatic and straight chain hydrocarbons in some selected surface water. *Water Qual. Expo. Health, 2,* 65-74.

Nwabueze, A. A. (2011). Levels of Some Heavy Metals in Tissues of Bonga Fish, *Ethmallosa fimbriata* (Bowdich, 1825) from Forcados River. *J. Appl. Environ. Biol. Sci., 1*(3), 44-47.

OEHHA: Air Toxics Hot Spots Program Risk Assessment Guidelines. Part IV. Technical Support Document for Exposure Assessment and Stochastic Analysis. (2000). Appendix.

Oikari, A. (2006). Caging techniques for field exposures of fish to chemical contaminants. *Aquat Toxicol, 78,* 370-381. http://dx.doi.org/10.1016/j.aquatox.2006.03.010

Olowoyo, D. N., Ajayi, O. O., Amoo, I. A., & Ayeisanmi, A. F. (2010). Seasonal variation of metal concentrations in catfish, blue crab and crayfish from Warri Coastal Water of Delta State, Nigeria. *Pakistan Journal of Nutrition, 9*(11), 1118-1121. http://dx.doi.org/10.3923/pjn.2010.1118.1121

Owen, C. A. (1981). *Copper deficiency and toxicity: acquired and inherited, in plants, animals, and man.* Noyes Publications, New Jersey. Priorities. Springer-Verlag, New York.

Sadiq, M. (1992). Toxic metal chemistry in marine environments. Marcel Dekker. New York.some Nigerian rivers. *Environ. Pollut.* (Series B)2, 87-95.

Sindayigaya, E., Cauwenbergh, R. V., Robberecht, H., & Deelstra, H. (1994). Copper, zinc, manganese, iron, lead, cadmium, mercury, and arsenic in fish from Lake Tanganyika, Burundi. *The Science of the Total Environment, 144,* 103-115. http://dx.doi.org/10.1016/0048-9697(94)90431-6

South Coast Air Quality Management District (SCAQMD). (1988). Multi-pathway health risk assessment input parameters guidance document.

Spacie, A., & Hamelink, J. L. (1982). Alternative models for describing the bioconcentration of organics in fish. *Environ. Toxicol. Chem., 1*, 309-320. http://dx.doi.org/10.1002/etc.5620010406

Sri Noegrohati. (2006). Bioaccumulation dynamic of heavy metals in Oreochromis Nilotycus. *MakalahDiterimatangal 2 maret.*

Stegeman, J. J., & Hahn, M. E. (1994). Biochemistry and molecular biology of monooxygenase: current perspective on forms, functions, and regulation of cytochrome P450 in aquatic species. In D. C. Malins, & G. K. Ostrander (Eds.), *Aquatic toxicology; Molecular, Biochemical and Cellular Perspectives* (pp. 87-206). Lewis Publishers, CRC press, Boca Raton.

Taylor, D. (1983). The significance of the accumulation of cadmium by aquatic organisms. *Ecotoxicol Environ Safety, 7*, 33-42. http://dx.doi.org/10.1016/0147-6513(83)90046-5

Timothy, D. J., Deborah, L. M., Wayne, L. F., Gerald, C., & Scott, B. B. (2005). Development of a short-term in situ caging methodology to assess long-term effects of industrial and municipal discharges on salmon smolts. *Ecotoxicol Environ Saf., 62*, 331-340. http://dx.doi.org/10.1016/j.ecoenv.2004.12.006

Tulasi, S. J., Reddy, P. U. M., & Ramano RAO, J. V. (1989). Effects of Lead on the spawning potential of the fresh water fish. *Anabas testudineus. Bull. Environ. Contam. Toxicol., 43*, 858-863. http://dx.doi.org/10.1007/BF01702056

U.S. EPA. (1993). Wildlife Exposure Factors Handbook. vol. I. EPA/600/R-93/187a.

US EPA. (1985). United States Environmental Protection Agency. Ambient water quality criteria for cadmium - 1984. Office of Water Regulations and Standards , Criteria and Standards Division. Washington, DC. EPA 440/5-84-032.

USEPA – Interpretation for the purpose of sediment Quality Assessment Status and need. (2000). Bioaccumulation testing and Bioaccumulation Analysis Workshop, Washington DC, 20460.

van der Oost, R., Jonny Beyer, J., & Vermeulen, N. P. E. (2003). Fish bioaccumulation and biomarkers in environmental risk assessment: a review. *Environ. Toxicol. Pharm., 13*, 57-149. http://dx.doi.org/10.1016/S1382-6689(02)00126-6

Wang, W. X., & Rainbow, P. S. (2006). Subcellular partitioning and the prediction of cadmium toxicity to aquatic organisms. *Environ. Chem., 3*, 395-399. http://dx.doi.org/10.1071/EN06055

Wong, P. T. S., Silverberg, B. A., Chau, Y. K., & Hodson. P. V. (1978). Lead and the aquatic biota. In J. O. Nriagu (Ed.), *The biogeochemistry of lead in the environment* (pp. 279-342). Part B. Biological effects. Elsevier/North Holland Biomedical Press, Amsterdam.

Zhou, R. B., Zhu, L. Z., & Kong, Q. X. (2007). Persistent Chlorinated Pesticides in fish species from Qiantang River in East China. *Chemosphere, 68*, 838-847. http://dx.doi.org/10.1016/j.chemosphere.2007.02.021

# 5

# On-site Sanitation Influence on Nitrate Occurrence in the Shallow Groundwater of Mahitsy City, Analamanga Region, Madagascar

Mamiseheno Rasolofonirina[1], Voahirana Ramaroson[1,2] & Raoelina Andriambololona[3]

[1] Department of Physics, Faculty of Sciences, University of Antananarivo, Madagascar

[2] Department of Isotope Hydrology, Institut National des Sciences et Techniques Nucléaires, Madagascar

[3] Institut National des Sciences et Techniques Nucléaires, Madagascar

Correspondence: Mamiseheno Rasolofonirina, Department of Physics, Faculty of Sciences, University of Antananarivo, BP 906 –Antananarivo 101, Madagascar.

**Abstract**

Nitrate contamination of groundwater has inclined to be a critical issue in areas where groundwater is the only available resource for water supply for drinking use purpose. In developing countries such as Madagascar, on-site sanitation can be a significant source of nitrate contamination of shallow groundwater, depending on the type of sub-surface layer and hydrogeological environment, the arrangements and behavior of sanitation, and the design of sanitation used for defecation. This study was carried out to investigate the nitrate occurrence in shallow groundwater of Mahitsy city, Analamanga Region of Madagascar, and to assess the on-site sanitation influence on nitrate concentration in drinking water well. Water samples were collected from dug wells in rainy and dry seasons.

The analytical results showed that the measured nitrate concentration was in the range of 1.5 mg/L and 580 mg/L with an average of 348 mg/L for all water samples. Thirteen out of fifteen samples had nitrate concentration exceeding the WHO guideline value (50mg/L). Data analysis indicated that nitrate concentration in dry season (average 409 mg/L) was greater as compared to rainy season (371 mg/L). However, the difference was not significant at the 0.05 level. Significant positive correlation (0.849, $p < 0.01$) was found between nitrate and chloride concentration with chloride/nitrogen ratio of about 1:2.23, suggesting the same source for nitrate and chloride. Nitrate concentrations of well waters were strongly correlated to distance between water wells and sanitation facilities (-0.466, $p = 0.08$), to water table level (-0.558, $p < 0.05$) and to age of water wells (0.655, $p < 0.01$).

**Keywords:** nitrate contamination, groundwater, on-site sanitation, Mahitsy

## 1. Introduction

Nitrate occurs naturally in soil and groundwater, but at low concentrations (less than 3 mg/L) in water unaffected by human- related activities according to Madison & Brunett (1984). It is part of the nitrogen cycle. Nitrogen is a key element for all living organisms and large amounts of inorganic nitrogen as well as other essential nutrients are needed specifically by plants for their sustainable high yields (Blumenthal, Baltensperger, Cassman, Mason, & Pavlista, 2008).

While used in fertilizers, nitrate can be released into groundwater when nitrogen fertilizer is applied indelicately. Plants take up the amount of nitrate they need for their growth and the exceeding nitrate stays in the soil, to further finds its way into groundwater through the downward movement of soil water into aquifer (Tredoux, Engelbrecht & Israel, 2009). That constitutes the main source of nitrate groundwater pollution known as diffuse source in the zone where fertilizers are used in excessive amounts and/or without proper management.

Other potential sources of nitrate contamination of groundwater include the use of animal manure and sanitation system leaching. Animals and human excreta increase the soil organic nitrogen, which is converted into nitrate by a series of bacterial transformations under aerobic conditions. It enhances the leaching of nitrate into groundwater (Assessing Risk to Groundwater from On-Site Sanitation [ARGOSS], 2002).This last phenomenon is well known in the rural villages and the urban cities in developing countries due to the inadequate use of sanitation facilities and the improper design of animal waste storage.

Whereas on-site sanitation is now recognized as one of the major sources of nitrate contamination to groundwater in the Southern African sub-continent such as South Africa, Botswana and Namibia, use of nitrogen fertilizer is the main anthropogenic source of nitrate in groundwater for the USA and the European countries (Tredoux, et al., 2009). Several authors reported that inadequate on-site sanitation design and the hydrogeological conditions of terrain are potential sources of groundwater pollution, including fecal contamination (ARGOSS, 2001; Dzwairo, Hoko, Love, Guzha, 2006; Pujari, Padmakar, Labhasetwar, Mahore, Ganguly, 2011). In Botswana, Staudt (2003) found that eleven out of thirty-one sampled boreholes in the Ramotswa area had nitrate concentrations exceeding the nitrate concentration guideline value of 45 mg/L for drinking water (World Health Organization [WHO], 2006). The pit latrines were identified as the major source of the groundwater pollution. In Zimbabwe, the investigation carried out by Dzwairo et al. (2006) depicted that on-site sanitation negatively impacted on the groundwater quality in the Kamangira village, resulting from the sandy nature of the soil and the pit latrine technology. In addition, Oday & Dugbantey (2003) showed in the case study of Ashanti region in Ghana that the distance between the water points and the sanitation facilities had influence on the groundwater pollution level: the greater the distance of groundwater source from the on-site facility, the lower the fecal and nitrate contamination level.

Although nitrate is harmless when ingested at low concentration, exposure to high level of nitrate can affect human health, particularly infants under the age of three months (WHO, 2011). Nitrate reduction occurs in the infant digestive system, leading to nitrite production. Nitrite is highly toxic, as it mediates the oxidation of hemoglobin in the blood to methemoglobin, causing the condition called methemoglobinemia. The methemoglobin cannot transport the oxygen and carbon dioxide to the tissues, resulting in dysfunction to the system and a bluish-tinge in the lips or other body parts well-known as "bleu-baby syndrome" in case of excessive nitrate exposure (Agency for Toxic Substances and Disease Registry [ATSDR], 2011).

In Madagascar, only about forty percent (39.6%) of the total population had access to safe drinking in 2005 with a rate of 31.2% for rural areas (Institut National de la Statistique [INSTAT], 2006). In 2010, slightly less than forty-five percent of the population of Madagascar had access to potable drinking water according to the demographic household survey (INSTAT, 2011). Over fifty percent (50.4%) of the population in the rural areas use groundwater as source of drinking water supply. In addition, only two percent (2%) of the rural population have modern latrines while around forty six percent (46.2%) do not use latrines at all, opting to defecate in nature for convenience (INSTAT, 2006).

Few data are available for nitrate occurrence in groundwater in Madagascar (Smedley, 2002). Groundwater contamination, including nitrate hazard is of concern insofar as the population do not have access to or do not use at all appropriate sanitation facilities because of cost constraint and/or culture barrier. The on-site livestock breeding appears as well to be a potential risk of groundwater pollution, regarding animal waste management practices.

In this sense, the purpose of the present work is to provide an overview of the nitrate occurrence in the shallow groundwater of Mahitsy city and to assess the on-site sanitation influence on nitrate concentration in drinking water well.

## 2. Study Area Description

### 2.1 Study Area Location and Characteristics

Mahitsy is located in the Highlands Plateau of central Madagascar, in the Ambohidratrimo district within the Analamanga region, at a distance of about 30 km northwestwards from the Capital city, Antananarivo. It stretches between longitudes 47° 20' 00" E and 47° 21' 25"E, and latitudes 18° 44'30" S and 18° 45' 25"S. The area elevations lie between 1250 m and 1280 m above mean sea level from the north-west boundary to south-east extremity of the city, respectively.

Most of the study area population is from lower to medium income categories with an average household size of 5 members. It consists of a mid-urban city, where small trading and farming make the population livings. The agglomeration is overcrowded, where housing is built in unorganized way and regardless of space for recreational use due to the lack of appropriate urban plan (Figure 1). Additionally, there is no sewage system in place and the domestic wastewater is discharged through open drains or is simply scattered around the yard. The local population mainly use pit latrines for sanitation purpose. Whilst some people in the north-western part of the city are using tap water, groundwater still constitutes the main source for drinking and domestic use in the area since the provision of water supply by the Mahitsy Commune Authorities cannot cover the whole population.

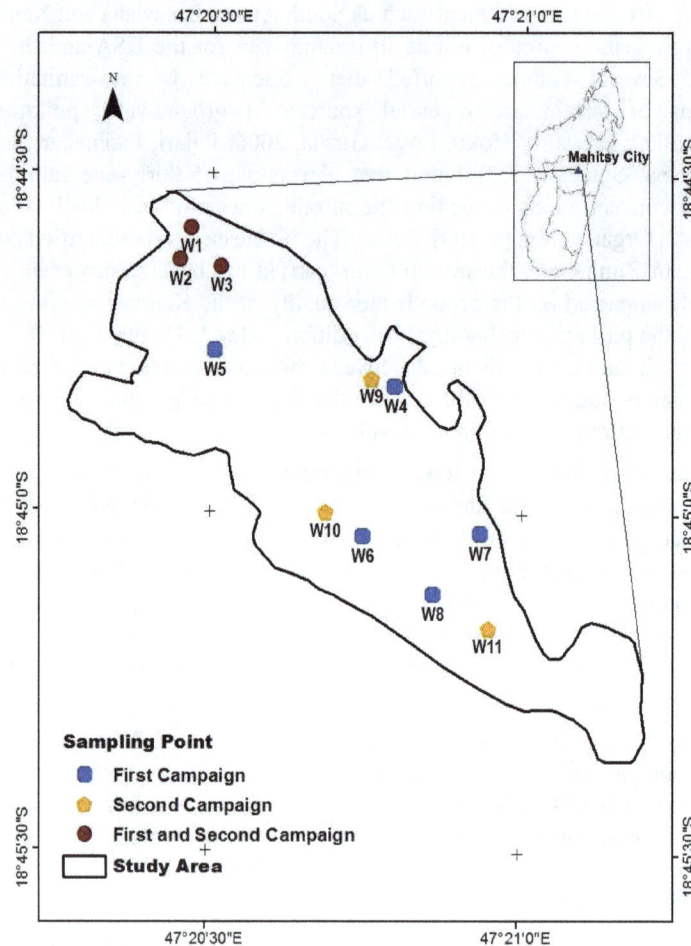

Figure 1. Sampling point location

## 2.2 Geological and Hydrogeological Overview

The study area lies on crystalline rocks of Precambrian age (Besaire, 1973). It is a gneissic-granitic basement, which consists of late Archaean, and Neoproterozoic gneisses, and granitoids (Goncalves, Nicollet & Lardeaux, 2002). The basement is deeply weathered and eroded, resulting in formation of thick red lateritic soils covering plains, dark brown soils deposited on volcanic rocks or grey alluvial soils in the valleys (Jeffrey, 2009). The weathering profile showing ridged lateritic layers at near surface level indicates that weathering affects the unsaturated zone and continues at depth down to the basement rocks, the layer base constituting the weathering front (Jeffrey, 2009).

Groundwater in the study area is mainly stored in weathered zones, including unconfined aquifer in weathered laterite and a semi-confined aquifer in fissure granitic basement, separated by clayey layer (Grillot, Blavoux, Rakotondrainibe, Raunet & Randrianarisoa, 1987; Dussarrat & Ralaimaro, 1993).There is interaction between these aquifers, as some infiltration from upper weathered laterite aquifer moves down through the interface clayey layer as a recharge to the deeper basement aquifer (Grillot & Ferry, 1990). Recharge to the two aquifers takes place preferentially within the interfluve zone (Grillot and Blavoux et al, 1990).

## 2.3 Climate, Temperature and Rainfall

The central highlands, where is located the study area experience a tropical mountain climate with mean temperature ranging from 16° C to 24° C. The rainy season occurs from November to April (warm summer), and the dry season from May to October (mild winter). Annual rainfall ranges between 1200 mm and 1400 mm. Heavy rains occurs during the rainy season mainly inside the period from December to March.

## 3. Methods

### 3.1 Sanitary Condition Surveying

Inspection of the physical installations was performed to identify any eventual water contamination source. For that, a survey questionnaire was developed to collect the distances between the water points and any potential source of well water pollution, including sanitation facilities, solid waste disposal area, animal manure storage site and/or livestock breeding location. The type of sampled well was also reported in the survey questionnaire.

### 3.2 Sampling

Two sampling campaigns were carried out in March 2005 and July 2005. A total of fifteen (15) water samples were collected, including eight (8) in the first campaign and seven (7) during the second campaign. Four out of eight wells (W1, W2, W3 and W7) were resampled within the second campaign. The water samples were collected from private traditional dug wells, which tap the upper unconfined aquifer. They are relatively protected, cased with bricks or stones, of about 1 m of diameter and was covered with a wooden cap. The best protected wells are installed inside a small shed equipped with door. In general, the groundwater table is shallow and is in the range of 2.2 m and 20 m under the land surface, depending on the well location topography.

Two 100 mL polyethylene (PE) bottles were filled with water sample for anion and cation measurements for each sampled well. Containers were rinsed with the water to be sampled prior to sample collection. All samples collected were filtered using cellulose nitrate membrane (0.2 μm) in the field. Samples for cation analysis were acidified with $HNO_3$ after filtration for preservation purposes(Barcelona, Gibb, Helfrich & Garske, 1985).

### 3.3 Field Measurement

Water quality parameters measured in the field were pH, Potential Redox (Eh), Electrical Conductivity (EC), Dissolved Oxygen (DO), Temperature (T), Total Dissolved Solids (TDS), Alkalinity and groundwater level. The first five parameters were measured using portable multimeter (Multi 340i model). The measurement of Total Dissolved Solids was done using Hach conductimeter (SensION 5). Alkalinity as the concentration of bicarbonate was determined by acidic titration ($H_2SO_4$) using digital titrator. Field measurements except for groundwater table were performed on aliquots of well waters immediately following collection. The groundwater level was measured by water level meter.

### 3.4 Analysis of Anions and Cations

Analysis of ions such as ammonium ($NH_4^+$), sodium ($Na^+$), potassium ($K^+$), magnesium ($Mg^{2+}$), calcium ($Ca^{2+}$), chloride ($Cl^-$), bromide ($Br^-$), nitrate ($NO_3^-$), and sulfate ($SO_4^{2-}$) were done using two separate Dionex Ion Chromatograph systems DX-120, one system for anion and one system for cation measurements. For the anion, the Chromatograph is equipped with AS 14A and for the cation, with a CS 12A analytical column. 20 mmole/L of $CH_3SO_3H$ and a combination of 8 mmole/L of $Na_2CO_3$ with 1 mmole/L of $NaHCO_3$ were used as eluent for cationic and anionic measurements, respectively. All water samples were measured in duplicate. The analytical results precision varied within 5%.

## 4. Results and Discussion

### 4.1 Physico-chemical and Chemical Parameters

The measured pH values ranged from 3.93 – 6.38 (averaging 5.1) in the first campaign to 4.14 – 5.87 (averaging 4.88) during the second campaign. All sampled waters were of acidic conditions and more likely to be corrosive.

Groundwater temperatures varied between 19.4$^0$C and 24.6$^0$C for all sampling points, averaging 22.6$^0$C and 21.2$^0$C in warm summer (first campaign) and mild winter (second campaign), respectively. Well water temperatures reflected seasonal air temperatures. They were warmer in March than in July.

The dissolved oxygen values ranged between 6.14 mg/L and 8.35 mg/L, averaging 6.81 mg/L for samples collected during the first campaign. They varied from 2.31 mg/L to 6.36 mg/L with an average of 3.76 mg/L within the second campaign. The second batch of water samples were under reducing conditions, as compared to the first batch.

As far as electrical conductivity was concerned, it varied in the range 118 μS/cm – 1734 μS/cm during the first sampling campaign and in the range13μS/cm – 1353 μS/cm in the second campaign. Samples collected during the first campaign were more charged with dissolved ions with a mean electrical conductivity value of 1096 μS/cm than those collected in scope of the second campaign (with an average of 904 μS/cm). Electrical conductivity of resampled wells recorded a decreasing trend, except for that of W1 well.

The measured Total Dissolved Solids (TDS) values ranged from 59 mg/L to 868 mg/L and from 7 mg/L to 813 mg/L for waters sampled during the first and second campaigns, respectively. The average values of TDS were 548 mg/l and 495 mg/L for the first and second lot of samples, respectively. Only two wells, namely W8 and W11, presented a TDS value less than 400 mg/L.

Bicarbonate measured during this investigation ranged from less than 1 mg/L to 46.7 mg/L, averaging 22.2 mg/L and 8.9 mg/L for the first and second sampling campaigns, respectively. The regression analysis between bicarbonate and pH values showed that bicarbonate decreased with a decrease of pH values (Figure 2).

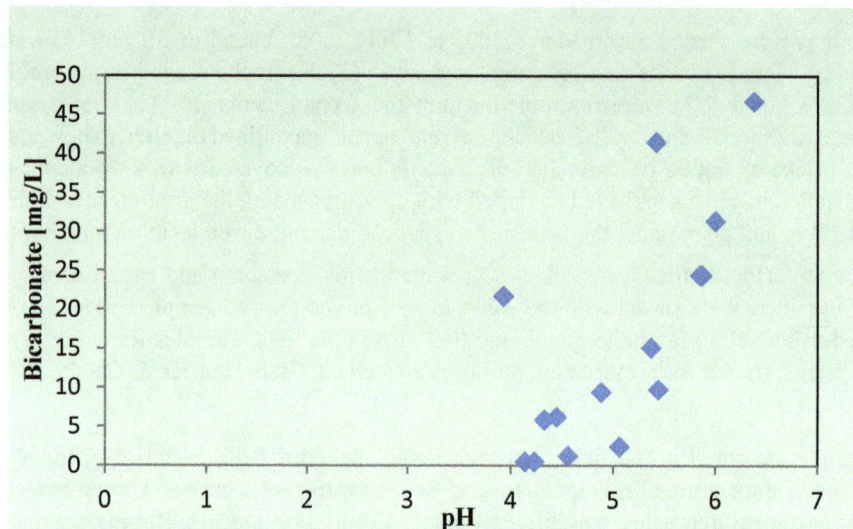

Figure 2. Relationship between chloride concentration and pH value for all well waters

Regarding chloride parameter, its concentrations were in the range of 6 mg/L – 205 mg/L with an average of 114 mg/L in the well waters sampled during the first campaign. Its contents varied from 0.2 mg/L to 195 mg/L (averaging 111 mg/L) for samples collected during the second campaign. Statistical analysis of data showed that there was no significant difference (p=0.935) between the concentrations of chloride measured within the two sampling campaigns. A plot of electrical conductivity and chloride data indicated correlation between the two parameters (Figure 3). Such correlation was stronger for samples collected during the first campaign (0.929, p < 0.01), as compared to those sampled within the second campaign (0.822, p < 0.05).

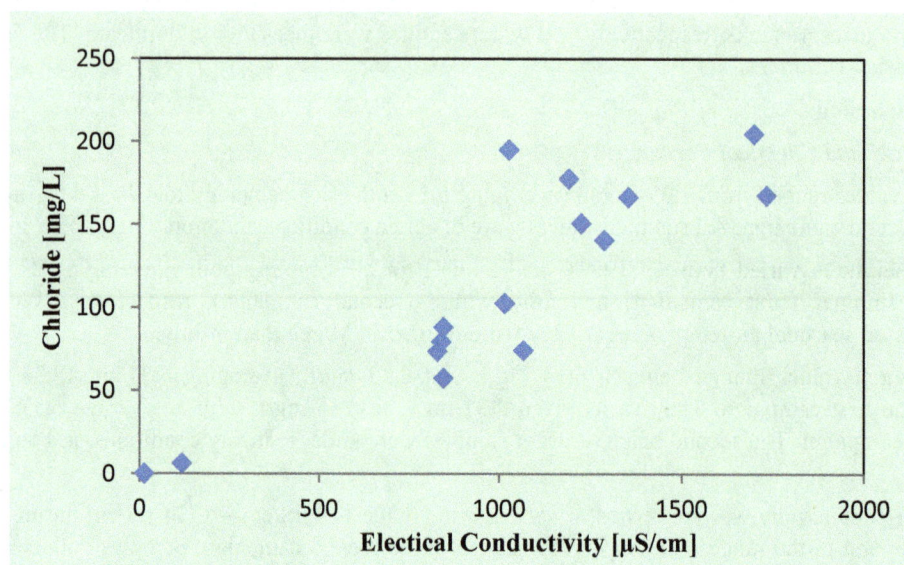

Figure 3. Relationship between chloride concentration and electrical conductivity for all well waters

*4.2 Nitrate Seasonal Variation*

Measured nitrate concentrations stretched between 1.5 mg/L and 580 mg/L for all water samples collected from wells in scope of the present investigation. Of the fifteen water samples analyzed from wells in Mahitsy city, only two samples (W8 and W11) did not exceed the maximum recommended nitrate concentration value of 50 mg/L nitrate for drinking water (WHO, 2006). All wells revealing nitrate concentration higher than the WHO recommended value were of nitrate contents exceeding 250 mg/L. Amongst them, W3 and W4 samples presented nitrate concentration values greater than 500 mg/L.

Regarding nitrate seasonal variation, water samples collected within the first batch had nitrate concentration values varying from 16 mg/L to 580 mg/L. Well waters sampled during the second collection batch presented nitrate content in the range of 1.5 mg/L – 556 mg/L. The mean value of nitrate concentration of drinking water from wells W1, W2, W3 and W7 sampled in the first campaign was less (371 mg/L) than that of the same samples but collected during the second campaign (409 mg/L).Statistical analysis of data demonstrated that there was no significant difference between the mean values of the first and second set of samples, as the p recorded value was 0.192 (greater than 0.05).Although the observed variation of mean values (from 371 mg/L to 409 mg/L) was not statistically significant, it presented a trend of nitrate seasonal variation. The obtained results recorded a slight increase of nitrate concentration during dry season, which was accompanied by a decrease of water level from 0.4 m to1.3 m. The dilution occurring from recharge of upper groundwater by rain infiltration could explain the reduced nitrate concentration in rainy season, since the recharge travel time to saturated zone can be of days or weeks in thin weathered basement (ARGOS, R68692). In addition, measured dissolved oxygen for the samples of interest (W1, W2, W3 and W7) varied from 2.57 mg/L to 5.22 mg/L during dry season, suggesting a condition favorable to the accumulation of nitrate to high level. Denitrification process occurs only when dissolved oxygen concentration falls below 2 mg/L in the groundwater (Rissmann, 2011).

A plot of nitrate concentration versus electrical conductivity for all samples (Figure 4) indicated that there was a good relationship between them (0.899, $p < 0.01$), suggesting that nitrate plays a major role in the electrical conductivity variation: the greater the nitrate concentration, the higher its influence on electrical conductivity.

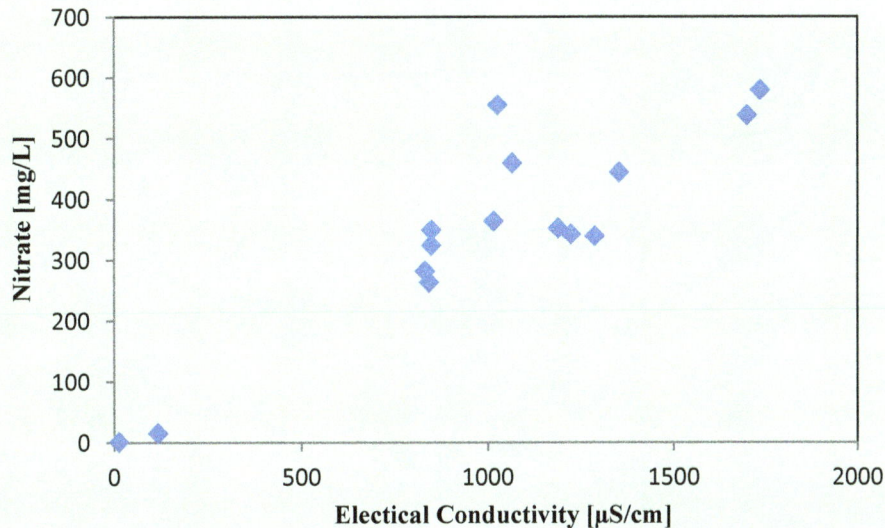

Figure 4. Relationship between nitrate concentration and electrical conductivity for all well waters

Considering separately nitrate concentration data for each sampling campaign, Figure 5 showed that nitrate concentration of the first batch of samples presented better relationship (0.960, $p < 0.01$) with electrical conductivity, as compared to the second batch of samples (0.872, $p < 0.05$).

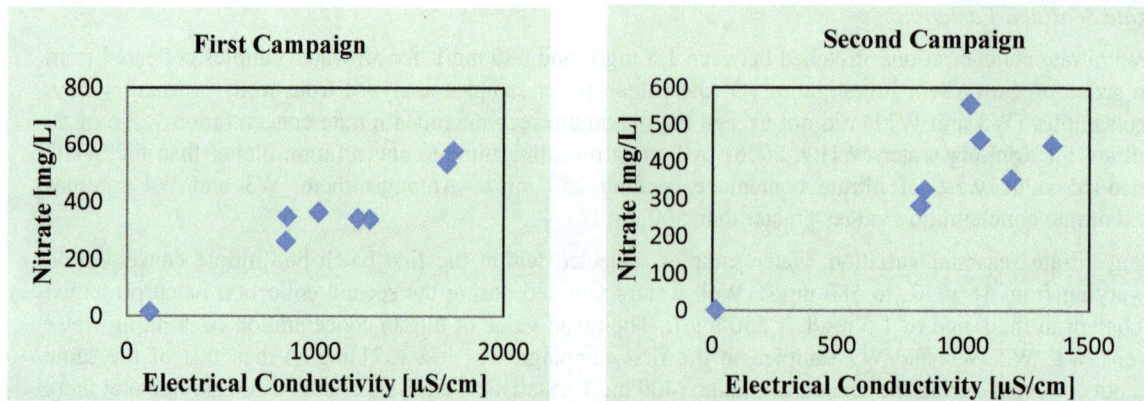

Figure 5. Relationship between nitrate concentration and electrical conductivity for sampled well waters during the first and second campaigns, respectively

The slight difference implied that the mechanism of nitrate groundwater contamination did not present the same magnitude of influence on the electrical conductivity values of samples collected within the two campaigns.

### 4.3 Variation of the Nitrate Concentrations along the Groundwater Flowpath

To investigate the variation of the nitrate concentrations along the groundwater flowpath, the graph between the nitrate concentration and the elevation (assuming the groundwater flows from high to low elevation) was plotted as shown in Figure 6. The graph showed scattered plotted points. However, two trends can be distinguished: (i) a clear increase of the nitrate concentration along the groundwater flowpath (blue oval), indicating a stronger occurrence of a nitrification process at shallower depth while approaching the discharge zone (ii) a rather obvious stagnant nitrate concentration along the groundwater flowpath (red oval), suggesting a rapid flushing of nitrate along the groundwater flowpath.

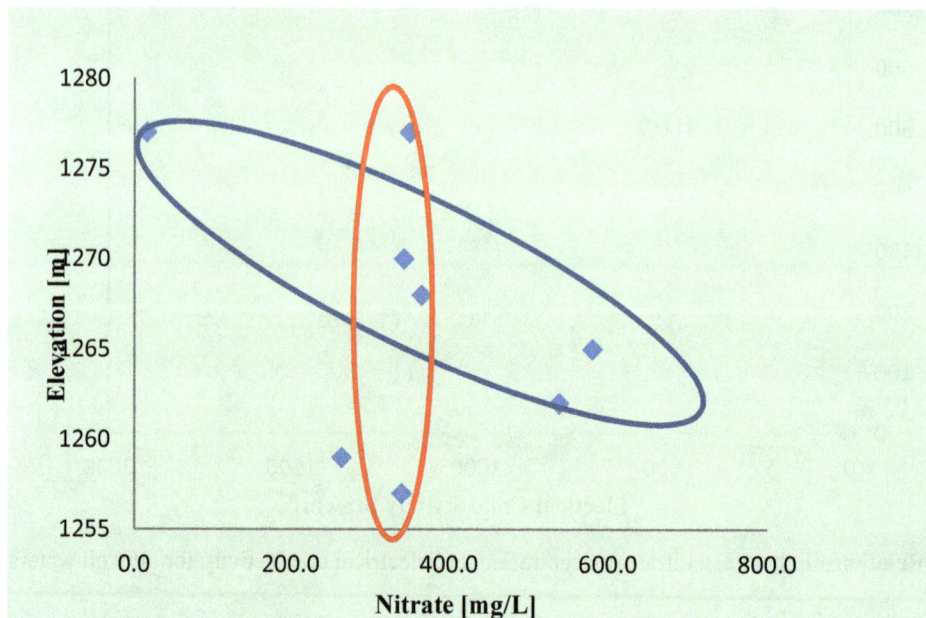

Figure 6. Graph between nitrate concentration and elevation in the study area

### 4.4 On-site Sanitation Effect on Nitrate Concentration

The sampled wells were located within the range of 3 – 20 m from the sanitation facilities. Population in the study area used mainly dry pit latrines where were stored the wastes. The pit latrines were not sealed and the liquid part of the wastes infiltrated into the soil.

According to Assessing Risk to Groundwater from On-Site Sanitation (ARGOSS) scientific review (2002), nitrate and chloride were found to be the major chemical contaminants generated from on-site sanitation.

ARGOSS (2001) report showed that released nitrogen through human excreta is of about 4 kg/year/person. Nitrification occurs in the unsaturated zone under an aerobic condition and converts a significant percentage of organic nitrogen into nitrate. Human waste also contains a large amount of chloride. Each person discharges on average 4 g/day of chloride through urine, feces and sweat (ARGOSS, 2001).

Figure 7 indicated that the nitrate and chloride concentrations were correlated, suggesting that the concentration rise was related with chloride/nitrogen ratio of around 1:2.23. In addition, the bivariate correlation test showed a significant correlation between nitrate and chloride concentration with a Pearson correlation coefficient of 0.849 (p-value less than 0.01). That fact implied that nitrate and chloride in the shallow groundwater could be the same source. They could be released from the human excreta in pit latrines, as the ratio of chloride to nitrogen in human excreta was approximately 1:2 (ARGOSS, 2002).

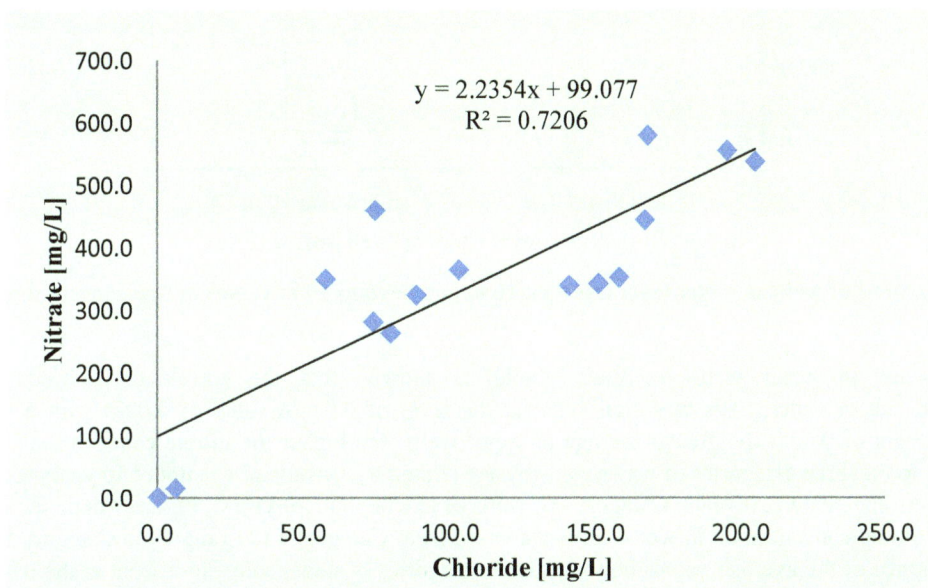

$$y = 2.2354x + 99.077$$
$$R^2 = 0.7206$$

Figure 7. Nitrate versus chloride concentration plot with points for all well waters sampled during the first and second campaigns

Regarding sanitation facility arrangement, a plot of nitrate concentration versus pit latrine location showed that there was a negative correlation between them with a Pearson correlation coefficient of -0.466, but it was not significant as the p-value was 0.08.

Considering the water table level parameter, the statistical correlation indicated that there was a significant correlation between nitrate concentration and water table level, as Pearson correlation coefficient was estimated at -0.558 with a p-value of 0.031(significant at the 0.05 level). The nitrate concentration tended to increase as the water table level decreased. Boxplot diagram of water wells of water level less than 10 m, and those of water level beyond 10 m (Figure 8) represented that the nitrate concentration average values were 415 mg/L and 213 mg/L for the first and second categories, respectively. The ANOVA test indicated that the difference between the average values was statistically significant (p-value less than 0.05).

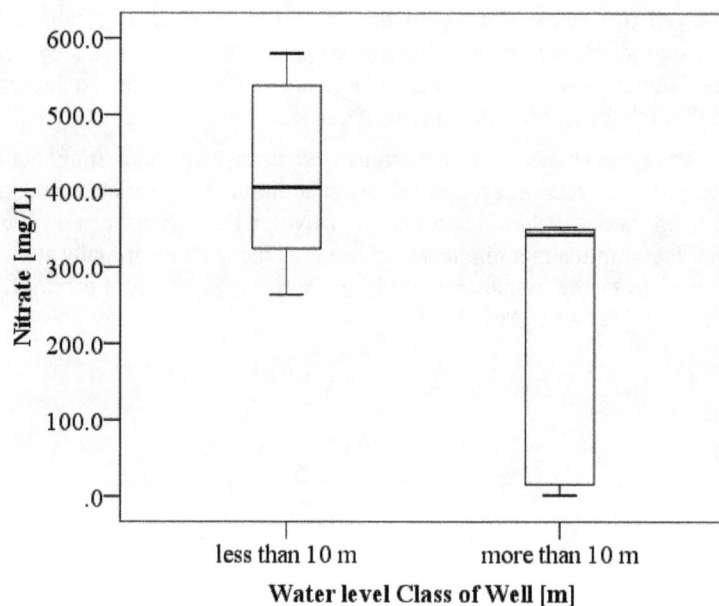

Figure 8. Boxplots of wells of water level less than 10 m and beyond 10 m versus nitrate concentration

Considering the age of water wells, correlation analysis showed that the correlation between nitrate concentration and age of water wells was significant at the level of 0.01 (p-value of 0.008) with a Pearson correlation coefficient of 0.655: the greater the age of water wells, the higher the nitrate concentration of well waters. Referring to the three categories of water wells by age (Category I: wells of age under 10 years; Category II: wells of age between 10 and 20 years; Category III: wells of age beyond 20 years), Figure 9 depicted that the average value of nitrate concentration in well waters increased from Category I (213 mg/L) to Category III (500 mg/L). The difference of the average values between three categories is statistically significant at the 0.05 level (p-value equal to 0.012).

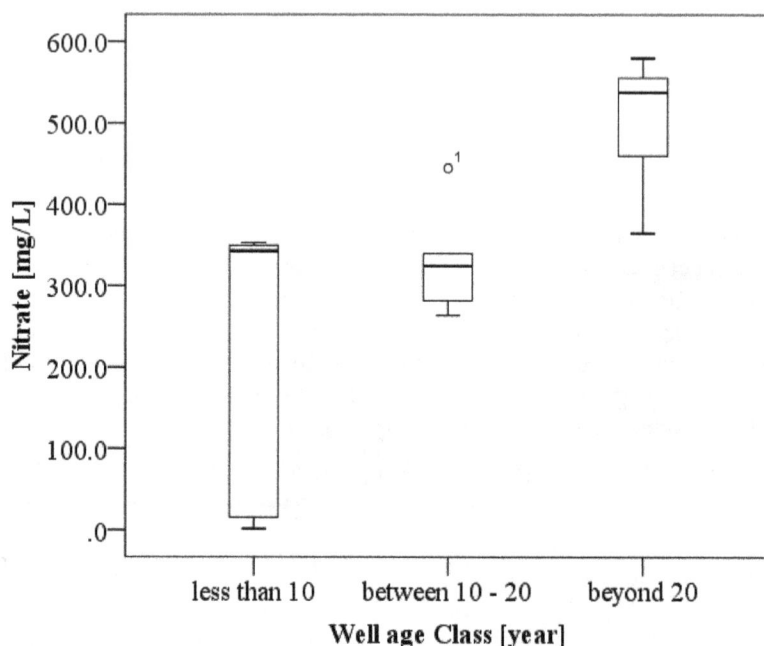

Figure 9. Boxplots of Category I, Category II and Category III of wells versus nitrate concentration

## 5. Conclusion

Very high nitrate concentrations were measured in most of sampled water wells in Mahitsy city, varying from 1.5 mg/L - 580 mg/L. Of the fifteen collected samples, thirteen had nitrate concentration exceeding the WHO recommended value (50 mg/L) for drinking water. For seasonal variation, investigation on set of resampled wells (W1, W2, W3 and W7) showed that the average nitrate concentration in rainy (371 mg/L) season was less than in dry season (409 mg/L). The reduced nitrate concentration in wet season could be due to the dilution resulting from recharge infiltration into the shallow groundwater. Additionally, content of dissolved oxygen of well waters (greater than 2 mg/L) could contribute to the increase of nitrate concentration during the dry season.

The on-site sanitation had influence on nitrate concentration, since there was a strong relationship between nitrate and chloride. In addition, data analysis resulted in the following findings:

- Nitrate concentration of shallow groundwater in the study area was negatively correlated to the distance between pit latrines and water point facilities: the closer the pit latrines, the higher the nitrate concentration.

- There was also a significant relationship between the nitrate concentration and the water table level: the deeper the groundwater table, the lower the nitrate concentration.

- Significant positive correlation was found between nitrate concentration and water well age as well: the older the water wells, the higher the nitrate concentration.

The upper groundwater in Mahitsy city will pose health issue when used for drinking purpose as far as nitrate concentration was concerned. The Mahitsy Commune authorities should reinforce the public water system, which provides drinking water supplies from mountain spring.

## Acknowledgements

The authors would like to thank Dr. Lucienne Randriamanivo, Head of X-Ray Fluorescence Department, INSTN-Madagascar and her team. This research was funded by the International Foundation for Science (IFS).

## References

Agency for Toxic Substances and Disease Registry (ATSDR). (2011). *Nitrates and Nitrites. CAS # 84145-82-4, 14797-65-0.* Atlanta, GA: U.S. Department of Public Health and Human Services, Public Health Service. Retrieved from http://www.atsdr.cdc.gov/toxfaqs/tfacts204.pdf.

ARGOSS. (2001).Guidelines for assessing the risk of groundwater from on-site sanitation. *British Geological Survey Commissioned report CR/01/142*

ARGOSS. (2002). Assessing Risk to Groundwater from On-site Sanitation: Scientific Review and Case Studies. *British Geological Survey Commissioned Report, CR/02/079N.*

Barcelona, M. J., Gibb, J. P., Helfrich, J. A., & Garske, E. E. (1985). Practical Guide for Ground-Water sampling. Illinois State Water Survey. *ISWS Contract Report 374.*

Besarie, H. (1973). Précis de géologie malgache. *Annales géologiques de Madagascar, 36*, 109-134.

Blumenthal, J. M., Baltensperger, D. D., Cassman, K. G., Mason, S. C., & Pavlista A. D. (2008). Importance and Effect of Nitrogen on Crop Quality and Health. In J. L. Hatfield, & R. F. Follett (Eds.), *Nitrogen in the Environment: Sources, Problems, and Management* (2nd ed.). Amsterdam: Elsevier. Retrieved from http://digitalcommons.unl.edu/agronomyfacpub/200.

Dzwairo, B., Hsoko, Z., Love, D., & Guzha, E. (2006). Assessment of the impacts of pit latrines on groundwater quality in rural areas: A case study from Marondera district, Zimbabwe. *Physics and Chemistry of the Earth, 31*, 779-788. http://dx.doi.org/10.1016/j.pce.2006.08.031

Goncalves, P., Nicollet, C., & Lardeaux, J. M. (2002). Finite strain pattern in Andriamena unit (North Central Madagascar): evidence for late Neoproterozoic-Cambrian thrusting during continental convergence. *Precambrian Research.*

Grillot, J. C., Blavoux, B., Rakotondrainibe, J. H., Raunet, M., & Randrianarisoa, N. (1987). A propos des aquifères d'altérites sur les hauts plateaux cristallophylliens de Madagascar. *Comptes Rendus de L'Académie des Sciences Paris Hydrogéologie 305(Série II): 1471-1476.*

Grillot, J.-C., & Ferry, L. (1990). Approche des échanges surface-souterrain en milieu cristallin altéré aquifère. *Hydrol. Continent. 5*(1), 3-12.

Institut National de la Statistique [INSTAT]. (2006). Enquête Périodique auprès des Ménages 2005. *Rapport*

*Principal*. Madagascar, 150-153.

Institut National de la Statistique [INSTAT]. (2011). Enquête Périodique auprès des Ménages 2010. *Rapport Principal*. Madagascar, 193-194.

Jeffrey, D. (2009). Hydrogeological mapping of north-central of Madagascar using limited data. Retrieved from http://nora.nerc.ac.uk/id/eprint/9062.

Madison, R. J., & Brunett, J. O. (1984). Overview of the occurrence of nitrate in ground water of the United States. *United States Geological Survey Water-Supply Paper 2275*. National Water Summary 1984 – Hydrologic Events Selected Water-Quality Trends and Ground-water, *93-105*.

Odai, S. N., & Dugbantey, D. D. (2003). Towards pollution reduction in peri-urban water supply: a case study of Ashanti region in Ghana. Diffuse Pollution Conference, Dublin 2003.

Pujari, P. R., Padmakar, C., Labhasetwar, P. K., Mahore, P., & Ganguly, A. K. (2011). Assessment of the impact of on-site sanitation systems on groundwater pollution in two diverse geological settings – A case study from India. *Environ Monit. Assess*. Retrieved from http://dx.doi.org/10.1007/s10661-011-1965-2.

Rissmann, C. (2011). Regional Mapping of Groundwater Denitrification Potential and Aquifer Sensitivity Technical Report. *Environment Southland. Publication No 2011-12*.

Smedley, P. L. (2002). Groundwater Quality: Madagascar. *British Geological Survey. NERC 2002*. Retrieved from http://www.bgs.ac.uk/sadcreports/madagascar2002smedleybgsgroundwaterquality.pdf

Staudt, M. (2003).*Environmental hydrogeology of Ramotswa*. Environmental Geology Division, Department of Geological Survey, Lobatse, Botswana, 65-66.

Tredoux, G., Engelbrecht, P., & Israel, S. (2009). Nitrate in groundwater: Why is it a hazard and how to control it? *Water Research Commission Report No.TT 410/09*.Republic of South of Africa. Retrieved from http://www.wrc.org.za/Knowledge Hub Documents/Research Reports/TT 410 Groundwater.pdf.

World Health Organization (WHO). (2006). *International Standards for drinking water* (3rd ed.). Geneva, pp. 417-420.

World Health Organization (WHO). (2011). Nitrate and Nitrite in Drinking-water. Background document for development of WHO Guidelines for Drinking-water Quality. Geneva, 10-12.

# 6

# The Protective Effects of Vitamin E and Zinc Supplementation Against Lithium-Induced Brain Toxicity of Male Albino Rats

Ahmed Th. Ibrahim[2], Marwa A. Magdy[2], Emad A. Ahmed[1] & Hossam M. Omar[1]

[1] Department of Zoology, Faculty of Science, Assiut University, Assiut, Egypt

[2] Department of Zoology, Faculty of Science, Assiut University, New Valley branch Assiut, Egypt

Correspondence: Hossam M. Omar, Department of Zoology, Faculty of Science, Assiut University, Assiut, Egypt. E-mail: hossameldin.mo@gmail.com

**Abstract**

Lithium (Li) therapy has widely used in the treatment of bipolar disorder. Consequently, consciousness of the side effects and pathogenesis of this metal is needed for such treatments. Recently, information on the interaction of Li with oxidative markers and organs toxicity attend the researchers over the world. In the present study we have tried to evaluate the influence of oral administration of LiCl for 4 weeks on the oxidative stress marker and histological structure of brain in male rats. Fifty adult male albino rats weighing $135\pm15$ gm was categorized into 5 groups (10 rats each). Group I worked as negative control, group II administrated with LiCl (0.20 mg/kg bw) in drinking water, group III, IV and V were administrated with Zn (10 mg/kg bw), VE (100 mg/kg bw) and their combination twice a week besides the daily administration of LiCl for 4 weeks, respectively. Rats after anesthesia with ether killed for collocation of brain for histopathological and biochemical analysis. Data obtained showed a significant increase in LPO, NO, GSH and Li content and the activities of SOD, CAT and AChE with demylination of the nerve fibers and degeneration of neurons in brain of LiCl treated rats. Co-treatment of rats with Zn or VE results in a significant decrease in LPO, NO, GSH content in the activities of SOD, CAT and AChE with less or normal structure of the brain. However, co-treatment with combination of Zn and VE caused a significant increase in SOD, CAT and AChE activities with normal histological structure.In conclusion, the data from the present study show that Zn and VE and their interaction are effective in protection against Li-induced brain toxicity in rat with priority for the combination.

**Keywords:** Lithium, vitamin E, brain, rats, oxidative stress, acetylcholinestrase (AChE)

## 1. Introduction

Lithium is a toxic alkaline metal that occur in the environment as industrial pollution and a therapeutic use. Moreover, it accumulated in algae, marine animals, vegetables, rock and tobacoo leaves (Schrauzer, 2002). Lithium is not essential element it used in therapeutic psychiatric diseases, in particular bipolar disorders like depression (Schou, 2001; Aral & Vecchio-Sadus, 2008). For therapeutic use, the dose of Li carbonate usually varies from 7-25 mg/kg per day (Allaguri *et al.*, 2006). For it is slowly movement from extracellular compartment to intracellular space it may require 6-10-days to reach steady blood concentration that desired for therapeutic responses (Groleau, 1994).

Clinical trial suggests that Li stops the progression of amyotrophic lateral sclerosis and inhibits a number of kinases and phosphatases that in turn affects many systems including inflammation, metabolism, receptor sensitivity, and adenyl cyclase (Young, 2009). However, prolonged Li therapy causes neuromuscular disorders (Sansone, 1985) and neuronal apoptosis in rat brain that effect on acetylcholine esterase (ACE) (Martins *et al.*, 2008). ACE is enzymes that terminate the neurotransmission at cholinergic synapses by splitting the neurotransmitter ACh to chloline and acetate (Tripathi & Srivastava, 2008). Acetylcholine plays an important role in sending signals from one neuron to the next when it is released from vesicles in the axon terminus, across the synapse, and onto receptors in the dendrites of the next neuron (Habila *et al.*, 2012).

Oxidative stress is one of the important mechanisms of toxic effects of Li (Oktem *et al.*, 2005). In fact, part of the adverse effects of Li seems to result from excessive formation of ROS andinhibition of antioxidant enzyme activities (Oktem *et al.*, 2005; Allagui *et al.*, 2007). Vitamin E (VE) as antioxidant is the primary membrane bound lipid-soluble, chain-breaking antioxidant that protects cell membranes against oxidative stress (Soylu *et al.*, 2006). VE prevents formaldehyde-induced tissue damage in rats (Gulec *et al.*, 2006) and endotoxin-induced oxidative stress in rat tissues (Kheir-Eldin *et al.*, 2001).

The role of zinc (Zn) is very important in antioxidant defense mechanism as well as in regeneration of damaged cells (Nuzhat & Mahboob, 2012). It is an essential trace mineral with important anti-inflammatory function (An *et al.*, 2005), antiapoptotic (Powell *et al.*, 2000), and antioxidant (Holland *et al.*, 1995). The role of Zn in antioxidant defense mechanism includes the protection due to redox active transition metals such as copper and iron, and the protection of -SH groups of protein from oxidative damage. The chronic antioxidant effects of Zn result in the induction of metellothionein synthesis that act as scavengers of toxic metals (Chvapil *et al.*, 1976), protection against VE depletion(Parsad *et al.*, 1988), induction of cell-proliferation and inhibition of NADPH oxidases (Oteiza *et al.*, 2000).

According to the aforementioned findings, the present work was aimed to study the protective effect of Zn, VE or combination of Zn and VE against Li induced brain toxicity in rats through measurement of oxidative stress markers and observation of histopathological changes.

## 2. Material and Methods

**Chemicals:** Lithium chloride (LiCl), zinc sulphate, VE (α-tochopherol), N, N diphenyl-p-phenylenediamine, superoxide dismutase, epinephrine, thiobarbituric acid (TBA), naphthylethylenediaminedihydrochloride, 5,5 dithiobis (2-nitrobenzoic acid (DTNB), triton-X100, sulfanilamide and acetylthiocholine (ATC), were obtained from Sigma Chemical Co. (St. Louis, MO, USA. All other chemicals and reagents were of the highest purity commercially available.

**Animals:** Fifty adult male albino rats (135±15 gm) were purchased from the Animal House of the Faculty of Medicine, Assuit University, Assuit, Egypt. Rats were housed in cages and were kept in a room temperature ($30^{\pm}3^{\circ}$C) with normal 12 h light/12 h dark cycle. They were allowed to acclimatize for one week before the experiments.

**Animal groups and treatment:**

Rats were divided into 5 groups of 10 rats each.

**Group I:** served as a control group.

**Group II:** received a dose of LiCl (0.20 mg/Kg bw) daily for 4 weeks in drinking water.

**Group III:** received a dose of LiCl (0.20 mg/Kg bw) and zinc sulphate (10 mg/kg bw) daily for 4 weeks in drinking water.

**Group IV:** received a dose of LiCl (0.20 mg/Kg bw) daily with VE (100 mg/kg bw) injected intraperitoneally twice a week for 4 weeks.

**Group V:** received a dose of LiCl (0.20 mg/Kg bw) with zinc sulphate (10 mg/kg bw) daily for 4 weeks, in drinking water with VE (100 mg/kg bw) injected intraperitoneally twice a week for 4 weeks.

**Collection and preparation brain cytosol:**

Animals of the different groups were killed after anesthesia with ether. The brain quickly removed and washed in (0.1 M) phosphate buffer (pH 7.4) and then stored at -20 °C for biochemical studies. Pieces of brain were fixed immediately in 10% neutral buffered formalin for histological studies. All experiments followed protocols approved by the Institutional Animal Care and Life Committee, Assiut University. 10% homogenate of brain was prepared by homogenization of 0.25 gm of tissue in 2.5 ml (0.1 M) phosphate buffer (pH 7.4) using homogenizer (IKA Yellow line DI 18 Disperser, Germany). The homogenates were centrifuged at 6,000 rpm for 1 hour at 4 °C and the supernatant cytosols were kept frozen at -20 °C for the subsequent biochemical assays.

**Biochemical measurements:**

Total protein concentration was determined by the method of Lowry *et al.* (1951). LPO products as TBARS were determined according to the method of Ohkawa *et al.* (1979). Nitric oxide (NO) was measured as nitrite concentration colorimetrically using the method of Ding *et al.* (1988). GSH was determined using the method of Beutler *et al.* (1963).Activity of SOD was determined according to its ability to inhibit the autoxidation of epinephrine at alkaline medium according to the method of Misra and Fridovich (1972). Activity of CAT was determined by the procedure of Luck (1963), basing on its ability to decompose hydrogen peroxide. Activity of AChE was estimated by method of Ellman *et al.* (1961).

Lithium, zinc and copperconcentrations in the samples were determined by ICP-MS (Thermo Fisher Scientific

(Bremen) GmbH) in central lab of Faculty of Science in New Valley. Standard solutions of multi-elements were prepared from commercial stock standard solutions at concentrations of 100 mg/L double deionised water. Working standard solutions were prepared by dilution of stock standard solution with the addition of hydrochloric acid, so that the acid concentration in working standard solutions matched the acid concentration in digested solutions.

For the histological part fixed tissues were processed routinely for paraffin embedding technique. Embedded tissues were sectioned at 5 μm and stained with hematoxylin and eosin (H & E) according to Drury and Wallington (1980).

**Statistical analysis:** Data were subjected to mean ± STD Err. The differences between means were done by using The Tukey-HSD test. Range test was used as a post-hoc test to compare between means at $p<0.05$. These analyses were carried out using Statistical Package for Social Sciences (SPSS) for windows, version 16.

## 3. Results

**Figs (1a, b)** show the level of LPO as TBARS and NO as nitrite in brain tissue. As compared to control rats, TBARS and nitrite level were significantly elevated in brain of Li group. In the same figure, data revealed that Zn, VE and the combination recovered TBARS and nitrite level significantly in brain tissue in comparison with Li group. **Fig (1 c, d)** showed the level of SOD and CAT activities in the brain tissue. Both enzyme activities were elevated in the brain of Li treated rats. Co-treatment of rats with Zn or VE alone caused reduction in these activities in comparison with Li treated rats, however the combination of Zn and VE not shown any effects. **Fig (1e)** show the level of GSH in brain tissue, as compared to control rats, GSH level elevated in brain of Li and Li, Zn treated groups. Also, the same fig, revealed that the treatment with VE and the combination have no significant effect in brain tissue. **Fig (1f)** showed the level of AChE in brain tissue, as compared to control rats, AChE showed no significant change in Li, Li, Zn and Li, VE, but it was highly significant increase in Li, Zn and VE treated groups.

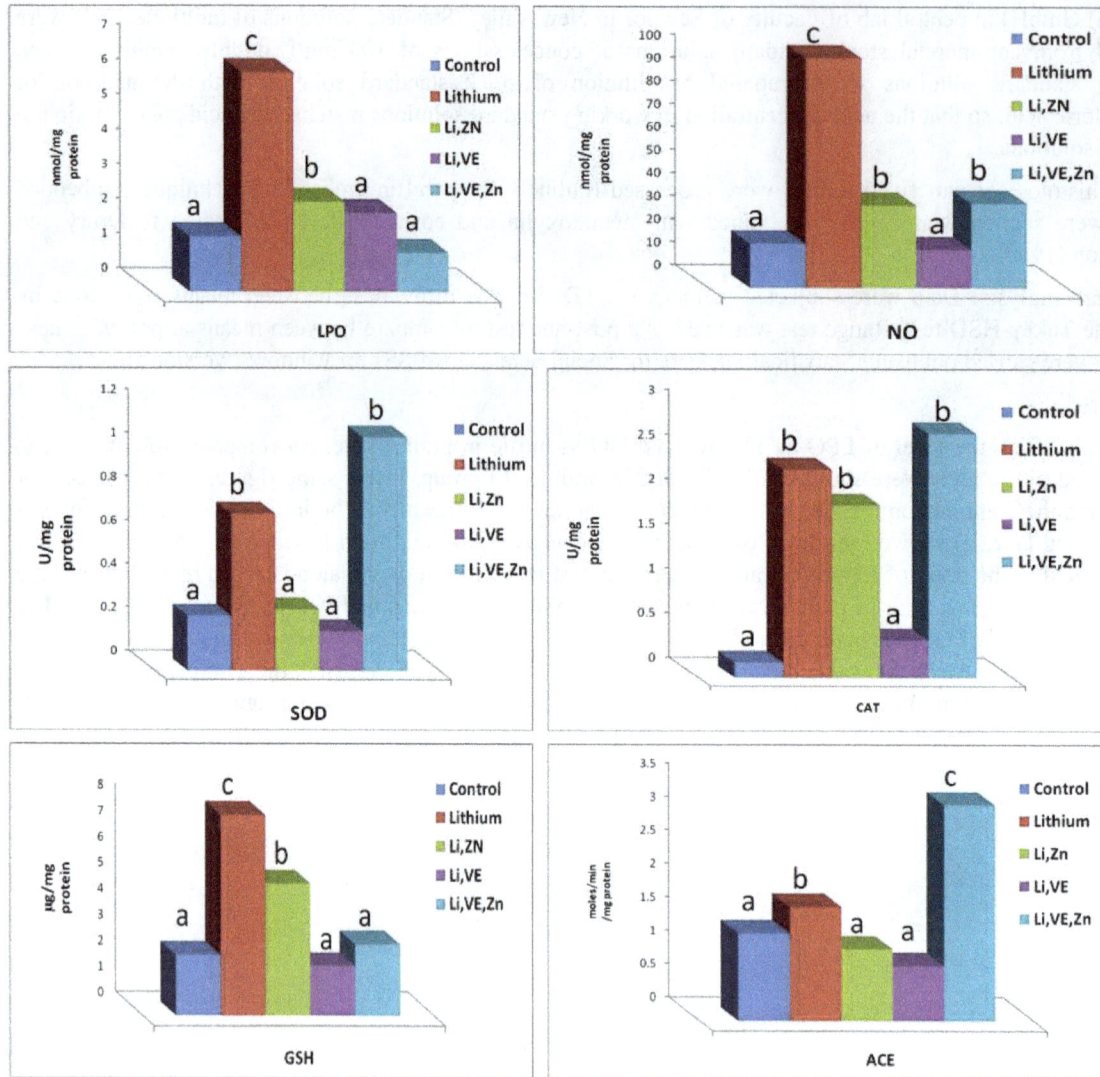

Figure 1. ShowLPO level (A), NO level (B), SOD activity (C), CAT activity (D), GSH level (E) and ACE level (F). Results presented as mean ± SE, different letter means significant different at p<0.05 between different group, where n=6

## Analysis of some trace elements

The values of Li, Zn and Cu as ppm in brain tissue in normal and different treated rats are presented in Fig (2). As compared to control rats, Li level was significantly increased in Li treated group. Co-treatment with combination of Zn & VE caused decrease in Li level in comparison to Li treated group and gives better result than Zn or VE alone.Level of Zn was significantly increased in Zn and Zn and VE treated groups, but still normal in Li and VE treated group. Cu level was significantly increased in interaction of Li, Zn and VE treated group, but still normal in Li, VE and Zn treated group

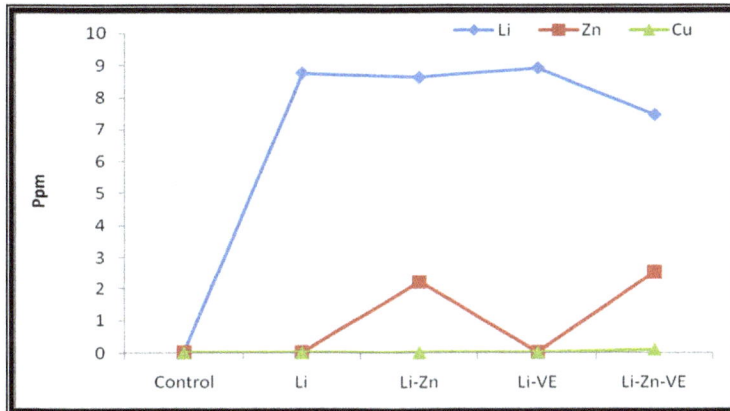

Figure 2. Show the some trace elements analysis in brain of normal and different treated rats

**Fig. 3:** Photomicrographs of brain sections: **A** control rat showing normal histological appearance of neurons in the white matter, **B** control rat showing normal histological appearance of neurons in the white matter, **C** brain or rat exposed to Li showing marked demylination in the nerve fibers **(arrows)**, **D** rat exposed to Li showing degeneration of the neurons **(arrow)**, **E** rat exposed to Li and treated with Zn showing perivascular edema **(arrow)** and perineural edema **(quadrate)**, **F** rat exposed to Li and treated with Zn showing mild degeneration changes in the neuronal cell body, **G** rat exposed to Li andtreated with VE showing perivascular edema **(arrow)**, perineural edema **(quadrate)** and mild degeneration changes in the neurons, **H** rat exposed to Li andco-treated with VE showing mild degeneration changes in the neurons, **I** rat exposed to Liand co-treated with combination of Zn and VE showing perivascular edema **(arrow) and J** rat exposed to Li and co-treated with combination of Zn and VE showing perivascular edema **(arrow) (H&E) (400X)**.

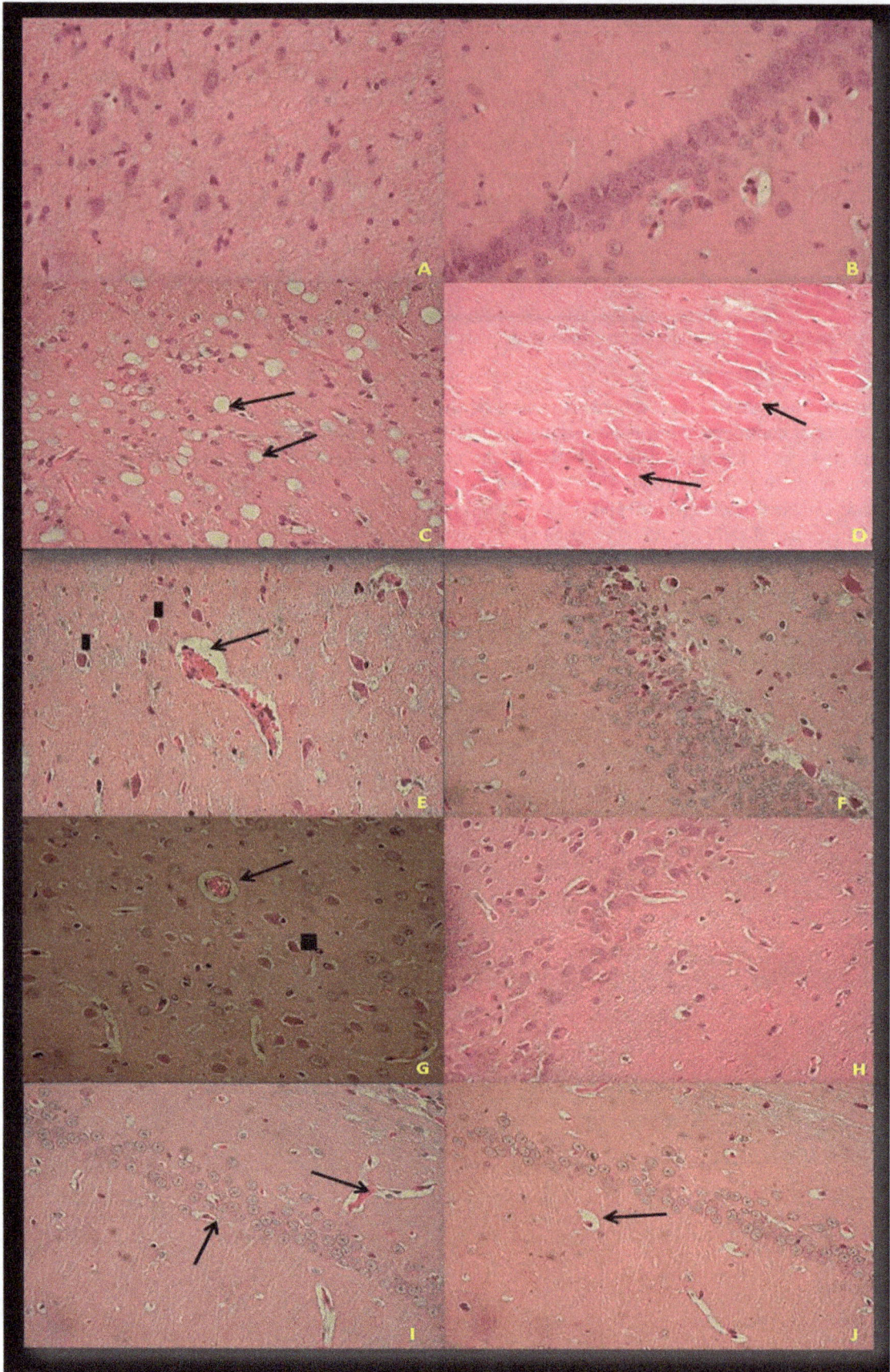

Figure 3.

## 4. Discussion

The present study showed a significant increase in LPO level in brain tissue of Li treated rats. Similar result was obtained by Buhalla *et al.* (2007) who found a significant increase in LPO level in brain tissue after Li treatment. Peroxidation of polyunsaturated fatty acids leads to degradation of phospholipids and cellular deterioration (Abou-Donia, 1981). The present result showed that Zn, VE and their interaction countered the LPO produced by Li. Similar result was observed by Buhalla *et al.* (2007) in the brain of rats treated with Li carbonate added to diet for two months. In this aspect, Chan *et al.* (1998) have demonstrated that Zn is involved in destruction of free radicals through Zn-metallothioneins which may serve as an efficient antagonist in inhibiting LPO in the brain tissue. Moreover, Zn causes inhibition of endogenous LPO to stabilize biomembranes (Dhawan & Goel, 1995).

NO is a messenger molecule with different functions in the body including long term potentiation, learning and memory (Floyd and Hensley, 2002). Inducible nitric oxide synthatase (iNOS) produces high amount of NO as a major contributor in the toxicity and disease pathway (Gouda *et al.*, 2010).In the present study, Li treatment elevated NO levels and caused degenerative changes in brain. Similarly, Harvey *et al.* (1994)found a significant increase in NO level in brain tissue after Li treatment. Also, strong positive immunoreactions for iNOS in cerebellar cortex with degenerated neurons and dilated congested capillaries of Li treated rats were observed by Bashandy (2013). Moreover, a significant increase in expression of iNOS in rat cerebellum under stress was detected by Gouda *et al.* (2010).In the current study, NO levels of rats treated with Zn, VE or their combination significantly decreased with improvement in the histological structure as compared with Li treated rats. In comparison, Bashandy (2013) found that neurons in the cerebellar medulla retained their normal appearance but still some degenerated neurons and slightly congested capillary in brain of rats treated Li and selenium. This could be attributed to the properties of Zn like selenium has antioxidant properties which provide protection from ROS induce cell damage (Chen and Berry, 2003).The protective effect Zn and VE may be due to it is role in regulation of redox status under physiological conditions (Reddy *et al.*, 2009) and reduction of LPO and NO (Savaskan *et al.*, 2003).

The present data showed that Li–induced the activities of SOD and CAT in the brain tissues of rats treated with Li. This altered of the two antioxidant balance in the brain by administration of Li may perturb the brain cell normal functioning, because balance between SOD and CAT are relevant for cell function (Savolainen, 1978). Co-treatment of rats with Zn along Li results in decline in the activities of SOD and CAT in comparison with Li treated alone. Several studies on the antioxidant property of Zn were reported (Sidhu, *et al.*, 2005 and 2006; Buhalla *et al.*, 2007),due toZn plays an important role as structural element of non-mitochondrial form of SOD (Choi, 1993). GSH is the most abundant low molecular weight thiol involved in antioxidant defense in animal cells. In the present study, the level of GSH in the brain tissue was increased by Li treatment. Similar results was obtained by Nanda *et al.* (1996) and Cui *et al.* (2007) who found a significant increase in GSH level in the brain of Li treated rat. However, Joshi *et al.* (2013) found a significant decreased in GSH level in different organs of rat treated with Li carbonate for 21 days. The increased levels of GSH in Li-treated rats may be due to increased detoxification capacity of the brain, most of the GSH in the brain is localized in glial cells rather than neurons, suggesting that Li affects the glial cells (Meister, 1984). Moreover, in brain, astrocytes play a central role in the metabolism of GSH (Takuma *et al.*, 2004). Co-treatment of rats with Zn or VE results in decline the level of GSH in the brain. This effect of Zn or VE could be returned to the antioxidative properties of Zn and VE as evident by decreasing the LPO levels and SOD and CAT activities in the present study.

Activity of AChE in brain homogenate showed a significant increase in comparison with control group and co-treatment of rats with Zn, VE or combination of Zn and VE elevated the reduction in activity of AChE. Jope (1979) found that LiCl treatment stimulates cholinergic activity in certain brain regions which may play a role in the therapeutic effect of LiCl in neuropsychiatric disorders. Also, Zn may act as a neuromodulator of excitatory or inhibitory processes (Vera-Gil *et al.*, 2003). Short-term orally supplementation of *Sonchusasper*, is traditionally used as a folk medicine to treat mental disorders, elevated brain antioxidant enzymes and inhibited ACE activity (Kumar *et al.*, 1994).

Generally, the levels of elements reflected dietary concentrations of these elements (Reinstein *et al.*, 1984). In the present study, concentration of Zn and Cu was increased in brain of rats treated with Zn or combination of Zn and VE. Autopsy studies of adults revealed that the cerebellum retains more Li than other organs, followed by the cerebrum and the kidneys (Schrauzer, 2002). Onosaka and Cherian (1982) returned this increase due in part to increase binding of Cu by metallothionein, which increases when the concentration of Zn increases. Also, in the present study, Li level was not detected in tissues of normal rats, however, it increased in tissues of rats treated with Li alone and in the rats co-treated with Zn, VE or combination of Zn and VE.

In conclusion, the data from the present study showed that Zn and VE and their interaction are effective in protection against Li- induced brain toxicity in rat. The effect of Zn may be attributed to formation Zn-metallothionein. In addition, Zn metallothionein and VE are free radical scavenger.

## References

Abou-Donia, M. B. (1981).Organophosphorous ester-induced delayed neurotoxicity. *Ann Rev Pharmacol Toxicol, 21*(1), 511-548. http://dx.doi.org/10.1146/annurev.pa.21.040181.002455

Allagui, M. S., Hfaiedh, N., Vincent, C., Guermazi, F., Murat, J.C., Croute, F., & El-Feki, A. (2006).Changes in growth rate and thyroid- and sex-hormones blood levels in rats under sub-chroniclithium treatment. *Hum Exp Toxicol, 25*(5), 243-250. http://dx.doi.org/10.1191/0960327106ht620oa

An, W. L., Pei, J. J., & Cowburn, R. F. (2005). Zinc-induced anti-apoptotic effects in SH-SY5Y neuroblastoma cells via the extracellular signal-regulated kinase ½.*Mol. Brain Res, 135*(1-2), 40-47. http://dx.doi.org/10.1016/j.molbrainres.2004.11.010

Aral, H., & Vecchio-Sadus, A. (2008).Toxicity of lithium to humans and the environment-A literature review. *Ecotoxicology and Environmental Safety, 70*, 349-356. http://dx.doi.org/10.1016/j.ecoenv.2008.02.026

Bashandy, M. A. (2013). Effect of lithium on the cerebellum of adult male albino rat and the possible protective role of selenium (Histological, Histochemical and immunohistochemical study). *Journal of American Science, 9*(11), 167-176. http://dx.doi.org/10.4172/2157-7099.1000263

Beutler, E., Duron, O., & Kelly, B. M. (1963). Improved method for the determination of blood glutathione. *Journal of Laboratory and Clinical Medicine, 61*, 882-888.

Bhalla, P., Chdaha, V. D., Dhar, R., & Dhawan, D. K. (2007). Neuroprotective effects of zinc on antioxidant defense system in lithium treated rat brain. *India Journal of Expermintal Biology, 27*(5), 595-607. http://dx.doi.org/10.1007/s10571-007-9146-0

Carney, S., Rayson, B., & Morgan, T. (1970). The effect of lithium on the permeability response induced in the collecting duct by antidiuretic hormone. *1'fluegers Arch, 373*(2), 105-112. http://dx.doi.org/10.1007/bf00584848

Chan, S., Gerson, B., & Subramanium, S. (1998). The role of copper, molybdenum, selenium and zinc in nutrition and health. *Clin Lab Med, 30*(2), 195.http://dx.doi.org/10.1079/bjn19730025

Chen, J., & Berry, J. (2003).Selenium and selenoproteins in the brain disease. *J. Neurochem, 86*(1), 1-12. http://dx.doi.org/10.1046/j.1471-4159.2003.01854.x

Choi, B. H. (1993). Oxygen, antioxidants and brain dysfunction. *Yonsei Medical J, 34*(1), 1. http://dx.doi.org/10.3349/ymj.1993.34.1.1

Chvapil, M., Janet, C., Ludwig, I., Sipes, G., & Ronald, L. (1976). Inhibition of NADPH oxidation and related drug oxidation in liver microsomes by zinc. *Biochem. Pharmacol.*, 1787-1791. http://dx.doi.org/10.1016/0006-2952(76)90417-2

Cui, J., Shao, L., Young, L. T., & Wang, J. F. (2007). Role of glutathione in neuroprotective effects of mood stabilizing drugs lithium and valproate. *Neuroscience, 144*(4), 1447-1453. http://dx.doi.org/10.1016/j.neuroscience.2006.11.010

Dhawan, D., & Goel, A. (1995).Further evidence of zinc as a hepatoprotective agent in rat liver toxicity. *ExptMolPathol, 6*, 110-117. http://dx.doi.org/10.1006/exmp.1995.1035

Ding, A. H., Nathan, C., & Steuehr, D. J. (1988). Release of reactive nitrogen intermediates and reactive oxygen intermediates from mouse peritoneal macrophages. *Journal of Immunology, 141*, 2407-2412. http://dx.doi.org/10.4049/jimmunol.171.10.5447

Drury, A. A., & Wallington, E. A. (1980). *Carletons histological technique* (5th ed.). Oxford University press, New York, Toronto. http://dx.doi.org/10.1017/s0038713400111546

Ellman, G. L., Courtney, K. D., Andres, V. J. R., & Featherstone, R. M. (1961). A new and rapid colorimetric determination of acetylcholinesterase activity. *Journal of Biochemical Pharmacology, 7*, 88-95. http://dx.doi.org/10.1016/0006-2952(61)90145-9

Floyd, R., & Hensley, K. (2002).Nitrones as neuro-protectants and anti-aging drugs. *Ann. N.Y. Acad. Sci, 959*, 321-329. http://dx.doi.org/10.1111/j.1749-6632.2002.tb02103.x

Gouda, S., Naim, M., El-Aal, H., & Mahmoud, S. (2010). Effect of alpha-phenyl-n-tertbutylnitrone on aging of the cerebellum of male albino rats (Histological and Immuno-histochemical study). *Egypt J. Histol, 33*(3), 495-507. http://dx.doi.org/10.4172/2157-7099.1000245

Grolea, G. (1994). Lithium toxicity. *Emerg. Med. Clin. N. Am, 12,* 511-531. http://dx.doi.org/10.1046/j.1442-2026.2000.00107.x

Gulec, M., Gurel, A., & Armutcu, F. (2006). Vitamin E protects against oxidative damage caused by formaldehyde in the liver and plasma of rats. *Mol Cell Biochem, 290,* 61-67. http://dx.doi.org/10.1007/s11010-006-9165-z

Habila, N., Inuwa, H. M., Aimola, I. A., Udeh, M. U., & Haruna, E. (2012). Pathogenic mechanism of *Trypanosomaevansi* infection. *Res Vet Sci, 93,* 13-17. http://dx.doi.org/10.1016/j.rvsc.2011.08.011

Harvey, B. H., Carstens, M. E., & Taljaard, J. J. (1994). Evidence that lithium induces a glutamatergic: nitric oxide-mediated response in rat brain. *Neurochem. Res., 19*(4), 469-474. http://dx.doi.org/10.1007/bf00967326

Holland, D. R., Hausrath, A. C., Juers, D., Matthews, B. W., & Christianson, D. W. (1995). Structural analysis of zinc substitutions in the active site of thermolysin. *Protein Sci., 4*(10), 1955-1965. http://dx.doi.org/10.1002/pro.5560041001

Jope, R. S. (1979). Effect of lithium treatment *in vitro* and *in vivo* on acetylcholine metabolism in rat brain. *Journal Neurochemistry, 33,* 487-495. http://dx.doi.org/10.1111/j.1471-4159.1979.tb05179.x

Joshi, D. K., Chauhan, D. S., Pathak, A. K., Mishra, S., Choudhary, M., Singh, V. P., & Tripathi, S. (2013). Lithium potentiate oxidative burden and reduced antioxidant status in different rat organs system. *International Journal of Toxicological and Pharmacological Research, 5*(1), 9-14. http://dx.doi.org/10.1007/s12291-008-0072-9

Kheir-Eldin, A. A., Motawi, T. K., Gad, M. Z., & Abd-ElGawad, H. M. (2001). Protective effect of vitamin E, β-carotene and N-acetylcysteine from the brain oxidative stress induced in rats by lipopolysaccharide. *Int J Biochem. Cell B, 33*(5), 475-482. http://dx.doi.org/10.1016/s1357-2725(01)00032-2

Kumar, U., Dunlop, D. M., & Richardson, J. S. (1994). The acute neurotoxic effect of β-amyloid on mature cultures of rat hippocampal neurons is attenuated by the anti-oxidant U-78517 F. *Intl J neurosci, 79*(3-4), 185-190. http://dx.doi.org/10.3109/00207459408986079

Luck, H. (1963). Catalase. In Bergmer H-U (Ed.), *Methods of enzymatic analysis* (pp. 885-888). Academic Press, New York. http://dx.doi.org/10.1016/b978-0-12-395630-9.50158-4

Martins, M. R., Petronilho, F. C., Gomes, K. M, Dal-Pizzol, F., Streck, E. L., & Quevedo, J. (2008). Antipsychotic-induced oxidative stress in rat brain. *Neurotox Res., 13,* 63-69. http://dx.doi.org/10.1007/bf03033368

Meister, A. (1984). New aspects of glutathione biochemistry and transport selective alteration of glutathione metabolism. *Nutr Rev, 42,* 397-410. http://dx.doi.org/10.1111/j.1753-4887.1984.tb02277.x

Misra, H. P., & Froidovich, I. (1972). The role of superoxide anion in the autoxidation of epinephrine and a simple assay for superoxide dismutase. *Journal of Biological Chemistry, 247,* 3170-3175. http://dx.doi.org/10.1016/0003-2697(78)90010-6

Nanda, D., Tolputt, J., & Collard, K. J. (1996).Changes in brain glutathione levels during postnatal development in the rat. *Brain Res. Dev., 94,* 238-241. http://dx.doi.org/10.1016/s0165-3806(96)80016-2

Ohkawa, H., Ohishi, N., & Yagi, K. (1979). Assay for lipid peroxidation in animal tissue by thiobarbaturic acid reaction. *Journal of Analytical Biochemistry, 95,* 351-358. http://dx.doi.org/10.1016/0003-2697(79)90738-3

Oktem, F., Ozguner, F., Sulak, O., Olgar, S., Akturk, O., Yilmaz, H. R., & Altuntas, I. (2005). Lithium-induced renal toxicity in rats: protection by a novel antioxidant caffeic acid phenethyl ester. *Mol Cell Biochem, 277*(1-2), 109-115. http://dx.doi.org/10.1007/s11010-005-5426-5

Onosaka, S., & Cherian, M. G. (1982).The induced synthesis of metallothionein in various tissues of rats in response to metals. II. Influence of zinc status and specific effect of pancreatic metal lothionein. *Toxicology, 23*(1), 11-20. http://dx.doi.org/10.1016/0300-483x(82)90037-3

Oteiza, P., Clegg, M. S., Paolazago, M., & Keen, C. L. (2000). Zinc deficiency induces oxidative stress and AP-1 activation in 3T3 cells. *Radical Biol. Med, 28*(7), 1091-1099. http://dx.doi.org/10.1016/s0891-5849(00)00200-8

Parsad, A. S. (1988). Zinc in growth and spectrum of human zinc deficiency. *J. Am. Coll. Nutr, 28*(7), 1091-1099. http://dx.doi.org/10.1016/s0891-5849(00)00200-8

Reddy, G. K., Sailaja, P., & Krishnaiah, C. (2009). Protective effects of selenium on fluoride induced alterations in certain enzymes in brain of mice. *Journal of Enviro Biol., 36*(3), 679-683. http://dx.doi.org/10.3724/sp.j.1035.2010.00679

Reinstein, N. H., Bolã–Nnerdal, K. C. N., & Hurley, L. S. (1984).Zinc-Copper interaction in the pregnant rat fetal outcome and maternal and fatal zinc, copper and iron. *Journal of the American College of Nutrition, 45*(8), 167-169.

Savaskan, N. E., Brauer, A. U., Kuhbacher, M., Eyupoglu, I. Y., Kyriakopoulos, N., & Behne, N. (2003). Selenium deficiency increases susceptibility to glutamate-induce excitotoxicity. *FASEB. J, 17*, 112-114. http://dx.doi.org/10.1096/fj.02-0067fje

Savolainen, H. (1978). Superoxide dismutase and glutathione peroxidase activities in rat brain. *Res Commv Chem Pathol Pharmacol, 28*(2), 89-96. http://dx.doi.org/10.1096/fj.02-0067fje

Schrauzer, G. N. (2002). Lithium: Occurrence, dietary intakes, nutritional essentiality. *Journal of the American College of Nutrition, 21*(1), 14-21. http://dx.doi.org/10.1080/07315724.2002.10719188

Sidhu, P., Garg, M. L., & Dhawan, D. K. (2004). Protective role of zinc in nickel induced hepatotoxicity in rats. *ChemBiol Interact, 150*(2), 199-209. http://dx.doi.org/10.1016/j.cbi.2004.09.012

Sidhu, P., Garg, M. L., & Dhawan, D. K. (2005). Protective effects of zinc on oxidative stress enzymes in liver of protein-deficient rats. *Drug Chem Toxicol, 28*(2), 211-230. http://dx.doi.org/10.1081/dct-200052551

Sidhu, P., Garg, M. L., & Dhawan, D. K. (2006). Zinc protects rat liver histo-architecture from detrimental effects of nickel. *Biometals, 19*(3), 301-313. http://dx.doi.org/10.1007/s10534-005-0857-8

Soylu, A. R., Aydogdu, N., Basaran, U. N., Altaner, S., Tarcin, O., Gedik, N., ... Kaymak, K. (2006). Antioxidants vitamin E and C attenuate hepatic fibrosis in biliary-obstructed rats. *World J Gastroenterol, 40*, 101. http://dx.doi.org/10.1016/s0168-8278(04)90331-5

Takuma, K., Baba, A., & Matsuda, T. (2004). Astrocyte apoptosis: implications for neuroprotection. *ProgNeurobiol, 72*(2), 111-127. http://dx.doi.org/10.1016/j.pneurobio.2004.02.001

Tripathi, A., & Srivastava, U. C. (2008). Acetylcholinesterase: A versatile enzyme of nervous system. *Ann Neurosci., 15*(4), 106-111. http://dx.doi.org/10.5214/ans.0972.7531.2008.150403

Vera-Gil, A., Pérez-Castejón, M. C., Lahoz, J. M., Aisa, M. P., Serrano, R. P., & Pes, N. (2003). 65Zn uptake in the rat cerebellum and brainstem. *Histol Histopathol, 35*(5), 457-462. http://dx.doi.org/10.1023/b:hijo.0000045944.07844.bd

Young, W. (2009). Review of lithium effects on brain and blood. *Cell Transplantation, 18*, 1-100. http://dx.doi.org/10.3727/096368909x471251

# 7

# Indoor Radon Characteristics in Canadian Arctic Regions

Jing Chen[1]

[1] Radiation Protection Bureau, Health Canada, Ottawa, Canada

Correspondence: Jing Chen, Radiation Protection Bureau, Health Canada, 775 Brookfield Road, Ottawa K1A 1C1, Ontario, Canada. E-mail: Jing.chen@hc-sc.gc.ca

**Abstract**

Radon is a naturally occurring radioactive gas generated by the decay of uranium bearing minerals in rocks and soils. Exposure to indoor radon has been identified as the second leading cause of lung cancer after tobacco smoking. In an indoor environment, there are many factors affecting indoor radon concentrations. Those factors could be different in the Arctic regions than the rest of Canada. Based on the results from recently completed Canadian residential radon survey, this technical note assessed indoor radon characteristics and associated radiation doses in Canadian Arctic regions and compared them to the average radon characteristics in Canada. In Arctic health regions the percentage of homes above 200 Bq/m$^3$ varied from 0% in Nunavut to 19.6% in Yukon Territory. On average, indoor radon characteristics in the Canadian Arctic regions are similar to the overall indoor radon characteristics in Canada. Although there are no significant differences in indoor radon exposure between the Canadian Arctic and rest of Canada, the average lung cancer incidence rate in the Arctic health regions is a factor of 1.6 higher than the national average lung cancer rate. The higher lung cancer rate in Canadian Arctic is likely due to the higher smoking rate in the northern communities.

**Keywords:** radon, indoor, lung cancer, tobacco smoking

## 1. Introduction

Radon is a naturally occurring radioactive gas generated by the decay of uranium bearing minerals in rocks and soils. Radon is invisible, odourless and tasteless and emits ionizing radiation as it decays. As a gas, radon can move freely through the soil enabling it to escape to the atmosphere or seep into dwellings. In the open air, the amount of radon gas is very small and does not pose a health risk. However, in enclosed or poorly ventilated spaces, radon can accumulate to high levels. As radon breaks down it forms radioactive particles called radon decay products or radon progeny. Radon gas and radon progeny in the air can be breathed into the lungs where they breakdown further and emit ionizing radiation in the form of alpha particles. Alpha particles release small bursts of energy which are absorbed by nearby lung tissue and result in lung cell death or damage. When lung cells are damaged, they have the potential to result in cancer when they reproduce. The only known health effect associated with long-term exposure to elevated radon levels in indoor air is an increased risk of developing lung cancer. Several large joint analyses of residential radon exposure and lung cancer incidence in Europe (Darby et al., 2005, 2006), North America (Krewski et al., 2005, 2006) and China (Lubin et al., 2004) provided strong evidence that exposure to indoor radon can increase the risk of lung cancer in the general population. Radon and its decay products have been identified as the second leading cause of lung cancer after tobacco smoking (WHO 2009, IARC 2004, 2012a, 2012b). Based on new scientific information and a broad public consultation, the Government of Canada revised the guideline for exposure to radon in indoor air from 800 to 200 Bq/m$^3$ in 2007 (Health Canada 2007). The new guideline recommends that remedial measures should be undertaken in a dwelling whenever the average annual radon concentration exceeds 200 Bq/m$^3$ in the normal occupancy area.

Indoors, radon concentrations can vary widely, depending on the type of rock underlying the dwelling, house type and structure, building materials used in the construction, ventilation, and other environmental factors including local weather conditions (UNSCEAR, 2009). In 2009, Health Canada launched a national residential radon survey to gain a better understanding of the distribution of radon concentrations in homes across Canada (Health Canada 2012). The survey was completed in 2011 with radon measurements in about 14,000 homes in all administrative areas defined by the provincial and territorial ministries of health across Canada.

The factors that affect indoor radon in the Canadian Arctic regions may be different than those in the rest of

Canada. The objective of this technical note was to assess indoor radon characteristics and associated radiation doses in Canadian Arctic regions and compare them to the average radon characteristics in Canada. Since radon and its decay products have been identified as the second known cause of lung cancer after tobacco smoking, lung cancer incidence rates and smoking rates in the Arctic regions were briefly discussed.

## 2. Methods

In this study, radon characteristics were determined for the administrative units as health regions. The same units were used in the national residential radon survey (Health Canada 2012). Canada is currently divided into 123 administrative health regions which are defined by the provincial and territorial health ministries (Statistics Canada 2011). Five health regions are completely or partially in the Arctic (north of 60° in latitude). They are, from east to west: Labrador-Grenfell Regional Integrated Health Authority (1014H) in Newfoundland and Labrador, Nunavik (2417F) in Quebec, Nunavut (6201F), Northwest Territories (6101E), and Yukon Territory (6001E), as shown in Figure 1 (for details, please visit Statistics Canada website, Statistics Canada 2011).

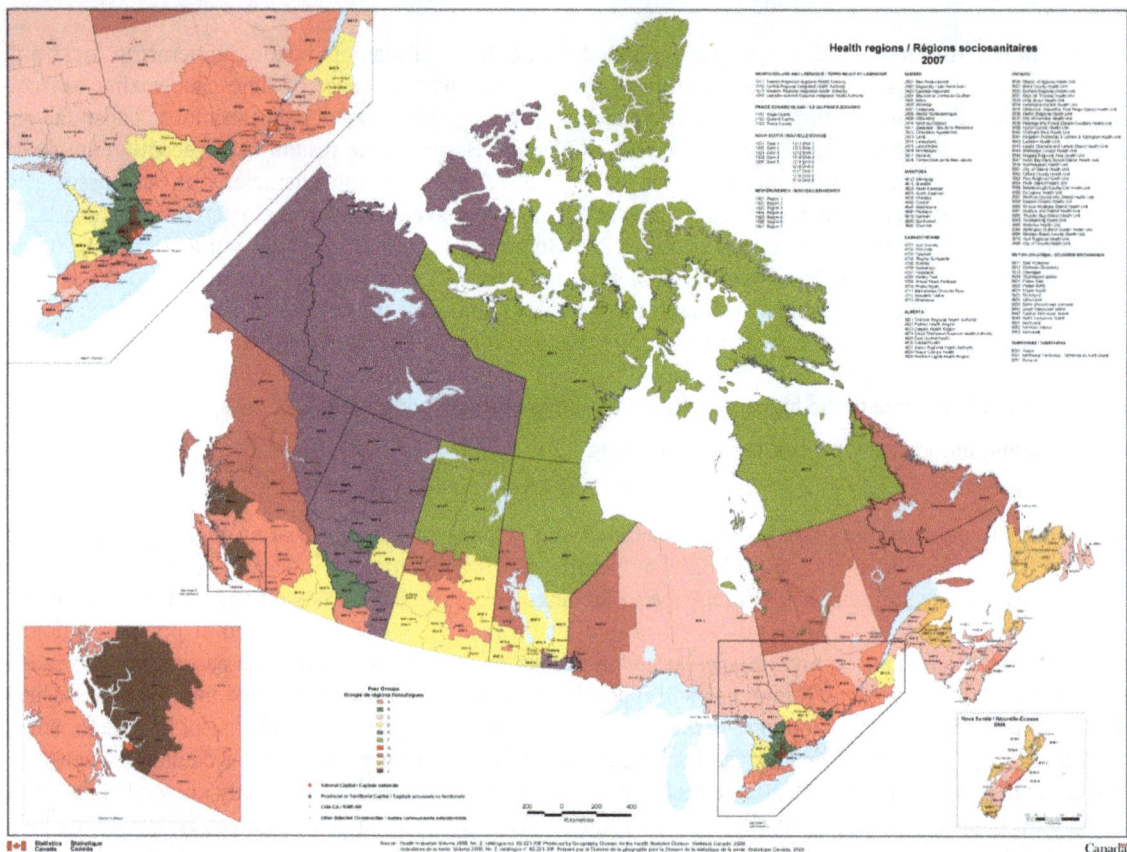

Figure 1. Health regions and peer groups in Canada (for detailed view, visit Statistics Canada Website) (Statistics Canada 2011)

Results of indoor radon measurements were taken from the summary report of the national radon survey (Health Canada 2012). The survey was started in 2009 and conducted over 2 years during the heating seasons (October to April) of 2009/10 and 2010/11. Long-term radon measurements were performed in all the homes surveyed. The observed radon concentrations in Canadian homes follow a log-normal distribution (Chen et al. 2012). Because of the wide distribution in radon concentrations, a central estimate alone, such as arithmetic mean (AM) or geometric mean (GM), will not be able to represent the distribution. Therefore, the percentage of homes above 200 Bq/m$^3$, the Health Canada current radon guideline (Health Canada 2007), was chosen to characterise the radon concentration level in a given health region. The average characteristics for the Arctic regions and for Canada are population weighted averages. Populations as of July 2012 were used (Statistics Canada 2013c).

In the survey, single family homes were randomly selected in each health region. The survey was targeted to have about 100 radon measurements in a health region. As indicated in a previous study on sample size required for a community radon survey (Chen et al. 2008), a large sample size, such as one to several thousand samples in a typical community of several tens of thousands of homes, can definitely provide high quality results with a very small uncertainty. However, from a cost-effectiveness point of view, a random sampling with a sample size of one to several hundred can serve the survey purpose well with an acceptable uncertainty of less than 25% for distribution parameters (such as AM and GM) as well as the percentages of homes above the Canadian radon guideline (Chen et al. 2008).

Once indoor radon concentrations are characterised, associated radon inhalation doses can be estimated. Based on the United Nations Scientific Committee on the Effects of Atomic Radiation (2009) formula, the annual effective dose due to inhalation of indoor radon and radon progeny for a population in a given area, $E_{Rn}$, was assessed

$$E_{Rn}(nSv) = C_{Rn} \times 0.4 \times 7000 \times 9 \tag{1}$$

where $C_{Rn}$ is the arithmetic mean (AM) radon concentration in the units Becquerel per cubic metre (Bq/ $m^{-3}$). The typical value of 0.4 was used as the equilibrium factor for radon indoors. A recommended value of 9 nSv (Bq/$m^3$ per hr.) was used to convert the radon equilibrium-equivalent concentration to the population effective dose, assuming an 80% home occupancy time (i.e. 7,000 hours). The population dose due to indoor radon exposure is proportional to the arithmetic mean indoor radon concentration in an area (UNSCEAR, 2009).

## 3. Results

Table 1 summarises the characteristics of indoor radon in the five Arctic health regions. Due to very limited radon data available in the North, over-sampling (twice of the average sampling in other health regions) was required for the Arctic regions. However, the recruitment in Nunavik was extremely difficult where only 9 participants were identified. In total, the five Arctic regions represent about 5% of homes tested for radon in the current survey (Health Canada 2012) even though those regions represent only 0.5% of Canadian population.

Table 1. Characteristics of exposure to indoor radon in Canadian Arctic regions

| Health Region | Population | Num. Homes tested | % homes > 200Bq/$m^3$ | GM Bq/$m^3$ | GSD | AM Bq/$m^3$ | Annual Effective Dose mSv |
|---|---|---|---|---|---|---|---|
| Labrador-Grenfell Regional Integrated Health Authority | 37,545 | 201 | 3.0 | 20.5 | 2.65 | 37.6 | 0.9 |
| Nunavik | 10,937 | 9 | 11.1 | 20.8 | 3.52 | 45.9 | 1.2 |
| Nunavut | 30,799 | 78 | 0.0 | 9.3 | 1.48 | 10.3 | 0.3 |
| Northwest Territories | 43,198 | 185 | 5.4 | 38.6 | 3.01 | 69.5 | 1.8 |
| Yukon Territory | 32,276 | 225 | 19.6 | 87.2 | 3.06 | 175 | 4.4 |
| Canadian Arctic Region | 154,755 | 698 | 7.1 | 37.2 | 2.66 | 70.3 | 1.8 |
| Canada | 32,576,074 | 13807 | 6.9 | 41.9 | 2.77 | 72.2 | 1.8 |

Radon characteristics varied widely, which was similar to the results for all health regions across Canada where the percentage of homes above 200 Bq/$m^3$ varied widely from 0% to 44%. In Arctic health regions the percentage of homes above 200 Bq/$m^3$ varied from 0% in Nunavut to 19.6% in Yukon Territory. Nunavut is a unique territory in that many homes are built on stilts because of the permafrost. This architectural factor means many homes in Nunavut will not suffer from infiltration of any radon that is able to find its way to the surface of the earth. However, the population weighted average percentage of homes above 200 Bq/$m^3$ in the Arctic health regions, was 7.1%, which was comparable to the Canadian national average of 6.9%.

The observed radon concentrations in Canadian homes follow a log-normal distribution (Chen et al. 2012). This was also true for the Arctic health regions. A log-normal distribution is characterised by two parameters, the geometric mean (GM) and geometric standard deviation (GSD). The parameters (GM and GSD) varied significantly from one Arctic health region to the other. However, the population weighted GM and GSD for the Canadian North were 37.2 Bq/$m^3$ and 2.66, respectively which were similar to the Canadian national average of

GM=41.9 Bq/m$^3$ and GSD=2.77 (Chen et al. 2012).

The annual effective doses due to indoor radon (based on Eq. 1) were lower in Nunavut (0.3 mSv) and higher in Yukon Territory (4.4 mSv); the difference was more than a factor of 15. However, the population weighted annual effective dose due to indoor radon exposure in the Canadian Arctic region was the same as the national average of 1.8 mSv.

## 4. Discussion

The only known health effect associated with long-term exposure to elevated radon levels in indoor air is an increased lifetime risk of developing lung cancer. The risk of developing lung cancer from radon depends on the level of radon and how long people are exposed to those levels. The survey results showed no significant difference of average radon characteristics for homes in the Canadian Arctic and the rest of Canada. The population weighted annual effective dose due to indoor radon exposure in the Canadian Arctic is the same as the national average, i.e. 1.8 mSv. Although average indoor radon in Canadian Arctic regions did not differ significantly from the Canadian average, it is of interest to examine lung cancer incidence rates and compare those in the North with the statistics for the rest of Canada. Since radon and its decay products have been identified as the second leading cause of lung cancer after tobacco smoking, a look at the smoking statistics would also be of public interest.

Canadian cancer incidence data are available on the website of Statistics Canada for the most recent years up to 2007 (Statistics Canada 2013a). Canadian smoking statistics can be found in Table 105-0501 from the Statistics Canada website (Statistics Canada 2013b). For smoking, the most relevant indicator to the present study was the percentage of current daily smokers in a given administrative unit. Even though long term smoking history is known to be more relevant to lung cancer development, this study used smoking statistics available from 2003 to 2011 on the website of Statistics Canada (2013b). Table 2 summarized indoor radon exposures, lung cancer incidence (averaged from 1996 to 2007) and current daily smoking rates (averaged from 2003 to 2011) for the five Arctic health regions. For comparison, Canadian average statistics were also included in Table 2.

Table 2. Characteristics of indoor radon exposure, lung cancer incidence (averaged from 1996 to 2007) and current daily smoking rate (averaged from 2003 to 2011) in the Canadian Arctic regions

| Health Region Name | % homes > 200Bq/m$^3$ | Annual Effective Dose mSv | Lung Cancer Incidence per 100,000 (1996 - 2007) | Current daily smoker % (2003 - 2011) |
|---|---|---|---|---|
| Labrador-Grenfell Regional Integrated Health Authority | 3.0 | 0.9 | 34.7 | 24.8 |
| Nunavik | 11.1 | 1.2 | 74.1 | - [a] |
| Nunavut | 0.0 | 0.3 | 250.5 | 51.9 |
| Northwest Territories | 5.4 | 1.8 | 70.4 | 28.8 |
| Yukon Territory | 19.6 | 4.4 | 56.8 | 25.6 |
| **Canadian Arctic Region** | **7.1** | **1.8** | **95.0** | **34.0** |
| **Canada** | **6.9** | **1.8** | **58.1** | **16.4** |

[a]: smoking data for Nunavik were marked as "not applicable" in Statistics Canada Tables (Statistics Canada 2013b).

As shown in Table 2, among the five health regions in the North, lung cancer incidence rates varied from 35 per 100,000 population in Labrador-Grenfell to 250 per 100,000 population in Nunavut. Among the five Arctic regions, Nunavut has the highest lung cancer incidence rate with lowest radon level indoors. The lung cancer incidence rate in Nunavut is more than 4 times of Canadian average while the estimated annual effective dose from radon exposure in Nunavut is less than 20% of the Canadian average radon dose. As can be seen in Table 2, Nunavut has the highest smoking rate among the Arctic regions (smoking statistics were not available for Nunavik according to Statistics Canada). The smoking rate in Nunavut is more than 3 times of the national average. Since tobacco smoking is the primary leading cause of lung cancer (IARC 2004, 2012a), the much higher than average smoking rate could likely be the main contributing factor to the much higher than average lung cancer incidence rate in Nunavut.

## 5. Conclusions

Even though local geology, climate and housing can be different in the Canadian Arctic region compared to other areas of Canada, on average, indoor radon characteristics in the Arctic region appeared to agree very well with the overall indoor radon characteristics in Canada. As expected, the average radon concentrations, such as percentages of homes above 200 Bq/m$^3$ vary widely from one health region to the other. Among the five Arctic health regions, indoor radon exposure is lower in Nunavut and higher in Yukon Territory.

Although there are no significant differences in indoor radon exposure between the Canadian Arctic and rest of Canada, the average lung cancer incidence rate in the Arctic health regions is a factor of 1.6 higher than the national average lung cancer rate. The higher lung cancer rate in Canadian Arctic is likely due to the higher smoking rate in the northern communities.

It is well known that tobacco smoking is the primary cause of lung cancer (IACR 2004, 2012a). As summarised in the Radon Handbook issued by the World Health Organization (2009), indoor exposure to radon and radon progeny contributes to about 3 to 14% of all lung cancers depending on the average radon concentration in a community. Due to the high smoking rate in the Arctic communities, radon awareness and protective actions to reduce indoor radon exposure should be integrated with tobacco control programs in order to effectively improve the healthy living in the northern communities.

## References

Chen, J., Moir, D., & Whyte, J. (2012). Canadian population risk of radon induced lung cancer – a reassessment based on recent cross Canadian radon survey. *Radiat. Prot. Dosim., 152*, 9-13. http://dx.doi.org/10.1093/rpd/ncs147

Chen, J., Tracy, B. L., Zielinski, J. M., & Moir, D. (2008). Determining the sample size required for a community radon survey. *Health Physics, 94*, 362-365. http://dx.doi.org/10.1097/01.HP.0000298226.47660.e5

Darby, S. et al. (2005). Radon in homes and risk of lung cancer: collaborative analysis of individual data from 13 European case-control studies. *BMJ, 330*(7485), 223-227. http://dx.doi.org/10.1136/bmj.38308.477650.63

Darby, S. et al. (2006). Residential radon and lung cancer: detailed results of a collaborative analysis of individual data on 7148 subjects with lung cancer and 14208 subjects without lung cancer from 13 epidemiologic studies in Europe. *Scand J Work Environ Health, 32*(Suppl1), 1-83.

Health Canada. (2007). *Government Canada Radon Guideline.* Retrieved from http://www.hc-sc.gc.ca/ewh-semt/radiation/radon/guidelines_lignes_directrice-eng.php

Health Canada. (2012). *Cross-Canada Survey of Radon Concentrations in Homes.* Retrieved from http://www.hc-sc.gc.ca/ewh-semt/alt_formats/pdf/radiation/radon/survey-sondage-eng.pdf

International Agency for Research on Cancer. (2004). IARC monographs on the evaluation of carcinogenic risks to humans. Volume 83, Tobacco smoke and involuntary smoking. Lyon.

International Agency for Research on Cancer. (2012a). Tobacco smoking. IARC monographs – 100E. Lyon.

International Agency for Research on Cancer. (2012b). Radiation – a review of human carcinogens. IARC monographs – 100D. Lyon.

Krewski, D. et al. (2005). Residential radon and risk of lung cancer: a combined analysis of 7 North American case-control studies. *Epidemiology, 16*, 137-145. http://dx.doi.org/10.1097/01.ede.0000152522.80261.e3

Krewski, D. et al. (2006). A combined analysis of North American case-control studies of residential radon and lung cancer. *J Toxicol Environ Health A, 69*, 533-597. http://dx.doi.org/10.1080/15287390500260945

Lubin, J. H. et al. (2004). Risk of lung cancer and residential radon in China: pooled results of two studies. *Int J Cancer, 109*, 132-137. http://dx.doi.org/10.1002/ijc.11683

Statistics Canada. (2011). *2011 Health regions and peer groups.* Retrieved from http://www.statcan.gc.ca/pub/82-583-x/2011001/article/11587-eng.pdf

Statistics Canada. (2013a). Table 103-0404 Cancer incidence, by selected sites of cancer and sex, three-year average, Canada, provinces, territories and health regions (2011 boundaries).

Statistics Canada. (2013b). Table 105-0501 Health indicator profile, annual estimates, by age group and sex, Canada, provinces, territories, health regions (2011 boundaries) and peer groups.

Statistics Canada. (2013c). Population of census metropolitan areas. Population as of July 1$^{st}$ 2012.

The United Nations Scientific Committee on the Effects of Atomic Radiation. (2009). *UNSCEAR 2006 Report, Volume II, Annex E: Sources-to-effects assessment for radon in homes and workplaces.* United Nations, New                               York.                               Retrieved                               from http://www.unscear.org/docs/reports/2006/09-81160_Report_Annex_E_2006_Web.pdf

The World Health Organization. (2009). WHO Handbook on Indoor radon. Retrieved from http://whqlibdoc.who.int/publications/2009/9789241547673_eng.pdf

# 8

# Urbanization and Major Ion Hydrogeochemistry of the Shallow Aquifer at the Effurun - Warri Metropolis, Nigeria

Irwin Anthony Akpoborie, Kizito Ejiro Aweto & Oghenero Ohwoghere-Asuma

[1] Department of Geology, Delta State University, Abraka, Nigeria

Correspondence: Irwin Anthony Akpoborie, Delta State University, Nigeria. E-mail: tony.akpoborie@gmail.com

## Abstract

Results from chemical analyses of forty dug well water samples in the Effurun-Warri metropolis show that mean pH is 7.1 and mean TDS is 193 mg/l. Representative mean levels of cation occurrence include Ca, Mg, Na and K at 6.13mg/l, 4.09 mg/l, 4.89 mg/l and 3.37 mg/l respectively. Mean anion concentration for bicarbonate was 8.20mg/l, 1.27mg/l for sulphate and 23.74mg/l for chloride. Physical and chemical parameters are thus well below WHO and Standard Organization of Nigeria drinking-water quality standards. Piper diagram plots of the data indicate that ground water is predominantly Ca+Mg+Na Chloride facie and that mixing and ion exchange processes control the dominant cation in space and thus at each specific locality. Leachates from the many, widely distributed and unregulated landfills and dumpsites have been identified as possibly the principal sources of major ion loading to groundwater. The ubiquitous onsite sewage treatment soak away pits also contribute major ions to groundwater. These two sources are thus accountable for any observed local spikes in groundwater chloride content rather than sea water intrusion as had been previously suggested.

Keywords: urban water, leachates, major ions, dump sites, ion exchange, Niger Delta, Benin Formation

## 1. Introduction

The quality of ground water and surface water in the Effurun - Warri metropolis, the densely populated hub of the oil and gas industry in the western Niger Delta, Figure 1, Figure 2 has attracted considerable research interest because in the absence of reliable formal public water supply systems the majority of an estimated population of up to one million residents (Babatola & Uriri, 2013), commerce and industry rely on self-supplies from dug wells and shallow boreholes. Thus, potential contamination of shallow groundwater with heavy metals from four primary sources have been suggested: oil and gas and related industry activities (Aremu, Olawuyi, Metshitsuka, Sridhar, & Oluwande ,2002; Nduka & Orisakwe, 2007, 2009; Emonyan, Akporhonor & Akpoborie, 2008; Etchie, Etchie & Adewuyi, 2011), leachates from unregulated garbage dumps (Akudo, Ozulu & Osogbue , 2010), road wash and storm water runoff (Egboh, Nwajei & Adaikpoh, 2000) and soils (Iwegbue, Nwajei, Ogala, & Overah, 2010). In addition, Olobaniyi and Owoyemi (2004; 2007) have also suggested possible chloride enrichment of the underlying aquifer by recharge from the tidal Warri River and its tributary creeks and argue that heavy ground water abstractions in parts of the city are potentially inducing sea water intrusion from the Atlantic Ocean. The influences of all these factors on the major element geochemistry and the quality of groundwater are not well understood.

Furthermore, published research on the chemistry of water from the area has usually been devoid of geolocated data points and this has hitherto severely limited the identification and closer examination of any inherent spatial and possible temporal trends in the occurrence of chemical constituents in groundwater. Thus the objective of this paper is to provide a description of the major element geochemistry of groundwater using geo-referenced data and against which future evaluations and trends in this highly vulnerable and rapidly expanding urban environment may be compared. The results reported herein constitute part of a groundwater evaluation study, partial results from which have been reported by Akpoborie, Uriri and Efobo (2014).

### 1.1 Study Area

#### 1.1.1 Climate, Physiography and Drainage

The Effurun-Warri metropolis lies roughly between latitude $5^0 30'N - 5^0 45N$ and longitude $5^0 15'E - 5^0 50'E$, Figure 2 and enjoys a hot ($23^0C - 37^0C$) and humid (Relative Humidity, 50 - 70 per cent) equatorial climate with a dry season that extends from about November to February, and a wet season that begins in March, peaks in July and October. 30-year mean annual rainfall is 3000mm (Adejuwon, 2011).

The area rests mainly on the Sombreiro –Warri Deltaic Plain (SWP) one of the distinguishable physiographic landforms resulting from Recent and modern delta top deposition in the Niger Delta (Short & Stauble, 1967; Allen,1965) the others being the Brackish Water and Mangrove Swamps (BMS) and the Fresh Water Swamps (FWS). The city is bound on its western edge by the BMS, swamp terrain that has so far limited city expansion in that direction.

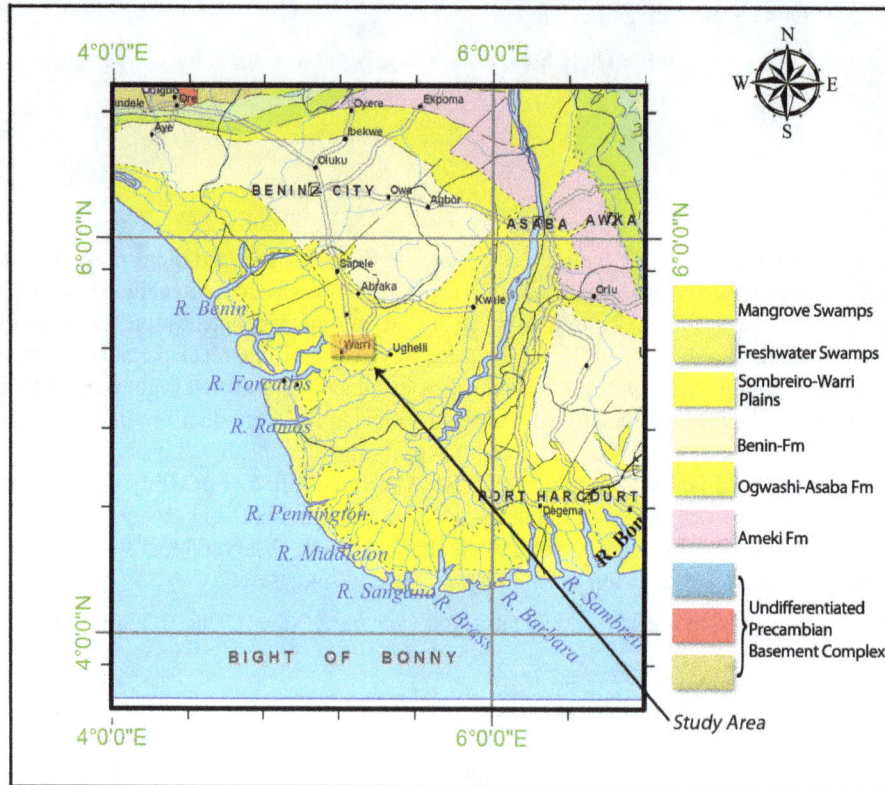

Figure1. Geological map of the western Niger Delta showing location of the Effurun-Warri Metropolis (Adapted from NGSA, 2004)

The Sombreiro –Warri Deltaic Plain is low lying, with an almost imperceptible seaward gradient and its lower parts merge with the water table and turn into swamps as evident in the Effurun-Warri metropolis. The area is circumscribed and drained by the tidal Warri River to the south and its tributary Ogunu/Edjeba Creek to the north, Figure 2. Mangroves line river banks especially in the western sector where they are sustained by weakly brackish water propagated landwards by a tidal regime.

1.1.2 Geology and Groundwater Conditions

The sedimentary fill of the late Quaternary delta top BMS and SWP physiographic terrains, Figure 1 has been described in the western Niger Delta by Akpoborie (2011) and in the eastern sector by Abam (2007) and Amajor (1991) and consists of an admixture of fine, to medium grained and coarse sands; silty clay and discontinuous thin clay layers. In the Effurun-Warri area, sandy layers are dominant up to 30m below ground at Osubi and Ejeba neighborhoods (Olobaniyi & Owoyemi, 2004), while further east Otobo, Aigbogun and Ifedili (2007) report the presence of up to 30m thickness of clay in the shallow aquifer at Aladja. In the western river port part of the city that is located on the boundary zone between the SWP and the BMS, Akpoborie (1996) describes two 100m deep boreholes that reveal a succession of black and gray colored silty clay and gravelly clay layers interbedded with coarse and medium grained sand beds. Because it has not been possible to distinguish between the younger Quaternary deposits and the Benin Formation proper in the subsurface, these delta top deposits are universally considered to be a continuation of the Benin Formation, Figure 1.

These shallow deposits are exploited everywhere in the Niger Delta with dug wells and shallow boreholes for domestic water supplies. Specifically in the Effurun-Warri area where public water supplies are inadequate, homeowners rely on groundwater from shallow dug wells and relatively inexpensive boreholes that are predominantly less than 40m deep. The dug wells and boreholes tap the medium to coarse – grained shallow aquifer sands. The permeability of the sands ranges from a reported $2.3x10^{-5}$cm/sec to $3.8x10^{-5}$cm/sec (Akpoborie, Ekakite & Adaikpo, 2000; Olobaniyi & Owoyemi, 2004; Niger Delta Development Commission, 2006). Deep regional groundwater flow in the Niger Delta region is reportedly in the west and south west direction (Ophori, 2007). In the Effurun –Warri area local flow in the shallow aquifer is controlled by a south westerly trending groundwater mound that extends from the Effurun GRA towards the neighborhood of Ekurede - Itsekiri and from which mound groundwater flows northwards, westwards, southwestwards and even eastwards (Akpoborie et al., 2014).

## 2. Methodology

Replicate water samples were collected from randomly located but evenly spread forty dug wells in the Effurun-Warri metropolis during a sampling program that was undertaken in mid - October, 2011. New, one litre size polyethylene bottles were used for collection of water samples. Prior to collection, the bottles were each washed with clean water and cleaning reagents then thoroughly rinsed with distilled, deionised water. After the pH, Total Dissolved Solids (TDS) and Electrical Conductivity (EC) were measured at point of collection, samples were sealed stored in ice chests and eventually transported to the laboratory within the hour of collection. The coordinates for selected sampling locations and associated codes are shown in Table 1 and Figure 2.

Electrical Conductivity and Total Dissolved Solids were measured *in situ* using the HACH Conductivity/TDS meters respectively. The pH was determined by means of a Schott Gerate model pH meter and the HACH Spectrophotometer was employed in determining the $NO_3$ ion using the cadmium reduction method. Na and K ion concentrations were obtained with a Jenway Clinical flame photometer. Sulphate content was determined by turbidimetry and Ca, Mg, $HCO_3$ and Cl with appropriate titrimetric methods as described by APHA (1992).

Table 1. Dug well water sampling locations at Effurun – Warri and sample codes used in the Piper and Stiff diagrams

| Coordinates | Street Address | Sample Code |
|---|---|---|
| N05$^0$30.9; E005$^0$49.13 | Boro Street, Oruhuwhorun | Orw |
| N05$^0$30.11;E00549.16 | Ero street, Oruhuwhorun | ORUWero |
| N05$^0$29.0;E005$^0$49.7 | Apostolic Church, Aladja | Aladja |
| N05$^0$29.5;E005$^0$45.5 | Umoji Compound, Aladja | AladjaUmoji |
| N05$^0$34.4;E005$^0$47.6 | Arubayi Str Eff. GRA | Ef GRA |
| N05$^0$31.14;E005$^0$46.10 | Inikoro Str Enerhen | enerhen |
| N05$^0$31.8;E005$^0$44.6 | Ekpen Ajamimogha | Ajm |
| N05$^0$34.8;E005$^0$46.5 | Army Barrack | ARM |
| N05$^0$34.6;E005$^0$41.9 | Tenumah Str, Ubeji | UBTem |
| N05$^0$35.6;E005$^0$33.9 | Ubeji health Centre | UBH |
| N05$^0$34.5;E005$^0$41.13 | Deeperlife Camp Rd | UBdpl |
| N05$^0$34.6;E005$^0$43.5 | St Gregory Hospital, Ubeji | UBStG |
| N05$^0$31.14;E005$^0$43.12 | Ekurede-Itsekiri | UB K-Itse |
| N05$^0$30.8;E005$^0$49.16 | Abuja street, DSC Township | DSCAbj |
| N05$^0$34.2;E005$^0$44.11 | Ekpan | Ekp |
| N05$^0$32.6;E005$^0$45.14 | Ugborikoko | Ogborik |
| N05$^0$31.9;E005$^0$44.12 | Agbassa | Agbassa |
| N05$^0$32.6;E005$^0$45.14 | Ugboroke close to dump site | Ugboroke |
| N05$^0$34.8;E005$^0$46.5 | Mammy Market | MammyMkt |
| N05$^0$30.8;E005$^0$42.6 | Ajamimogha | AJM2 |

| N05$^0$34.5;E005$^0$43.5 | Town Hall Ubeji | UBTH |
| N05$^0$34.12; E005$^0$49.10 | 70 Okuokoko | okuok |
| N05$^0$18.33; E005$^0$27.51 | Iyara3 | Iyara3 |
| N05$^0$34.6;E005$^0$47.8 | Dudu Crescent Eff. GRA | GRA |
| N05$^0$31.9; E005$^0$44.12 | Okere | OK |

## 3. Results and Discussion

### 3.1 Water Chemistry

### 3.1.1 Physical and Chemical Characteristics

Descriptive statistics of physical characteristics and major ion occurrence in ground water sampled at the different locations in the metropolis are shown in Table 2. Water ranges from mildly acidic, pH 5.2 at Iyara to alkaline, pH, 8.9 from a dug well near Ubeji Town Hall. Total Dissolved Solids (TDS) also varies widely from 545mg/l at Ubeji to 19mg/l at Abuja St., Steel Town, Aladja. Fifty seven per cent of all the samples showed a TDS of 200 mg/l or below while twenty five per cent have TDS of less than 100mg/l. The magnitude of range in spatial variation observed in all parameters suggests that mean values of parameters are of limited utility in describing shallow groundwater quality. However, on the basis of the data shown in Table 2, groundwater in the area is fresh and the major chemical constituents each occur at well below the limits specified in the WHO (2011) and SON (2007) drinking-water quality standards.

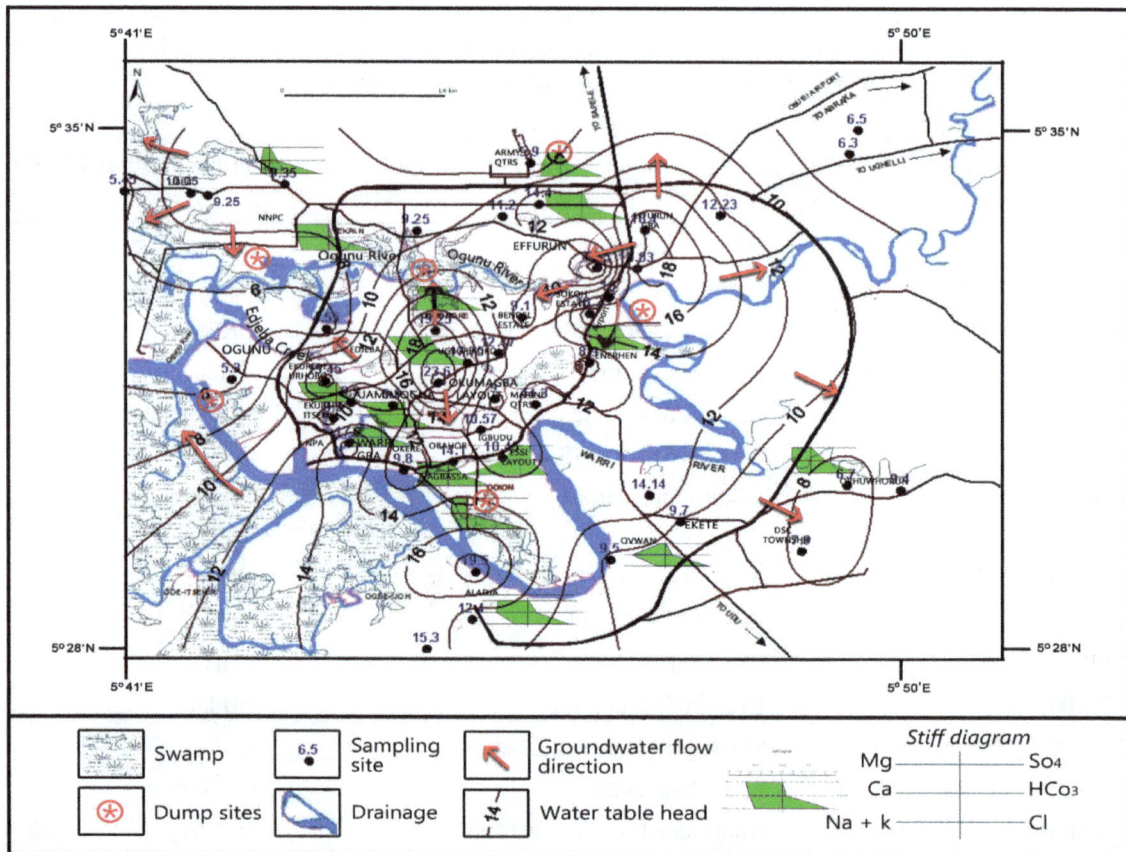

Figure 2. Map of Warri - Effurun area showing water table head distribution and Stiff pattern diagrams (Modified from Akpoborie, Uriri & Efobo, 2014)

These results appear to be in general agreement with previous studies undertaken in the metropolis, a selection of which include those by Akudo et al. (2010), Olobaniyi and Efe (2007) and Olobaniyi and Owoyemi (2004; 2007).

Table 2. Summary statistics of groundwater physical and chemical characteristics

| N =40 | Min | Max | Range | Mean | Std. Deviation | SON (2007) |
|---|---|---|---|---|---|---|
| pH | 5.2 | 8.9 | 3.7 | 7.1 | 1.04 | 6.5-9.2 |
| TDS (mg/l) | 19 | 545 | 526 | 193.20 | 138.22 | 500 |
| EC | 22 | 1091 | 1069 | 381.14 | 272.50 | 500 |
| Na (mg/l) | 0.04 | 23.67 | 23.63 | 4.89 | 5.39 | 200 |
| K (mg/l) | 0.05 | 16.43 | 16.38 | 3.37 | 3.91 | - |
| Ca (mg/l) | 0.25 | 29.03 | 28.78 | 6.13 | 5.97 | 75 |
| Mg (mg/l) | 0.07 | 18.70 | 18.63 | 4.09 | 4.21 | 50 |
| $HCO_3$ (mg/l) | 0.43 | 24.71 | 24.28 | 8.20 | 6.50 | 500 |
| $SO_4$ (mg/l) | 0.0 | 4.30 | 4.3 | 1.27 | 1.20 | 200 |
| Cl (mg/l) | 1.59 | 104.21 | 102.62 | 23.74 | 22.0 | 200 |
| $NO_3$ (mg/l) | 0.0 | 5.5 | 5.5 | 1.01 | 1.28 | 50 |

3.1.2 Major Ion Geochemistry

The Piper (1944) diagram plot of analyses Figure 3 contains only selected samples because many samples plotted in the same position in both ternary diagrams and the central diamond. In order to reduce clutter many of such samples were removed from the diagram in order to reduce clutter. The ions plot linearly on the upper right quadrant of the Piper diamond reflecting variability and a well developed mixing trend that is also clearly reflected in the cation ternary plot. As a result of this, there appear to be no dominant cation and ground water is predominantly Ca + Mg + Na Chloride facie.

Data from Akudo et al. (2010) who suggested a relationship between dumpsite leachate chemistry and nearby borehole water quality have also been plotted in the Piper diagram, Figure 3. The leachate sample from this data, red circle plots in the lower quadrant of the diamond while borehole water (blue circle) is in near linear alignment with other dug well data. In the cation ternary diagram both borehole water and leachate plot together at the end point of a mixing line that is indicative and may be interpreted as a source of sodium enrichment (Smith & Wahl, 2003).

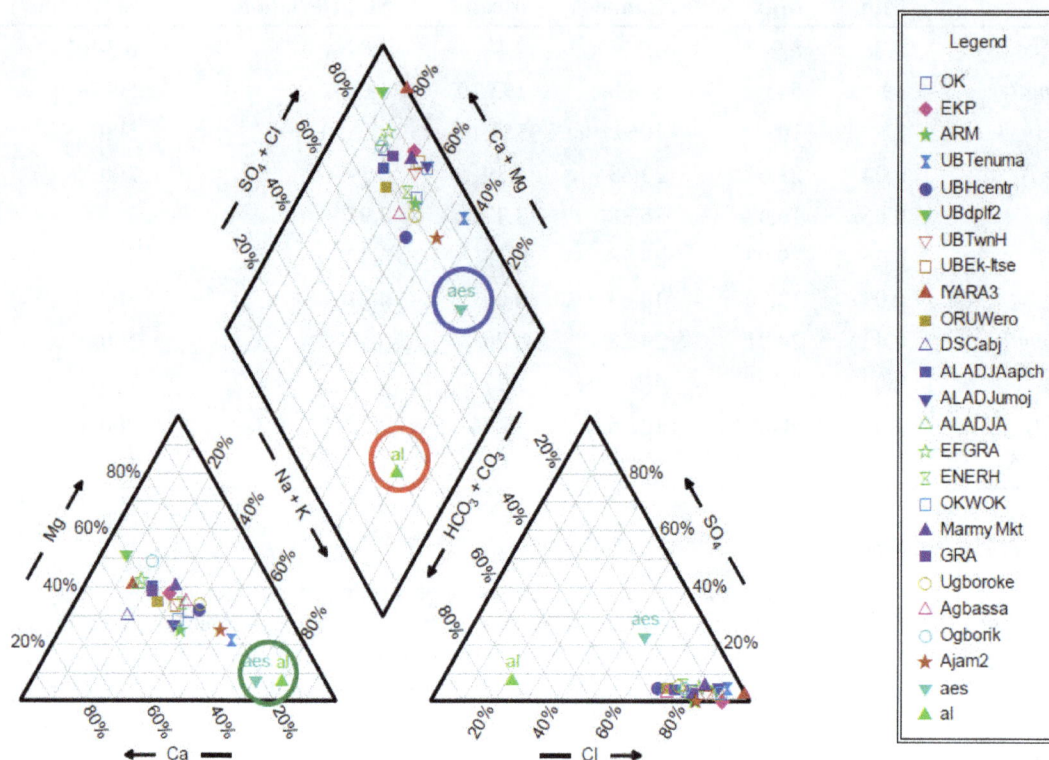

Figure 3. Piper diagram for dug well data and leachate data

Notes: al = leachate from waste dump at Esisi area; aes = groundwater from domestic borehole near the waste dump; all other sample codes explained in Table 1. Data for the leachate and borehole water obtained from Akudo et al. (2010)

Dumpsite and landfill leachate and sewage are indeed known sources of chloride, bicarbonate, calcium and magnesium loading to native groundwater (Panno et al., 2006; Uma, 2004; Hanchar 1991; Bradley et al, 1987, Baedecker & Back, 1979). Thus Akudo et al. (2010) report dumpsite leachate concentrations of up to 932mg/l, 2,208mg/l, 170mg/l and 396 mg/l for Na, $HCO_3$, $SO_4$ and Cl respectively from dumpsites in Warri, while Efe, Cheke and Ojoh (2013) show that up to twenty five huge and open garbage dumpsites and landfills are randomly located in the metropolis, Figure 2. This is in addition to the potentially tremendous amount of sewage generated by close to a million persons from numerous home and industrial onsite septic tank sewage treatment systems.

The shapes, and spatial distribution of Stiff (1951) diagrams derived from chemical analyses of water from selected locations, Table 3 and Figure 2, also suggest groundwater mixing and the resultant transitory nature of ground water chemistry within short distances in this urban setting. Ion exchange involving Na, Mg and Ca is also suggested by the interchange in dominance of these ions in space as groundwater moves in the area. Mixing is enhanced by existing groundwater gradients that control complex and multiple directions of groundwater movement.

Thus while Aweto and Akpoborie (2011) identify the importance of water – rock interactions in determining the chemistry of shallow groundwater in the SWP deposits that occur in a more rural setting at Orerokpe, anthropogenic factors associated with urbanization in the Effurun-Warri area contribute significantly to ultimate groundwater chemistry.

Table 3. Stiff Diagrams and cation sequences at selected locations at Effurun-Warri

| Stiff Diagram | Sample location/ code[*] | Cation sequence |
|---|---|---|
| | Okere /ok | Na+k>Ca>Mg |
| | Ekurede Itsekiri / | Mg>Ca>Na+k |
| | Ajamimogha/Ajam | Mg>Ca>Na+K |
| | Ubeji St Gregory's Hospital/ UbStG | Na+k>Ca>Mg |
| | Iyara3 /Iyara3 | Ca>Mg>Na+k |
| | Boro St, Oruworun/ Orw | Ca>Mg>Na+k |
| | Ubeji health cmtr /Ubhcentr | Na+k>Mg>Ca |
| | Leachate [**]from Esisi dumpsite/al | Na+k>Ca>Mg |
| | Esisi area borehohe [**]/aes | Na+k>Ca>Mg |

*Sample location code used in Table 1 and Piper Diagram, Figure 3

**Data obtained from Akudo et al. (2011)

However, potential direct recharge from an average annual 3000mm rainfall is high as indicated by reported groundwater level fluctuation in the wet/dry seasons of up to 5m (Akpoborie, Ekakite & Adaikpoh, 2000) in the SWP. This would enhance dilution of contaminating leachate and sewage as it mixes with native groundwater moving through the shallow aquifer and also possibly explain the low occurrence of the nitrate ion in groundwater, Table 2.

Therefore, evolution of groundwater chemistry in the shallow aquifer underlying the city appears to be driven by such factors as the infusion of dumpsite leachate that is an important source of bicarbonate, chloride and sodium ions into the system; direct recharge from mildly acidic and low pH rainwater (Efe, 2005, Olobaniyi & Efe, 2007), flood water and storm water from drains and gutters (Gobo, Amangabara & Agobie, 2014; Egboh, Nwajei, & Adaikpoh, 2000), untreated sewage and finally, the complex ground water flow patterns (Akpoborie et al., 2014) that exist in the area. Omo-Irabor, Olobaniyi, Oduyemi and Akunna (2008) have also stressed the influence of anthropogenic factors in the determination of water chemistry in the Niger Delta region.

Sea water intrusion into the aquifer underlying Warri-Effurun has also been suggested as an important process in perceived elevated chloride levels in shallow groundwater by Olobaniyi and Owoyemi (2004, 2007) and Olobaniyi and Efe (2007) who argue that sea water intrusion results from excessive ground water abstraction and also from recharging water from the tide influenced Warri River. However, Warri River and tributary creeks are perennial gaining rivers (Akpoborie, et al., 2014)) and there is no evidence of heavy groundwater withdrawals in any part of the city that has caused a reversal of existing gradients. Indeed, the Warri River and its tributary creeks carry predominantly fresh water (Aghoghovwia, 2011; Ogbeibu, Chukwurah & Oboh, 2011). Thus the presence of copious quantities of dump site-generated leachate and untreated sewage from ubiquitous septic tank soak- away pits in the metropolis suggests alternative and more credible sources of any and as has been shown, transitory chloride enrichment to shallow groundwater.

## 4. Conclusion

This study has shown that groundwater in the SWP deposits at the Effurun Warri metropolis has comparatively low mean TDS and major ions occur generally at levels that are well below SON and WHO drinking-water quality standards. However, because of rapid spatial and horizontal changes in major ion content, statistical mean values of major ion content should be used with caution.

Further, the evolution of groundwater chemistry in the shallow aquifer appears to be driven by urban induced factors that include infusion of dumpsite leachate that is an important source of bicarbonate, chloride and sodium ions into the system; direct recharge from acidic and low pH rainwater, flood water and storm water from drains and gutters, untreated sewage and finally, the complex ground water flow patterns that result from existing gradients and which enhance groundwater mixing.

## Acknowledgements

Partial funding for this study was provided by the Center for Research in Water and Environment (CREWE), Abraka nd for which the authors are grateful.

## References

Abam, T. K. S. (2007). N soil exploration and foundations in the Recent coastal areas of Nigeria. *Bulletin of Engineering Geology and the Environment, 53*, 13-9.

Adejuwon, O. A. (2012). Rainfall Seasonality in the Niger Delta Belt, Nigeria. *Journal of Geography and Regional Planning, 5*(2), 51-60. http://dx.doi.org/10.5897/JGRP11.096

Aghoghovwia, O. A. (2011). Physico-chemical characteristics of Warri River in the Niger Delta region of Nigeria. *Journal of Environmental Issues and Agriculture in Developing Countries, 3*(2), 40-46.

Akpoborie, I. A. (1996). Pollution from Oil in Delta State. In *Environmental Issues in Delta State* (pp. 49-59). Chevirol Resources Ltd., Ibadan.

Akpoborie, I. A. (2011). Aspects of the hydrology of the western Niger Delta wetlands: groundwater conditions in the Neogene (Recent) deposits of the Ndokwa area. *Africa Geoscience Review, 18*(3), 25-36.

Akpoborie, I. A., Ekakite, O. A., & Adaikpoh, E. O. (2000) The Quality of Groundwater from Dug Wells in Parts of the Western Niger Delta. *Knowledge Review, 2*(5), 72-75

Akpoborie, I. A., Uriri, A. E., & Efobo, O. (2014). Ground Water Conditions and Spatial Distribution of Lead and Cadmium in the Shallow Aquifer at Effurun- Warri Metropolis, Nigeria. *Environment and Pollution, 3*(3), 27-37. http://dx.doi.org/10.5539/ep.v3n3p27

Akudo, E. O., Ozulu, G. U., & Osogbue, L. C. (2010). Quality Assessment of groundwater in selected waste dump site areas in Warri. *Environmental Research Journal, 4*(4), 281-285. http://dx.doi.org/10.3923/erj.2010.281.285

Allen, J. R. L. (1965). Late Quaternary Niger Delta and Adjacent Areas: Sedimentary Environments and Lithofacies. *Bulletin American Association of Petroleum Geologists, 49*(5), 547-600.

Amajor, L. C. (1991). Aquifers in the Benin Formation (Miocene Recent), Eastern Niger Delta, Nigeria: Lithostratigraphy, Hydraulics, and Water Quality. *Environ Geol Water Sci., 17*(2), 85-101. http://dx.doi.org/10.1007/BF01701565

APHA. (1992). *Standard methods for the examination of water and waste water* (18th ed.). American Public Health Association, Washington, D.C.

Aremu, D. A., Olawuyi, J. F., Metshitsuka, S., Sridhar, M. K., & Oluwande, P. A. (2002). Heavy metal analysis of groundwater from Warri, Nigeria. *International Journal of Environmental Health Research, 12*(3), 261-267 http://dx.doi.org/10.1080/0960312021000001014

Aweto, K., & Akpoborie, I. A. (2011). Geo-Electric and Hydrogeochemical Mapping of Quaternary Deposits at Orerokpe in the Western Niger Delta. *J. Appl. Sci. Environ. Manage., 15*(2), 351-359.

Babatola, O., & Uriri, A. (2013). Assessment of Maternal Health Intervention Programme of Delta State, Nigeria:Application of the U.N Process Indicators. *Journal of Public Policy and Administration Research, 3*(9), 62-71.

Baedecker, M., & Back, W. (1979). Hydrogeological Processes and Chemical reactions at a landfill. *Groundwater, 17*(5), 429-437. http://dx.doi.org/10.1111/j.1745-6584.1979.tb03338.x

Efe, S. I. (2005). Urban effects on precipitation amount and rainwater quality in Warri metropolis, (Unpublished Ph.D. Thesis) Delta State University, Abraka, Nigeria.

Efe, S. I., Cheke, L. A., & Ojoh, C. O. (2013). Effects of Solid Waste on Urban Warming in Warri Metropolis, Nigeria. *Atmospheric and Climate Sciences, 3*, 6-12. http://dx.doi.org/10.4236/acs.2013.34A002

Egboh, S. H. O., Nwajei, G. E., & Adaikpoh, E. O. (2000). Selected Heavy metals concentration in sediments from major roads and gutters in Warri, Delta State, Nigeria. *Nig. J. Sc. Env., 2*, 105-111.

Emoyan, O. O., Akporhonor, E. E., & Akpoborie, I. A. (2008). Environmental Risk Assessment of River Ijana, Ekpan, Delta State. *Chem. Spec. and Bioavailability, 20*(1), 23-32. http://dx.doi.org/10.3184/095422908X295825

Etchie, T. O., Etchie, A. T., & Adewuyi, G. O. (2011). Source Identification of Chemical Contaminants in Environmental Media of a Rural Settlement. *Research Journal of Environmental Sciences, 5*, 730-740, http://dx.doi.org/10.3923/rjes.2011.730.740

Gobo, A. E., Amangabara, G. T., & Agobie, O. (2014). Impacts of Urban Land use changes on flood events in Warri, Delta State Nigeria. *Int. Journal of Engineering Research and Applications, 4*(9), 48-60.

Hanchar, D. W. (1991). Effects of septic-tank effluent on ground-water quality in northern Williamson County and southern Davidson County, Tennessee U. S. Geological Survey Water-Resources Investigations Report 91-4011.

Iwegbue, C. M., Nwajei, G. E., Ogala, J. E., & Overah, C. L. (2010). Determination of trace metal concentrations in soil profiles of municipal waste dumps in Nigeria. *Environ Geochem Health, 32*, 415-430 http://dx.doi.org/10.1007/s10653-010-9285-y

Nduka, J. K., & Orisakwe, O. E. (2007). Heavy Metal Levels and Physico – Chemical Quality of Potable Water Supply in Warri, Nigeria. *Annali di Chimica, 97*, 867-874. http://dx.doi.org/10.1002/adic.200790071

Nduka, J. K., & Orisakwe, O. E. (2009). Effect of Effluents from Warri Refinery Petrochemical Company WRPC on Water and Soil Qualities of "Contiguous Host" and "Impacted on Communities" of Delta State, Nigeria. *The Open Environmental Pollution & Toxicology Journal, 1*, 11-17. http://dx.doi.org/10.2174/1876397900901010011

Niger Delta Development Commission. (2006). Environmental Impact Assessment of the proposed Iyara Sand Fill Project, Document NDDC-005, NDDC Port Harcourt, Nigeria.

Nigeria Geological Survey Agency. (2004). Geological Map of Nigeria, NGSA, Abuja, Nigeria.

Ogbeibu, A. E., Chukwurah, A. E., & Oboh, I. P. (2013). Effects of Open Waste Dump-site on its Surrounding Surface Water Quality in Ekurede-Urhobo, Warri, Delta State, Nigeria. *Natural Environment, 1*(1), 1-16. http://dx.doi.org/ 10.12966/ne.06.01.2013

Olobaniyi, S. B., & Efe, S. I. (2007). Comparative assessment of rainwater and groundwater quality in an oil producing area of Nigeria: environmental and health implications. *Jnl Env Health Res, 6*(2), 111-117.

Olobaniyi, S. B., & Owoyemi, F. B. (2004). Quality of Groundwater in Deltaic Plain Sands Aquifer of Warri and Environs, Delta State, Nigeria. *Water Resources, 15*, 38-45.

Olobaniyi, S. B., & Owoyemi, F. B. (2007). Characterization by factor analysis of the chemical facies of groundwater in the Deltaic Plain Sands Aquifer of Warri, western Niger Delta, Nigeria. *African Journal of Science and Technology (AJST) Science and Engineering Series, 7*(1), 73-81.

Omo-Irabor, O. O., Olobaniyi, S, B., Oduyemi, K., & Akunna, J. (2008). Surface and groundwater water quality assessment using multivariate analytical methods: A case study of the Western Niger Delta, Nigeria. *Physics and Chemistry of the Earth, 33*, 666-673. http://dx.doi.org/10.1016/j.pce.2008.06.019

Ophori, D. U. (2007). A simulation of large scale groundwater flow in the Niger Delta, Nigeria. *Env. Geosciences, 14*(4), 181-195. http://dx.doi.org/10.1306/eg.05240707001

Otobo, E., Aigbogun, C. O., & Ifedili, S. O. (2007). Geoelectrical Evaluation of Waste Dump Sites at Warri and its Environ, Delta State, Nigeria. *J. Appl. Sci. Environ. Manage., 11*(2), 61-64.

Panno, S. V., Hackley, K. C., Hwang, H. H., Greenberg, S., Krapac, I. G., Landsberger, S., & O'Kelly, D. J. (2006). Characterization and identification of the sources of Na-Cl in ground water. *Ground Water, 44*(2),

176-187. http://dx.doi.org/10.1111/j.1745-6584.2005.00127.x

Piper, A. M. (1944). A graphic procedure in geochemical interpretation of water analyses. *Trans. Am. Geophys. Union, 25*, 914-923. http://dx.doi.org/10.1029/TR025i006p00914

Short, K. C., & Stauble, A. J. (1967). Outline of geology of Niger delta. *Bull. Amer. Assoc. Petr. Geol., 54*(5), 761-779.

Smith, S. J., & Wahl, K. L. (2003). Changes in streamflow and summary of major-ion chemistry and loads in the North Fork Red River Basin upstream from Lake Altu, northwestern Texas and western Oklahoma, 1945-1999, USGS Water Resources Investigation Report 03-1999.

Standards Organization of Nigeria. (2007). *Nigerian Standard for Drinking Water Quality NIS 554:2007.* Abuja, Nigeria.

Stiff, H. A. Jr. (1951). The interpretation of chemical water analysis by means of patterns. *Journal of Petroleum Technology, 3*(10), 15-17. http://dx.doi.org/10.2118/951376-G

Uma, K. O. (2004). Hydrogeology of the perched aquifer systems in the hilly terrains of Nsukka Town, Enugu State, S.E.Nigeria, *Water Resources, 14*, 85-92.

WHO. (2011). *Guidelines for drinking-water quality* (4th ed.). Retrieved from http://www.who.int

# 9

# A Comparison of Using Dominant Soil and Weighted Average of the Component Soils in Determining Global Crop Growth Suitability

T. Avellan[1], F. Zabel[1], B. Putzenlechner[1] & W. Mauser[1]

[1] Ludwig-Maximilians-Universität Munich, Dept. for Geography and Remote Sensing, Munich, Germany

Correspondence: Tamara Avellan, Ludwig-Maximilians-Universität Munich, Dept. for Geography and Remote Sensing, Luisenstr. 37, Munich 80333, Germany.

**Abstract**

Soil parameters represent key data input for crop suitability analysis. Soil databases are complex offering soil mapping units made up of various component soils. In the case of the Harmonized World Soil Database there can be up to 8 component soils per unit. In roughly 1/3 of soil mapping units, the additional component soils take up more than 50% of the pixel share value. The soil parameter value estimate, such as pH, salinity and organic carbon content, may differ between the value of the dominant soil component and the weighted average of the values of all component soil. Understanding the effect of these differences on crop model outputs may allow quantifying the error. In this study, we show the changes in crop suitability of 15 crops while using the parameter value estimates of the dominant soils versus a weighted average of the component soils. In the case of the latter, global crop suitability amounts to 54.5% of the earth's land surface–1% more than when using the values of just dominant soils. Intrinsic regional differences in the quality of the soil database influence the distribution of crop suitability classes especially in areas where share values of the dominant soil are low. The uncertainty range for the use of dominant versus component soils on the overall global crop suitability could be considered to be 1%, while that of each suitability class can amount to up to 4%.

**Keywords:** crop suitability, HWSD, quality control, dominant soil, mapping units, component soils

## 1. Introduction

Ensuring food security for the global population is already challenging in current times and will be even more, when population rises up to around 8.3 billion by 2030 (UNDP, 2008). Enhanced food production relies on three factors: increased yield, enhanced cropping intensity and the expansion of agricultural land (FAO, 2003). In 2009, the total amount of agricultural and permanent crops amounted to 2.5 billion ha which equals about 19% of the earth's land surface (Bontemps, Defourny, Van Bogaert, Arino, & Kalogirou, 2009). In the last four decades of the past century, 172 million ha of land have been added in developing countries (FAO, 2003). To ensure global food security, an additional 120 million ha of converted land are projected to be necessary until 2030 and an extra 5% will be necessary up to 2050 (Bruinsma, 2009). Most land is expected to be transformed in South America and Sub Saharan Africa (Fischer, 2000).

Models based on climate and soil inputs can help discern the areas where crops can grow optimally for given natural conditions. Fischer et al. (2002) showed that roughly 2.8 billion ha are to some degree suitable for rain-fed agriculture and Avellan, Zabel, and Mauser (2012) showed that about a quarter of the earth's land surface is suitable to highly suitable for the rain-fed growth of 15 major crops (Avellan, Zabel, & Mauser, 2012; Fischer, 2002). Both authors base their different models (Global agro-ecological zones versus fuzzy logic crop suitability) on global soil and climate databases. However, global soil databases are scarce and rely on patchy soil sampling. Few sets exist, such as the Harmonized World Soil Database (HWSD) (FAO/IIASA/ISRIC/ISSCAS/JRC, 2009) and the ISRIC-WISE derived soil properties on a 5 by 5 arc minute grid (Batjes, 2006). Global Climate Datasets are more varied. Past climate data can be obtained from interpolated station data (WorldClim), reanalysed forecasts (ERA) or hind-casted climate models (ECHAM, HadCM).

Avellan et al. (2012) showed that the quality of climate inputs is quite homogenous while global soil databases can differ widely. The choice of the database can have a strong effect on the amount and distribution of crop suitable areas, leading to a 10% difference between the two most common global soil datasets (Avellan et al.,

2012). Soil databases are immensely complex and the quality of the data is geographically diverse. For example, the HWSD is made up of four different input databases—each covering different areas of the world, using different sampling and compilation methods (FAO/IIASA/ISRIC/ISSCAS/JRC, 2009) (see Figure 1). Each pixel can contain up to 8 component soils which may, in sum, have a larger share within the pixel than the dominant soil class (see mock up example in Figure 2). When taking component soil classes into account, the soil parameter value estimate for each given pixel may be different than that of the dominant soil mapping unit (i.e. dominant soil value for pH is 8, but that of the weighted average of all component soils is 7.8).

In order to enhance modelling results a balance between the quantity and quality of the used input parameters has to be maintained. While more parameters might refine the modelling results, poor quality parameters might, in fact, be counterproductive. A careful analysis of both the quality of the data as well as their influence on final results might inform the choice of parameters. In Avellan et al. (2012), we started our crop suitability analysis using only the parameter value estimates of the dominant soil mapping unit of the topsoil (0-30 cm) on a pixel by pixel basis. In comparison, the Global Agro-ecological zones studies, used soil parameters from all component soils, top- and subsoils (0-30 cm and 30 cm and below), phases as well as management practices (IIASA/FAO, 2012). It is clear to the authors that other parameters relevant to soil databases such as subsoil parameters (30 cm and below), including drainage, granularity or acidity, as well as phases and management practices can have drastic effects on crop growth (Benjamin, Nielsen, & Vigil, 2003; Kirchhof et al., 2000; Van den Akker, Arvidsson, & Horn, 2003).

To our knowledge, the use of parameters in crop suitability models has not been substantiated by the analysis of the quality of the data. The inclusion of factors is defended by referring to standard works (i.e. FAO manuals (FAO, 1976, 2007) or similar) without questioning the validity of the usage. It is our intent to enhance model complexity in a step-by-step approach while showing the error margins incurred. Analogous to the well-known uncertainty ranges of climate models we wish to demonstrate a similar approach in the use of crop suitability estimations. Here, we assessed the influence of the area-weighted average of the additional component soils of the soil mapping units of the topsoil, on the amount and distribution of crop suitable areas.

## 2. Materials and Methods

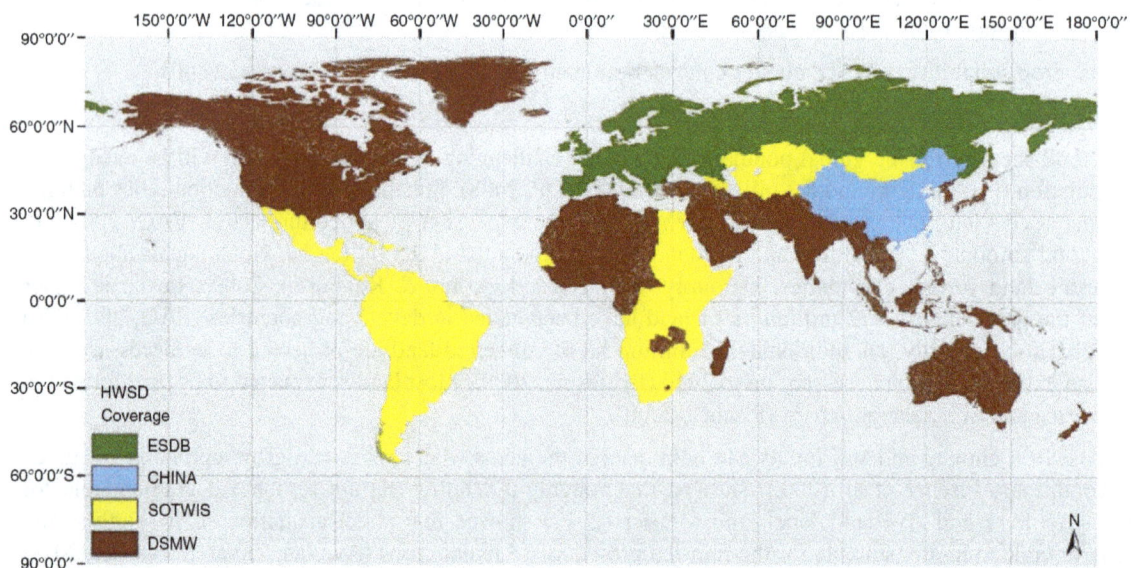

Figure 1. Distribution of the four underlying databases of the Harmonized World Soil Database (HWSD); European Soil Database (ESDB), Soil Map of China (CHINA), Soil and Terrain dataset (SOTWIS), Digital Soil Map of the World; adapted from (FAO, IIASA, ISRIC, ISS-CAS, & JRC, 2009)

*2.1 Datasets*

We used the following datasets and parameters at 30 arc seconds resolution (1 x 1 km at the equator):

Harmonized World Soil Database (HWSD, version 1.1): dominant and component soil mapping units of the topsoil (0-30cm) as input for eight parameter value estimates–textural class (USGS), coarse fragments volume (%), gypsum (%CaSO4), base saturation (%), pH, organic carbon (%), salinity (dS/m) and sodicity (%).

WorldClim dataset (Hijmans, Cameron, Parra, Jones, & Jarvis, 2005): mean annual temperature and mean annual precipitation

SRTM30 global digital elevation model (Farr et al., 2007; USGS, 2000): slope computed as percent rise.

Regions were defined for their economic relevance in global trade as a biophysical crop model was coupled to a Global Equilibrium Model in a subsequent step (Table 1).

Table 1. Coding of the regions

| Region code | Region name |
| --- | --- |
| AFR | Sub-saharan Africa |
| BEN | Belgium, Netherlands, Luxemburg |
| BRA | Brazil |
| CAN | Canada |
| CHI | China |
| FRA | France |
| FSU | Rest of former Soviet Republic |
| GBR | UK & Ireland |
| GER | Germany |
| IND | India |
| JPN | Japan |
| LAM | Rest of Latin America |
| MAI | Malaysia, Indonesia |
| MEA | Middle East, North Africa |
| MED | Spain, Portugal, Italy, Greece, Malta, Cyprus |
| MRC | Chile, Argentina, Uruguay, Paraguay |
| NAU | New Zealand, Australia |
| PAS1 | Guayanas |
| PAS2 | Iceland |
| PAS3 | Switzerland |
| PAS4 | Afghanistan, Pakistan |
| PAS5 | Mongolia |
| REU | Austria, Estonia, Latvia, Lithuania, Poland, Hungary, Slovakia, Slovenia, Czech Republic, Romania, Bulgaria |
| RUS1 | RUS1 (west) |
| RUS2 | RUS2 (east) |
| SCA | Finland, Sweden, Denmark |
| SEA | Kambodscha, Laos, Thailand, Vietnam, Myanmar, Bangaldesh |
| USA | United States of America |

*2.2 Dominant vs. Component Soil Areas and Soil Parameter Value Estimates*

Dominant soil is defined as the HWSD component soil with the largest share value irrespective of the fact that the other component soils together may have a larger share within one pixel. Soil parameter value estimates are the values each pixel has for a chosen parameter, i.e. pH, salinity, etc. In Figure 2 we have tried to show in a mock-up example how a pixel can be made up of several component soils and the effect the weighted average has on the parameter value estimate.

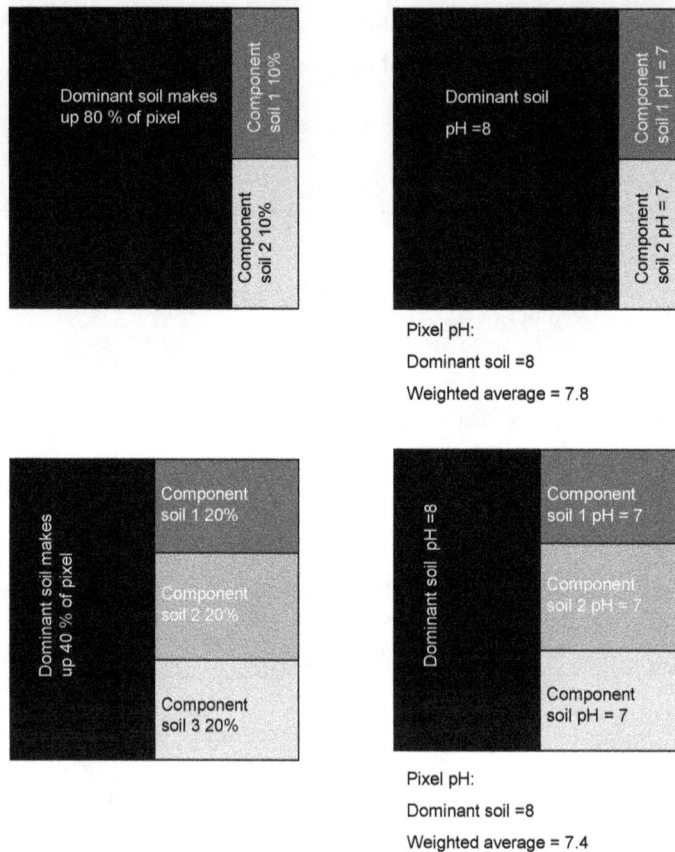

Figure 2. Mock-up examples of two pixels with different distributions of component soils (left); effect of using the weighted average on the overall parameter value estimate versus using that of the dominant soil (right)

We used GIS techniques to determine the area of prominence of dominant soils and compared it in size to that where component soils had higher percentages. We used Mondrian (version 1.2), an open source statistical analysis tool (University of Augsburg, 2012), to study the distribution of dominant soil units and component soil units. For the spatial representation of the soil units, a FORTRAN program was designed that allowed assigning the soil unit share to each pixel.

*2.3 Determination of Crop Suitable Areas*

We used the fuzzy logic approach as discussed in Avellan et al. (2012). Fuzzy classification methods define growth through membership functions and likelihoods (Burrough, MacMillan, & Deursen, 1992). The rationale behind this is that most soil parameters have a large error rate per se, due to sampling and handling errors, and crops are able to grow at various levels of these parameters (Rossiter, 1996). Thus strict Boolean classification systems may be too restrictive in growth ranges and areas. Fuzzy logic approaches have been used for a selected number of crops on limited study areas by other authors e.g. (Baja, Chapman, & Dragovich, 2002; Braimoh, Vlek, & Stein, 2004; Reshmidevi, Eldho, & Jana, 2009; Van Ranst, Tang, Groenemam, & Sinthurahat, 1996).

Raster-based soil, terrain and climate parameter values were matched on a sliding scale from 0 to 1 with their respective crop growth likelihoods as determined by (Sys, Van Ranst, Debaveye, & Beernaert, 1993) (Figure 3a). Subsequently, the most optimally matching crop was selected to be the most suitable for a given pixel. Each component soil was assigned one fuzzy value (Figure 3b). Depending on the number of component soils in each soil mapping unit, up to 8 fuzzy values per pixel were assigned. These were aggregated based on their weighted share value of the respective soil mapping unit. Component soils with high share values end up with a stronger influence on the final fuzzy value.

Crop growth abilities were then categorized into four subsets as defined by Sys et al. (1993) and (FAO, 1976). Fuzzy value between:

1) 0–0.4 Pixel not suitable for crop growth (N) (none).

2) 0.4–0.6 Pixel marginally suitable for crop growth (S3).

3) 0.6–0.8 Pixel suitable for crop growth (S2).

4) 0.8–1 Pixel highly suitable for crop growth (S1).

Pixels are subsequently transformed into land surfaces according to their location on the globe through a FORTRAN programme. The total land surface is considered except Antarctica.

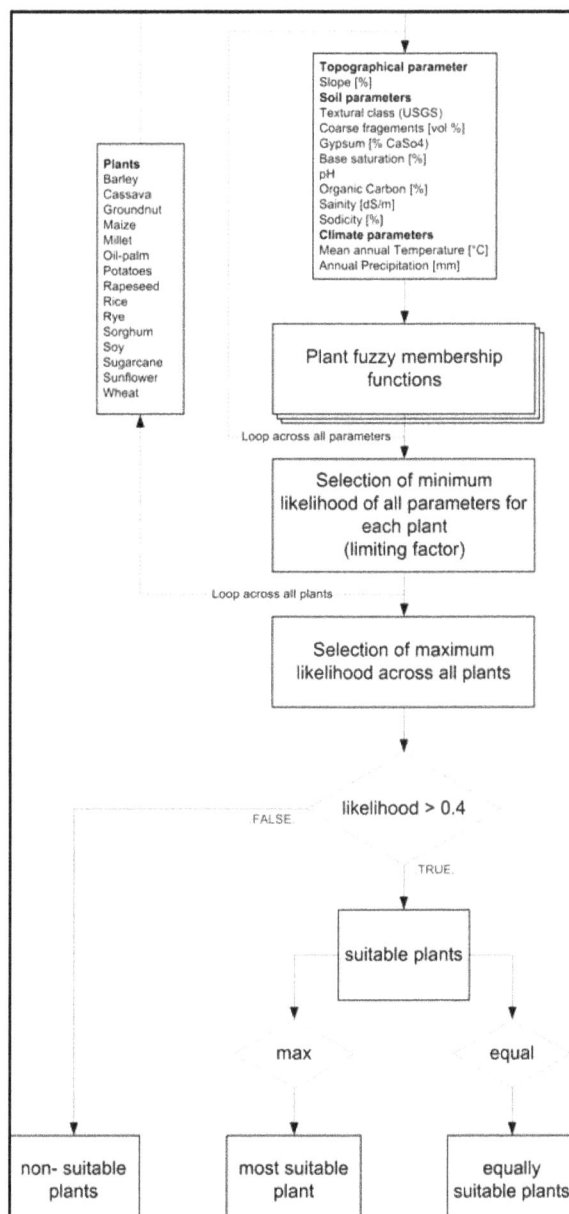

(a)

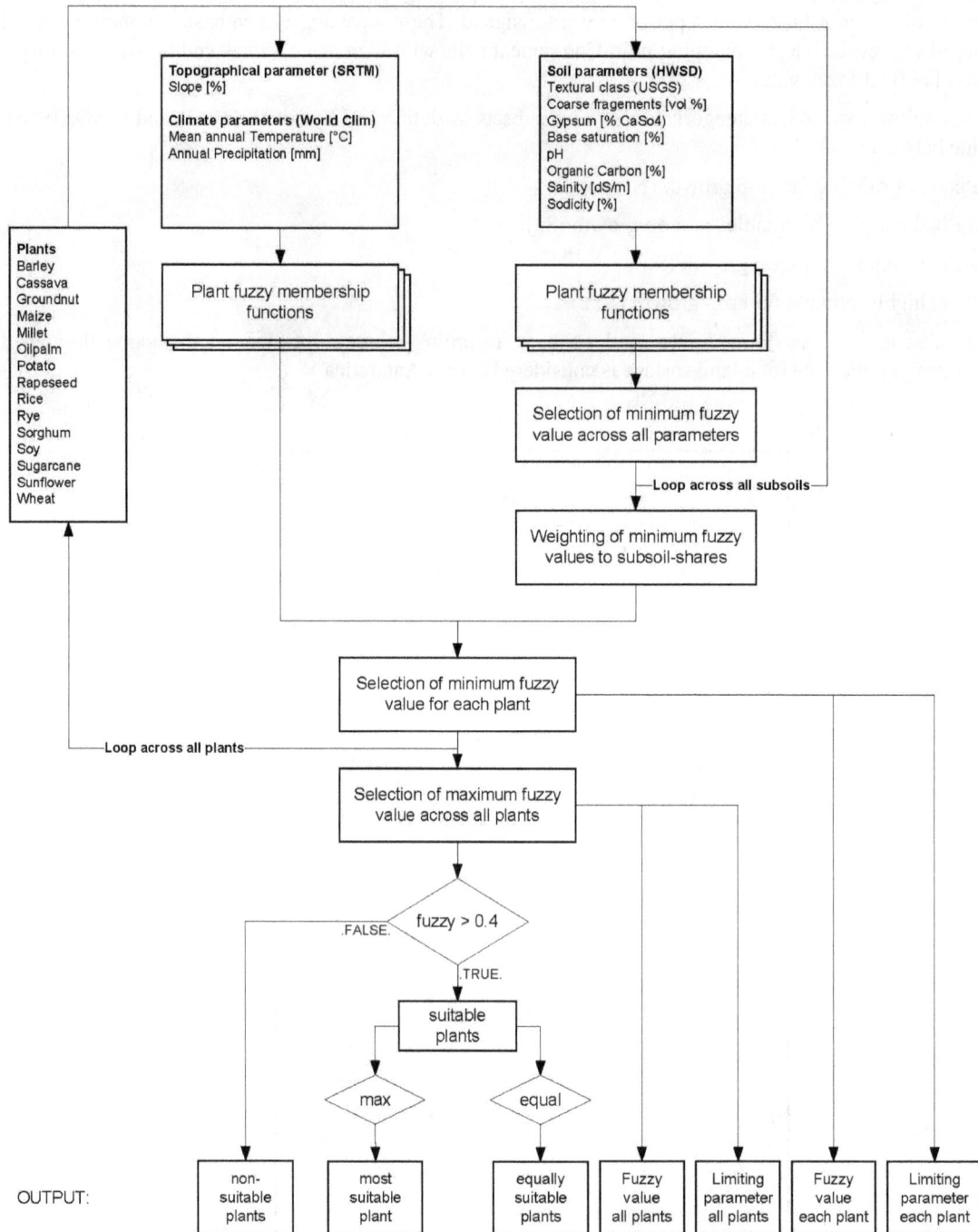

(b)

Figure 3. Overview of the methodology of fuzzy logic crop suitability analysis using just the parameter value estimates of a) the dominant soil (top) or b) of all component soils (bottom)

## 3. Results

### 3.1 Dominant vs. Component Soil Areas

In 64% of all pixel the dominant soil holds more than 50% of the pixel's share value. When looking at specific major soil groups, some only exist as dominant soil types (i.e. Is-Lithosols, Ns-Nitosols, U-Rankers and W-Planosols). Most soils comprise only two component soils in their soil mapping unit (i.e. dominant soil plus one additional component soil). Few cases exist where soil mapping units have 6 or more component soils. The share value of the dominant soil component is very high in most of northern Asia, Greenland, the North America and large parts of Africa. These are areas where the dominant soil defines the parameter value estimate (grey areas in Figure 4). In the case of China, due to the way the database was produced, only one-the dominant-soil exists. In the Middle East, Central Asia, the Pacific and Australia, share values of the dominant soil component were very low. These are areas where the other component soils play a larger role in determining the parameter value estimates of the given pixel (black areas in Figure 4, see also mock up example in Figure 2). South America exhibits mostly areas with intermediate share values (data not shown explicitly).

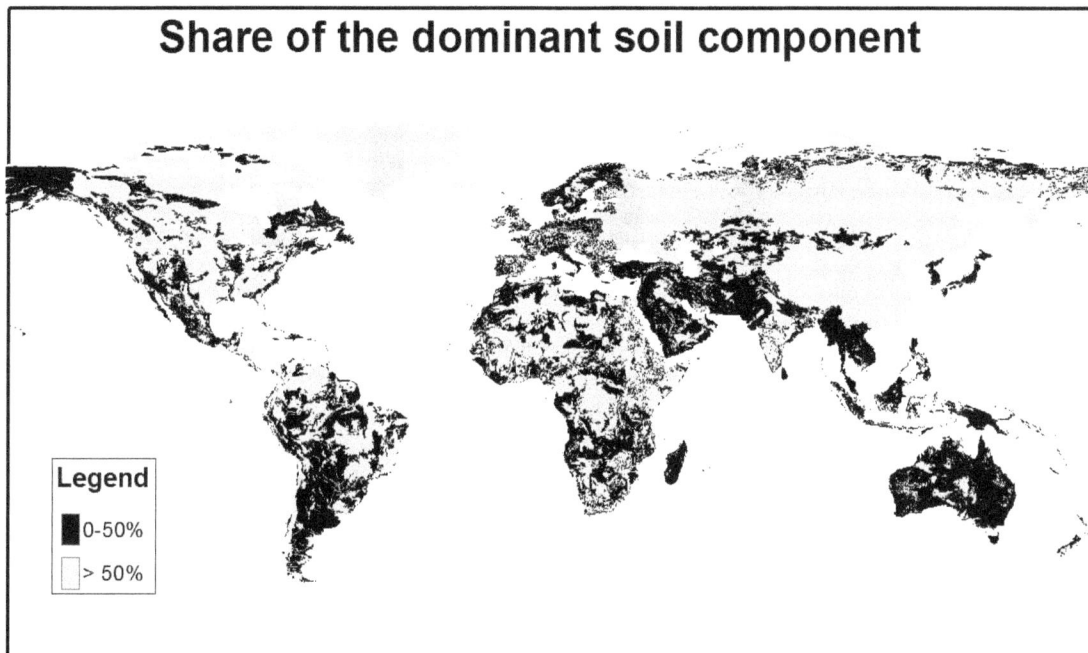

Figure 4. Analysis of shares and sequences of component soils. Grey areas represent soil mapping units where the share value of the dominant soil component holds more than 50%; Black areas are regions where the dominant soil component holds a share value of more than 50%

### 3.2 Determination of Crop Suitable Areas

While using the parameter value estimates of the dominant soil mapping units along with climate and terrain constraints, 9% of the earth's surface result in highly suitable (S1), 25% in suitable (S2) and 19% in marginally suitable (S3) areas (Figure 5). Barley (10.7%), wheat (5.6%), and oil palm (5.2%) are globally the most suitable crops (Figure 6) (Percentages of overall pixel, not of area).

While considering the parameter value estimates of all component soils in a given pixel, the area suitable for crop growth amounts to 54.5% of the earth's land surface excluding Antarctica. Roughly 4.5% can be categorised as highly suitable (S1); 27% and 23% can be classified as suitable (S2) and marginally suitable (S3), respectively (Figure 5). The most prominent crops were the same as when using dominant soils only, with adjustments in their overall percentages (barley-11.1%, wheat-6.5%, oil palm-5.9%) (Figure 6).

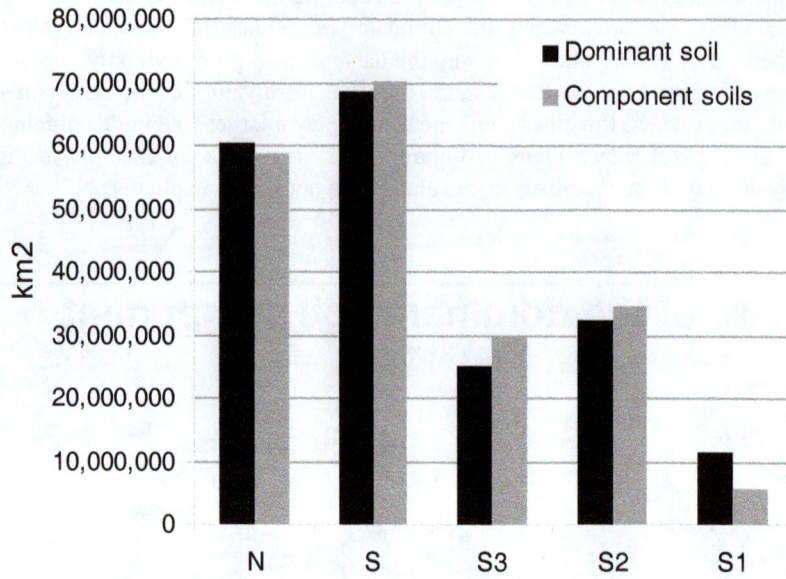

Figure 5. Amount of crop suitable areas while considering only dominant soils (black bars) or all component soils (grey bars). N–non suitable areas; S–sum of all suitable areas; S3–marginally suitable; S2–suitable; S1–highly suitable

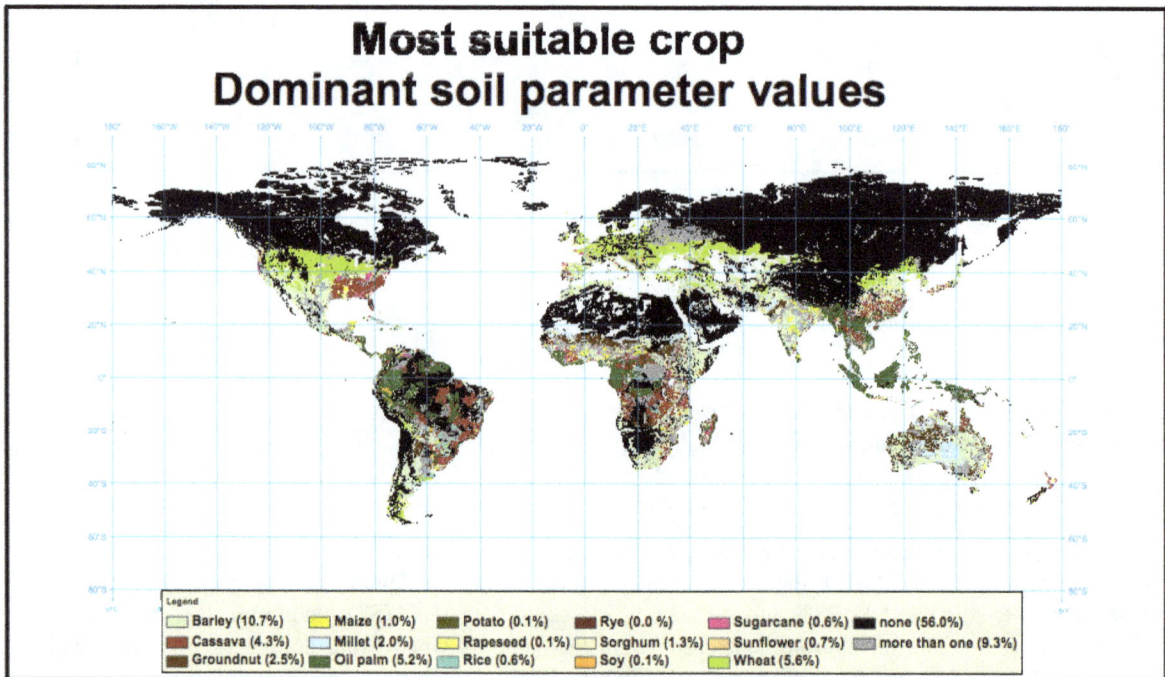

(a)

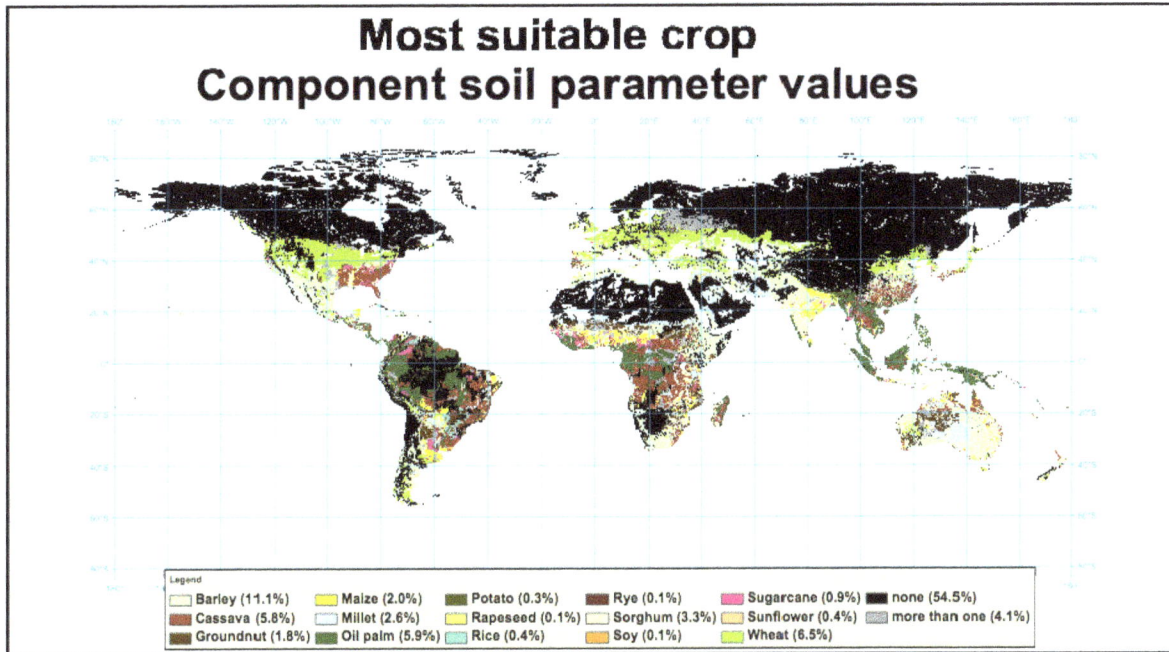

(b)

Figure 6. Distribution of the most suitable crop using all component soil parameter value estimates (top) or dominant soils (bottom) (% values represent the relative amount of pixel for that crop, not the relative area)

## 4. Discussion

In about 1/3 of the soil mapping units the share value of the dominant soil is less than 50% of the pixel. Its parameter value estimate, i.e. pH, salinity or organic carbon value, may not be the same as that of the weighted average of all component soils. In terms of crop suitability this translates in a 1% increase of crop suitable areas of the earth's land surface when using all soil components of the soil mapping unit of the topsoil. The global distribution of crops itself is marginally affected - the ranking of the top 5 crops with the highest amount of pixels remains equal. Changes in the distribution of the suitability classes are important. For instance, in the highly suitable areas, a reduction of 4.3% is observed when using component soils whereas the marginally suitable areas increased by 3.7% (Figure 5).

The model results reflect the qualitative differences of the underlying databases. The HWSD is an integrated patchwork of diverse datasets (see Figure 1). South and Central America, East Africa and parts of Central Asia are fed with the SOTWIS data which have the latest updates of soil samples (latest version of 2006). Europe and Russia is based on the datasets of the European Soil Database, a very comprehensive set (FAO/IIASA/ISRIC/ISSCAS/JRC, 2009). North America, West Africa, and large parts of Asia and Australia are still based on the outdated Digital Soil Map of the World (DSMW). Data for China was produced by assigning one soil class per pixel. No changes in crop suitability classes occurred for those areas which are composed of only one, the dominant, component soil, such as in the case of China. Changes in crop suitability classes were in general more prominent where dominant soil shares are below 50% such as in the Americas, Africa and Central Asia (Figure 7). For instance, while 867991 km$^2$ were assigned to be highly suitable when using the dominant soils in the Mercosur region (MRC), only 301704 km$^2$ were left in this category when using all component soils (Figure 7). Instead, 1300988 km$^2$ versus 1030206 km$^2$ were marginally suitable when using component soils or dominant soils only, respectively.

However, a linear relationship between data quality, more component soils and smaller share values in the dominant soils cannot be postulated. Some areas, in particular in the tropics, are predominantly composed of one dominant soil component with a share value of more than 50% but are based on "high quality" datasets (i.e. Amazon forest in Brazil based on SOTWIS, see grey areas in Figure 4). Other areas are based on "low quality" datasets, such as Australia on DSMW, and show large areas with several component soils and dominant soil shares below 50% (see black areas in Figure 4).

**Regional distribution of crop suitable area classes**

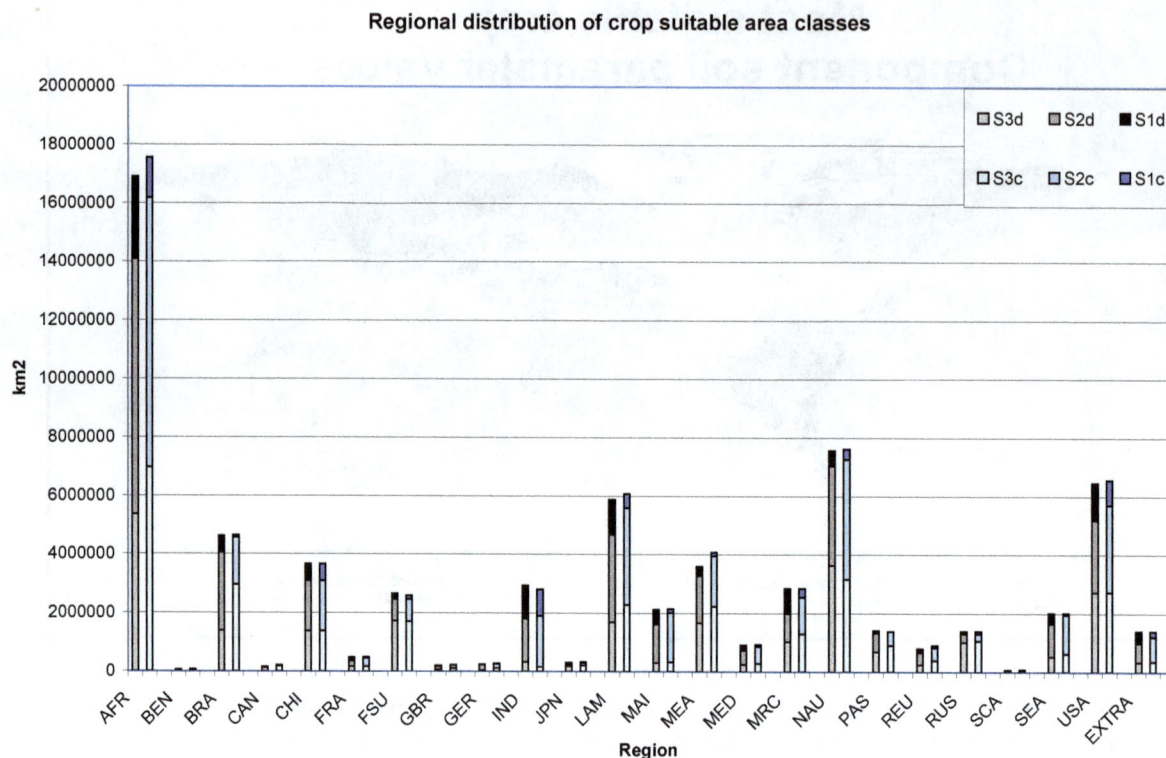

Figure 7. Region specific changes in crop suitability areas by categories using dominant soil parameter value estimates (d) or component soils (c). S3–marginally suitable, S2–suitable, S1–highly suitable

Now, how to make a choice of which dataset to use? The quality for all component soils is heterogenous; the effect on the extent and type of crop suitability minimal. The lack of consistent quality of global datasets is a known issue. A variety of research centres are working towards enhanced soil datasets and sampling, often in collaboration with many others such as in the Global Soil Initiative launched in 2011 (The Global Soil Partnership, 2011). In few cases of crop modelling some authors have undertaken extensive quality control of the underlying soil data and adapted it to their needs (Gijsman, Thornton, & Hoogenboom, 2007; Romero et al., 2012). This is very cumbersome and can only be carried out when sufficient expert staff is available for a specific target objective. However soil datasets are used widely by differing disciplines. We suggest explaining the inherent uncertainty attached to these datasets and lay open the error margin of their use. In this particular case, on the use of all component soils versus only the dominant soils we postulate that the error margin is of about 1% at a global scale.

It is clear to the authors that additional parameters can be used from the soil databases as well as a variety of other parameters such as refined climate datasets, in particular at the temporal scale. Knowledge on ethnicity, gender, management practices, adapted crops, irrigation, use of fertilizers and of the use of technology are all factors that influence the suitability of an area for agricultural purposes (FAO, 2007). Obtaining reliable data for these parameters may be even more challenging than for soil databases.

## 5. Conclusion

In this study, we intended to show the differences in model results when using all component soils for the analysis of crop suitability. This is important because it allows determining the level of uncertainty that modellers face when using current global soil databases. Including more parameters does not always mean better results. We showed that the distribution of the number of component soils of the HWSD is very heterogeneous on a geographical scale but is not linked to the quality of the underlying data subset. The error range for using either the dominant component soil versus all component soils could be considered to be 1%–the difference in crop suitable area between the two datasets. The margin of error varies according to the region and increases to up to 4% when looking at the individual suitability classes.

## References

Avellan, T., Zabel, F., & Mauser, W. (2012). The influence of input data quality in determining areas suitable for crop growth at the global scale–a comparative analysis of two soil and climate datasets. *Soil Use and Management, 28*(2), 249-265. http://dx.doi.org/10.1111/j.1475-2743.2012.00400.x

Baja, S., Chapman, D. M., & Dragovich, D. (2002). A Conceptual Model for Defining and Assessing Land Management Units Using a Fuzzy Modeling Approach in GIS Environment. *Environmental Management, 29*, 647-661. http://dx.doi.org/10.1007/s00267-001-0053-8

Batjes, N. (2006). *ISRIC-WISE derived soil properties on a 5 by 5 arc minute grid (version 1.1)* (No. 2006/02). Wageningen: ISRIC-World Soil Information.

Benjamin, J. G., Nielsen, D. C., & Vigil, M. F. (2003). Quantifying effects of soil conditions on plant growth and crop production. *Geoderma, 116*(1-2), 137-148. http://dx.doi.org/10.1016/S0016-7061(03)00098-3

Bontemps, S., Defourny, P., Van Bogaert, E., Arino, O., & Kalogirou, V. (2009). *GLOBCOVER 2009, Products Description and Validation Report*. ESA, Université catholique de Louvain.

Braimoh, A. K., Vlek, P. L. G., & Stein, A. (2004). Land Evaluation for Maize Based on Fuzzy Set and Interpolation. *Environmental Management, 33*, 226-238. http://dx.doi.org/10.1007/s00267-003-0171-6

Bruinsma, J. (2009). *The resource outlook to 2050: By how much do land, water and crop yields need to increase by 2050* (p. 33). ROME: FAO.

Burrough, P. A., MacMillan, R. A., & Deursen, W. (1992). Fuzzy classification methods for determining land suitability from soil profile observations and topography. *European Journal of Soil Science, 43*, 193-210.

FAO. (1976). *A framework for land evaluation*. Food and Agricultural Organization of the United Nations.

FAO. (2003). *World Agriculture: Towards 2015/2030: An FAO Perspective* (Jelle Bruinsma). Rome and London: FAO & Earthscan Publications Ltd.

FAO. (2007). *Land evaluation: Towards a revised framework*. Food and Agricultural Organization of the United Nations.

FAO/IIASA/ISRIC/ISSCAS/JRC. (2009). *Harmonized World Soil Database, Version 1.1*. FAO, Rome; IIASA, Laxenburg.

Farr, T. G., Rosen, P. A., Caro, E., Crippen, R., Duren, R., Hensley, S., ... Alsdorf, D. (2007). The Shuttle Radar Topography Mission. *Rev. Geophys., 45*, RG2004. http://dx.doi.org/10.1029/2005rg000183

Fischer, G. V. (2000). *Global Agro-Ecological Zones Assessment: Methodology and Results*. Laxenburg.

Fischer, G. V. (2002). *Global Agro-ecological Assessment for Agriculture in the 21st Century: Methodology and Results*. Rome, Laxenburg: Food and Agriculture Organization of the United Nations and International Institute for Applied Systems Analysis.

Gijsman, A. J., Thornton, P. K., & Hoogenboom, G. (2007). Using the WISE database to parameterize soil inputs for crop simulation models. *Computers and Electronics in Agriculture, 56*, 85-100. http://dx.doi.org/10.1016/j.compag.2007.01.001

Hijmans, R., Cameron, S., Parra, J., Jones, P., & Jarvis, A. (2005). WORLDCLIM–a set of global climate layers (climate grids). Retrieved from http://www.worldclim.org/

IIASA/FAO. (2012). Global Agro-ecological Zones (GAEZ v3.0). IIASA, Laxenburg, Austria and FAO, Rome, Italy. Retrieved from http://gaez.fao.org/Main.html#

Kirchhof, G., So, H. B., Adisarwanto, T., Utomo, W. H., Priyono, S., Prastowo, B., ... Sanidad, W. B. (2000). Growth and yield response of grain legumes to different soil management practices after rainfed lowland rice. *Soil and Tillage Research, 56*(1-2), 51-66. http://dx.doi.org/10.1016/S0167-1987(00)00122-7

Reshmidevi, T. V., Eldho, T. I., & Jana, R. (2009). A GIS-integrated fuzzy rule-based inference system for land suitability evaluation in agricultural watersheds. *Agricultural Systems, 101*, 101-109. http://dx.doi.org/10.1016/j.agsy.2009.04.001

Romero, C. C., Hoogenboom, G., Baigorria, G. A., Koo, J., Gijsman, A. J., & Wood, S. (2012). Reanalysis of a global soil database for crop and environmental modeling. *Environmental Modelling & Software, 35*, 163-170. http://dx.doi.org/10.1016/j.envsoft.2012.02.018

Rossiter, D. G. (1996). A theoretical framework for land evaluation. *Geoderma, 72,* 165-190. http://dx.doi.org/10.1016/0016-7061(96)00031-6

Sys, C. O., Van Ranst, E., Debaveye, J., & Beernaert, F. (1993). *Land evaluation: Part III Crop requirements* (Vols. 1-3). Bruxelles (Belgium): Administration Generale de la Cooperation au Developpement.

The Global Soil Partnership. (2011). Retrieved from http://www.fao.org/globalsoilpartnership/home/en/

UNDP. (2008). World Population Prospects: The 2008 Revision. Population Division of the Department of Economic and Social Affairs of the United Nations Secretariat.

University of Augsburg. (2012). *Mondrian.* Retrieved from http://stats.math.uni-augsburg.de/mondrian/

USGS. (2000). SRTM_30 Version 2.1. Retrieved from http://dds.cr.usgs.gov/srtm/version2_1/SRTM30/

Van den Akker, J. J., Arvidsson, J., & Horn, R. (2003). Introduction to the special issue on experiences with the impact and prevention of subsoil compaction in the European Union. *Soil and Tillage Research, 73*(1-2), 1-8. http://dx.doi.org/10.1016/S0167-1987(03)00094-1

Van Ranst, E., Tang, H., Groenemam, R., & Sinthurahat, S. (1996). Application of fuzzy logic to land suitability for rubber production in peninsular Thailand. *Geoderma, 70,* 1-19. http://dx.doi.org/10.1016/0016-7061(95)00061-5

# 10

# Interactive Web-Based Visualization for Lake Monitoring in Community-Based Participatory Research: *A Pilot Study Using a Commercial Vessel to Monitor Lake Nipissing*

Mark P. Wachowiak[1], Renata Wachowiak-Smolíková[1], Brandon T Dobbs[1,2], James Abbott[2] & Daniel Walters[2]

[1] Department of Computer Science and Mathematics, Nipissing University, Canada

[2] Department of Geography, Nipissing University, Canada

Correspondence: Mark P. Wachowiak, Department of Computer Science and Mathematics, Nipissing University, Canada. E-mail: markw@nipissingu.ca

**Abstract**

Environmental and limnological monitoring is of interest to government agencies, researchers, and the general public. In communities that rely on and are heavily affected by lakes and their watersheds, accessible and intuitive presentation of lake properties influences and aids decision-making, interventions, and the formulation of environmentally sound policies. In this paper, interactive web-based visualizations are employed as a mechanism to communicate environmental information collected from a commercial cruise vessel. A pilot study is presented for monitoring Lake Nipissing, a large culturally and environmentally important lake in northeastern Ontario, Canada. This example of community-based participatory research suggests that: (1) policy makers and researchers can quickly gain insight into what is happening in the lake through visualizations, which helps to direct subsequent, detailed investigations; and (2) through accessible, visual presentation, community members may be encouraged to become involved in contributing to environmental policies that directly affect them, thereby supporting environmental "citizen science".

**Keywords:** lake monitoring, visualization, citizen science, community based participatory research

## 1. Introduction

### 1.1 Problem

Community-based participatory research (CBPR), including "citizen science", has gained increasing prominence as a means of addressing challenges in data collection and transparency (Bonney et al., 2009; Newman et al., 2010). CBPR has been defined as research undertaken as a partnership between community members, academic researchers, and other organizations (Israel, Schulz, Parker, & Becker, 1998), whereas "citizen science" primarily refers to avocational research, where research goals are partly obtained by distribution of tasks to non-professionals or amateurs (e.g. the SETI@home or FOLDIT@home initiatives) (Hand, 2010). Both terms imply multiple stakeholders with differing goals and expectations. Furthermore, the research questions themselves can be posed through top-down (scientist-driven) or bottom-up (community-driven) processes (Danielsen et al., 2009). In addition to providing an increased amount of data (Crall et al., 2010; Kéry et al., 2010), stakeholder data collection can better reflect human-environment interactions (Kanjo & Landshoff, 2007; Weckel, Mack, Nagy, Christie, & Wincorn, 2010). Research involving stakeholders can also lead to increased public awareness of environmental issues (Kanjo & Landshoff, 2007) and increased participation in further research (Sullivan et al., 2009). However, the published literature largely focuses on the role that non-scientists can play within the collection of mostly quantitative data within limited parameters [but see (Viegas, Wattenberg, Van Ham, Kriss, & McKeon, 2007) for an example of unstructured collection of qualitative data]. The inclusion of qualitative data collection by non-scientists can also provide a better understanding of complex environmental processes, as well as the perceptions of stakeholders (Abbott & Campbell, 2009; St. Martin, 2001). Moreover, the literature is largely silent about how participatory science can improve policy dialogue beyond improving data quality and transparency.

Among the many research areas covered by citizen science and CBPR, environmental monitoring and ecology (Conrad & Hilchey, 2011; Hochachka et al., 2012; Tulloch, Possingham, Joseph, Szabo, & Martin, 2013; Voinov

& Gaddis, 2008) and the impact of environmental factors on public health (Eghbalnia et al., 2013; Israel et al., 2005; Miller et al., 2013) are prominent. In particular, environmental monitoring is complex, as it relies on multimodal data from sensors, images, and empirical observation. Furthermore, the large quantities of (often multi-dimensional) data acquired from various sensors motivate the need for accessible, interactive visualizations.

The current work focuses on CBPR research and visualization for lake monitoring, where data are collected using a sonde sensor affixed to a recreational cruise vessel, the *Chief Commanda II*. Although the concepts presented in this paper apply to aquatic monitoring in general, the specific target of this research, presented as a pilot study, is Lake Nipissing, the third-largest lake in Ontario, Canada. The lake, which drains into the Georgian Bay of Lake Huron, provides important economic, cultural and ecosystem services to residents and visitors. There is growing concern about declining fisheries, increased blue-green algae blooms, and the appearance of invasive aquatic species (Morgan, 2013; Hutchinson, Karst-Riddoch, & Köster, 2010; Filion, 2011). Because of its importance, the lake is the subject of several ongoing scientific studies of its biophysical properties. These studies involve measuring several parameters in different sections of the lake, generating large quantities of data that must be processed and analyzed. Applying the results of visualization and human-computer interaction studies, these data are subsequently presented through interactive web-based visualizations. The emphasis of the current work is therefore transparent data presentation and interaction.

Major stakeholders include environmental researchers, government policy-makers, First Nations communities, businesses – particularly the fisheries industry – and the general public in North Bay (pop. 54 000) and other communities that surround Lake Nipissing. Researchers, for instance, need to establish baseline environmental indicators, such as oxygen level and temperature. The public can gain an understanding of the data being collected and why they are collected, and how they may contribute potentially important qualitative data, such as unusual environmental events that may be overlooked by conventional surveys. Policy-makers are concerned about fish populations, particularly walleye (Morgan, 2013), as well as the potential effect of three dams on the French River, the primary outflow of the lake.

This research emphasizes the need to enhance data collection, to expand stakeholder participation in the research and policy process, and to disseminate results to the public through intuitive, interactive visualizations. To achieve these goals, an online platform was developed wherein sensor data can be displayed after acquisition. In turn, environmental data can be augmented by qualitative and quantitative input from other stakeholders. This two-way flow of information via the online platform serves several purposes: it allows stakeholders, on a real-time basis, to review the data, and to understand why they are collected; it complements standard data collection by providing additional data input; and it establishes common ground between scientists, policy makers and other stakeholders. The current study encompasses both CBPR, in that data are collected as a collaborative effort from the crew of a commercial cruise vessel, and citizen science, where not only government policy makers and academic researchers can access the data, but where all community members can make use of interactive visualizations and draw conclusions, or relate the information to personal experience.

*1.2 The Importance of Lake Monitoring*

Many large lakes are of vital significance to local communities, and convey their own important benefits and present their own unique challenges and problems. In the case of Lake Nipissing, the lake experiences considerable variability in biological productivity both within and between years; concerns have been raised regarding apparent declines in fish population and more frequent blue-green algae blooms (Morgan, 2013; Hutchinson et al., 2010). While this may be due in part to the inherent variability of the lake's ecosystem, growing levels of human activity within the region, as well as broader changes in water level and temperature associated with climate change, are also possible factors. In some cases, Lake Nipissing has been affected by human activity, such as modification of tributaries, eutrophication, and the introduction of invasive species (Filion, 2011).

*1.3 The Importance of Interactive Visualization in Environmental Monitoring and Citizen Science*

Accessible, web-based visualization technologies can stimulate scientific innovation for both researchers and members of the community who participate in the research and in environmental monitoring. Such visualizations encourage non-scientists to develop new questions, and scientists to address questions in new ways, given the availability of citizen participation (Goodchild, 2007). Specifically, interactive browser-based visualization and analysis allow participants and scientists to explore the data with fewer constraints, thereby enhancing innovation (Newman et al., 2012). For instance, visualization, visual communication tools, and immersive virtual experiences are encouraging community involvement in sustainability efforts aimed at assessing climate change effects (e.g. http://cirs.ubc.ca/research/research-area/visualization-tools-and-community-engagement).

Such highly interactive visualization techniques are increasingly used in geographical visualization (Dykes, MacEachren, & Kraak, 2005). Utilizing a variety of interactive methods allows data to be explored more thoroughly, thereby aiding understanding. For instance, multiple graphs and maps showing related data can be linked together such that an operation performed on one graph or map will be reflected on the others, creating a dynamic connection between the information and the visualizations (Theus, 2005). Especially for representation of geospatial data, maps and other visualizations often require simplification so that a high level understanding of fundamental concepts (e.g. datums, projection, resolution) is not necessary for participants to successfully use the site (Newman et al., 2010). This principle was employed in the present study, where the data collected from the *Chief Commanda II* are presented using multiple visualizations linked through an interactive map. Users can interact with a variety of different visualizations (described below) in such a way that they can explore different spatial, temporal, properties and other aspects using multiple visualization mechanisms. Such a "linked approach" is conducive to understanding and successful use by people who are not trained in geographic information systems or in spatial representation methods (Newman et al., 2010).

*1.4 Interactive Visualization of Multidimensional Transect Data*

Data were collected at regularly-sampled intervals, usually every minute, every three minutes, or every four minutes depending on the transect (a *Chief Commanda II* cruise where data were collected), and can therefore be considered as continuous. As cruise durations are generally less than three hours, and there may be hours or even days between each transect, the temporal dimension is described as discrete. Finally, the spatial coverage of each transect is described as continuous because the GPS collects positions at regular intervals. Due to the hybrid nature of the data, alternative graphing approaches are used: (1) An interactive overlay map for visualizing properties with high spatial resolution; (2) Parallel coordinate plots for communicating high dimensionality (i.e., the different properties collected by the sensors attached to the cruise vessel), providing both high temporal and spatial resolution; (3) Cloud lines, which are new, sophisticated scatter plots for displaying long-term trends in properties over time, allowing users to probe how properties change as a function of distance from a specific reference point; (4) Hovmöller (space-time) plots that provide a broad representation of changes in properties over time and space, targeted primarily to the research community and environmental organizations; and (5) Visualization of descriptive statistics.

1.4.1 Interactive Overlay Map

Easily understood maps can be overlaid with data to greatly aid understanding and ease of interpretation (Hochachka et al., 2012). The primary component of the linked visualizations is a map of Lake Nipissing overlaid with paths representing each transect taken by the *Chief Commanda II*. The cruise paths are by default coloured such that the colour at any particular coordinate reflects the value for a chosen attribute at that point along the path. The data can be coloured based on any chosen property, using one of several different colour maps. These include specialized colour maps, such as a greyscale colour map, colour maps for users with colour-deficient vision, and the standard pH scale, which can be adjusted to show small differences which normally occur in lake monitoring (e.g. setting 5 to to red/acidic, 8 to purple/alkaline). Groups of points can be selected on the map, and these spatial selections will be reflected throughout the rest of the visualization.

1.4.2 Parallel Coordinate Plots

Another facet of the linked visualization is a parallel coordinates plot (PCP), where $n$ parallel axes, each corresponding to a dimension of the data, are drawn. A single $n$-dimensional tuple is represented by a polyline (a piecewise connected line) that intersects each axis at the location corresponding to the tuple's value for that axis' dimension (Streit et al., 2006), allowing all dimensions of multidimensional data to be displayed simultaneously. The PCP used in the linked visualization assigns a parallel axis to each of the measurements collected in the study. PCP can be used for an arbitrary number of dimensions. It can also indicate when data fall outside the range of the sensor by showing these values below their corresponding coordinate axis (Siirtola & Räihäm, 2006).

Many enhancements have been made to the standard parallel coordinates technique. In the current study, a 3D plot was implemented so that every parallel axis was extended along the $z$-axis into a plane. The different *Chief Commanda II* transects are separated along the $z$-axis, clearly indicating the date of each collection (Johansson et al., 2014). Opacity controls were provided to reveal the density distribution of the graph, as polylines in dense areas are more visible than in less dense areas (Wegman & Luo, 1996). The order of the parallel axes can be changed to better observe relationships between two arbitrarily juxtaposed axes (Blaas et al., 2008).

The polylines can also be grouped into $k$ categories (Johansson et al., 2005), with each polyline coloured based on the cluster to which it belongs. The clustering can be performed based on any combination of $m$ properties, using some distance measure between the $m$-D data points (in the current study, Euclidean distances). Although many sophisticated clustering algorithms exist, the standard iterative $k$-means clustering approach was used in this work.

Groups of polylines can be selected directly on the plot by "brushing" (Edsall, 2002), wherein users draw a "brush line" over the plot to highlight polylines. Consistent with the goal of linked visualization, the selection ("brushed" polylines) is reflected throughout the other visualization components (i.e. the map and the statistics table). The polylines in the PCP are also coloured based on the property chosen for the map, with each polyline's colour matching that of its corresponding point on the map.

### 1.4.3 Cloud Lines

To represent time, a discrete measure due to the temporal distance between transects, a visualization based on cloud lines was employed (Krstajic et al., 2011). Cloud lines are essentially scatter plots where dependent variables (sonde properties) are each shown as single lines, as a function of the independent variable (time). Events (the sonde readings) along the time axis are represented as circular markers that are coloured according to their value, or whose radius and/or opacity reflects the value, resulting in lines of varying width and colour. The advantage of this technique is that it allows a large amount of data to be intuitively displayed with a relatively small amount of screen space. Initially proposed for discrete, episodic data (Krstajic et al., 2011), cloud lines were adapted to continuous data in the current study to convey properties represented in a straight line, whose colours vary over time according to their value. To represent the spatial dimension, the radius of each point in the cloud line represents the proximity (inverse distance) of that spatial coordinate from a user-specified point, where readings closer to the selected point appear larger. Cloud line plots for multiple properties can be shown simultaneously, and can be juxtaposed arbitrarily using a dragging mechanism.

### 1.4.4 Space-Time Plots

Hovmöller (space-time plots) present a standard image of how properties change over space ($y$-axis) and time ($x$-axis). The image colour indicates the value of the property currently being displayed (Buzzelli, Ramus, & Paerl, 2003). To make these diagrams more useful, space is represented as distance from a user-selected point on the map, allowing users to view changing properties based on specific geographic locations.

The time sparsity of the transects in this pilot study posed challenges for space-time interpolation, as the data are sparse in time but not in space (Wright & Goodchild, 1997). Therefore, the natural neighbour method based on Voronoi methods was employed (Gold & Condal, 1995; Wright & Goodchild, 1997).

### 1.4.5 Statistics

Two statistics tables and corresponding graphs are also available. The statistics tables show the mean, standard deviation, median, interquartile range, minimum, and maximum values of every property. The first table shows these statistics for the data points that are currently selected in the linked visualizations, and is updated dynamically whenever the selection changes on either the interactive map or the PCP. The second table displays statistics on a transect-by-transect basis, and is not dynamic. The available graphs are a multi-bar graph of the mean ± standard deviation for all selected properties, and a box-whisker plot of the minimum, first quartile, median, third quartile, and maximum for a selected property.

### 1.4.6 Colour Maps

For the coloured properties described in 1.4.1 to 1.4.4, the standard rainbow (blue = low, red = high) and grayscale (black = low, off-white = high) colour maps that vary colour across the visual spectrum are included, but the former are sometimes considered to be among the least useful. Recent visualization research has focused on diverging colour maps, where a transition between two chosen colours passes through an unsaturated colour (Moreland, 2009). Several of these diverging colour maps are implemented in the current system. These colour maps may also be more easily understood by users with colour-deficient vision (Moreland, 2009).

## 2. Methods

### 2.1 Study Area

The study area covered a large NE to SW sector of Lake Nipissing (see the map on Figure 1). Most data were collected in an area within approximately 20 km of North Bay, Ontario, but many routes also included the environmentally important Callander Bay area in the southeastern part of the lake. A limited amount of data is also available for the French River, which is the primary outflow for the lake, located in its southwestern sector.

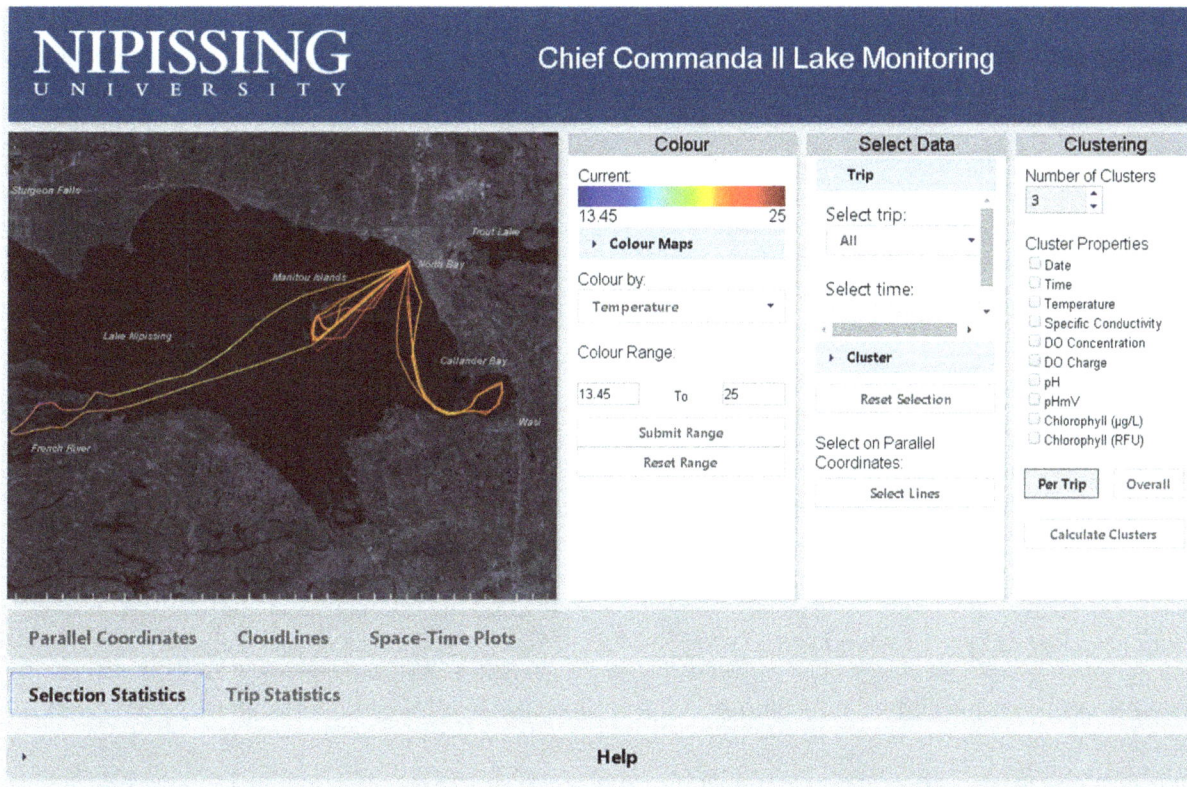

Figure 1. The main web-based graphical user interface, with map coloured by temperatures for all transects. The interactive map of Lake Nipissing is on the left. The expandable visualization and statistics panes are located below the main view pane

### 2.2 Data Collection

The data were collected as part of the *Chief Commanda II* Water Quality Monitoring Pilot Study. This project is one of several studies to gauge the health of the Lake Nipissing ecosystem. The *Chief Commanda II*, a cruise ship operating daily on Lake Nipissing between May and October, takes several different cruise routes through the lake. Due to the regularity and consistency of its routes, as well as the coverage of large parts of the lake, the cruise ship is an excellent tool for surveying the conditions of the lake over time and at multiple points. Examples of other community-based research from commercial vessels contributing to lake monitoring have been presented in the literature (Buzzelli et al., 2003). In the summer months between 2010 and 2012, the research team attached a YSI (YSI Incorporated, Yellow Springs, Ohio) 6600 V2 sonde to the ship to acquire and transmit readings from the lake at regular intervals during cruises. The sonde, simultaneously used for other studies, was calibrated biweekly. The measurements taken were: (1) temperature (°C); (2) specific conductivity (μs/cm); (3) dissolved oxygen (DO) concentration (mg/L); (4) DO charge; (5) pH; (6) pH charge (mV); and (7) chlorophyll (μg/L and RFU). Additionally, all transects from 2011 measured conductivity (mS/cm), all transects from August 2011 measured turbidity (NTU), and all transects from September 2011 measured total algae (cells/mL and RFU). These attributes were recorded along with the time and GPS coordinates.

### 2.3 Web Service Implementation

The interactive visualization web service was programmed in JavaScript, a lightweight language that runs on virtually all current web browsers, and employs the JavaScript-based WebGL application programming interface, which facilitates 3D graphics by utilizing the client's graphics hardware. Customized free Dygraphs libraries (http://dygraphs.com) were specially adapted for the cloud lines and box-whisker visualizations. Database connectivity and data retrieval were effected through the PHP language.

### 2.4 Visualization Implementation

Users have the ability to change colour maps for the map overlay, the 3D parallel coordinate plots, cloud lines, and Hovmöller plots. However, pH property colour defaults are set to the standard pH colour scale. Furthermore, the user can select a rectangular spatial region on the interactive map. The property values for this selection are then updated on the PCP, cloud lines, and statistics visualization. The user may also select individual transects for analysis. Although transects generally run less than three hours, a time slider is provided to allow users to track the progress of a single selected transect, and to assess how properties change during this transect. For the cloud lines, an interactive range selector (adapted from Dygraphs) located below the graphs allows zooming in on smaller date ranges, and hovering over a point on the graph displays the value of all properties at that point in time.

Due to the computational complexity of computing the Hovmöller plots and the limited system resources generally available for web-based visualization, Hovmöller plots were pre-computed for thirty-nine points close to meaningful observation sites across the lake. The site closest to the user-selected point was used for display. Up to four diagrams can be displayed simultaneously.

### 2.5 Human-Computer Interaction

To facilitate the community-based research aspect of this work, the web-based interface was designed to be as intuitive and interactive as possible. The diverse visualizations are linked through the interactive map "hub". This map allows users to select specific transects (either by clicking them on the display or by selecting them from a drop-down list), dates, and geographic locations.

The interface is divided into three components to reduce scrolling. Each component can be minimized to save screen space. The interactive map, colour map selection, and transect selection are located in separate panels in the top component. The second component houses the PCP, cloud lines, and space-time plots. A single visualization can be displayed in this panel, allowing the user to rapidly cycle through the visualization with tabs, resulting in a cleaner, easier-to-navigate interface. Users, especially non-researchers, may find it easier to concentrate on one representation at a time. Options that are unique to each visualization are displayed on the right side of the panel.

The bottom component features the statistical tables and plots. A tab interface separates statistics over all data collections with per transect representations. Finally, although the interface, visualization, and components were designed to be intuitive and self-teaching, "tool tips" (text that explains the function of a button, menu, or feature that appears when the pointing device/mouse hovers over it) and text in the interface are liberally employed to aid navigation. Because this is a citizen science initiative, a tutorial video is included that provides an overview of all of the features of the service with special attention given to the more advanced visualizations.

## 3. Results

The website is hosted on Nipissing University's server system, and is found at the URL http://visual.nipissingu.ca/Commanda2. All the features previously described in Section 2 can be found on this site (Figure 1). Three examples illustrating the use of the visualizations for exploratory data analysis follow.

### 3.1 Temperature and Dissolved Oxygen

Using the North Bay municipal dock as the reference point from which spatial distances were calculated, a cursory analysis of the Hovmöller plots for temperature and DO may lead to the interpretation of an inverse relationship between the two properties (Figure 2). Because of the low temporal resolution of these plots, the relationship is further examined with cloud lines visualizations, using the same reference point (Figure 3). Here, four properties of interest (temperature, DO concentration, specific conductivity, and pH) are displayed, with the temperature and DO concentration cloud lines moved to the top to facilitate interpretation. The larger radii denote closer proximity to the reference location. The apparent inverse relationship seen in the Hovmöller plots was also observed in the cloud lines, with possible hyponoxic (or even anoxic) conditions in parts of the lake.

However, the PCP plot shows that some polylines fall below the DO coordinate axis; consequently, those DO values are suspect (Siirtola & Räihäm, 2006). On the specific sonde used in this study, when the DO charge falls below 25, DO readings are suspect. When the cloud lines zoomed to show the last five transects (all with valid DO values), a positive relationship is seen on the August 22, 2011 transect, and a negative relationship on the September 4, 2011 trip. The visualization led to a closer statistical analysis being performed, the results of which are shown in Table 1. The correlation coefficients ($r$) confirm the visual results.

To take the analysis further, the PCP was clustered based on temperature and DO concentration with five clusters (Figure 4). Not only is an inverse relationship generally observed between these properties, but upon further examination with rotating the PCP in 3D and by brushing the time period covering the three clusters, it was seen that three of the five clusters were generated for the longest cruise towards the distant French River, suggesting wide spatial variability in DO concentration. By reducing the number of clusters to two, the French River cruise was clearly separated from the other cruises.

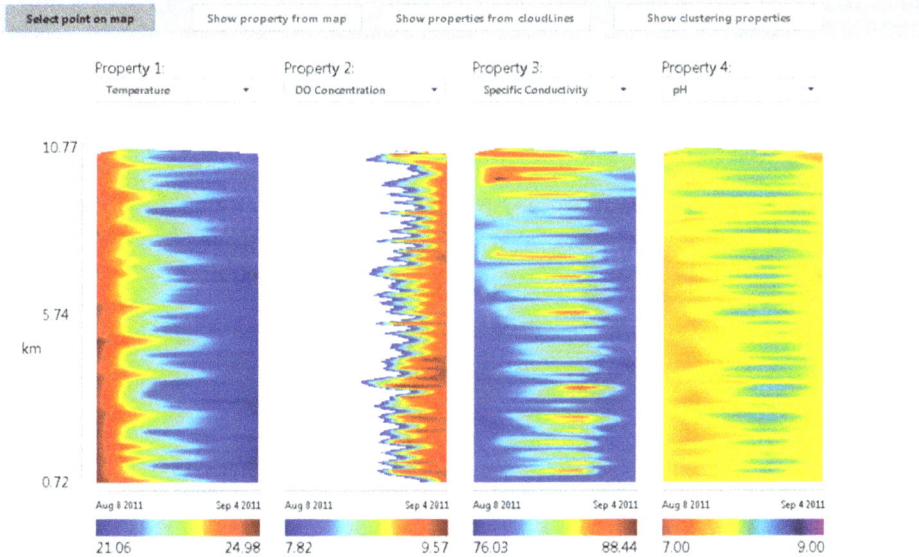

Figure 2. Hovmöller (space-time) plots of four properties. Invalid DO values are not displayed in the second panel

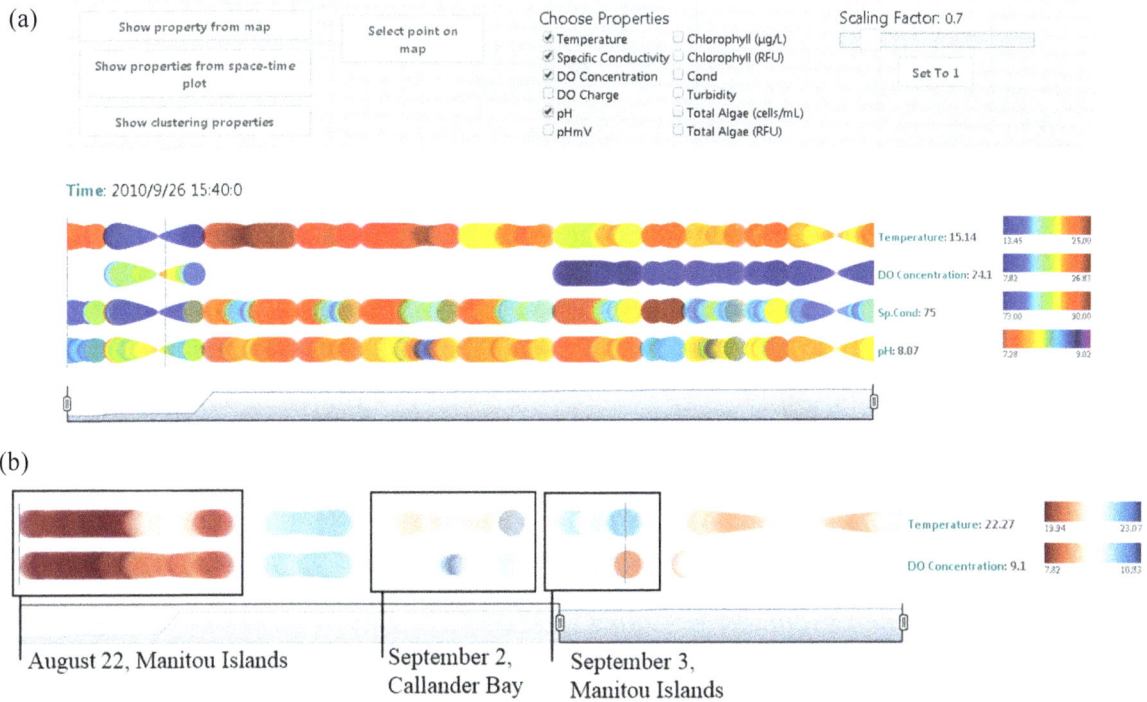

Figure 3. Cloud lines of: (a) four properties; the contrast of Temperature and DO concentration is seen by the two cloud line plots placed at the top of the panel and (b) Temperature and DO concentration zoomed by selecting transects in late August and early September 2011 to more clearly show relationships

Table 1. Correlation coefficients (*r*) and *p*-values for three pairs of properties. The Temperature/DO and Chlorophyll/Total Algae relationships were investigated after analyzing the visualizations of these properties

| Date | Extent | Temperature and DO | | Temperature and pH | | Chlorophyll and Total Algae | |
|---|---|---|---|---|---|---|---|
| | | *r* | *p* | *r* | *p* | *r* | *p* |
| Aug 30 2010 | Manitou Islands | N/A | N/A | 0.09 | 0.24 | N/A | N/A |
| Sep 26 2010 | French River | 0.49 | 0 | -0.28 | 0.00 | N/A | N/A |
| Aug 8 2011 | Manitou Islands | N/A | N/A | 0.81 | 0.00 | N/A | N/A |
| Aug 9 2011 | Manitou Islands | N/A | N/A | 0.59 | 0.00 | N/A | N/A |
| Aug 9 2011 | Callander Bay | N/A | N/A | 0.72 | 0.00 | N/A | N/A |
| Aug 11 2011 | Callander Bay | N/A | N/A | 0.12 | 0.36 | N/A | N/A |
| Aug 22 2011 | Manitou Islands | 0.73 | 0 | 0.77 | 0.00 | N/A | N/A |
| Sep 2 2011 | Manitou Islands | 0.07 | 0.57 | 0.10 | 0.43 | 0.07 | 0.55 |
| Sep 2 2011 | Callander Bay | 0.59 | 0.00 | 0.72 | 0.00 | 0.57 | 0.00 |
| Sep 3 2011 | Manitou Islands | -0.46 | 0.00 | -0.22 | 0.07 | 0.34 | 0.00 |
| Sep 4 2011 | French River | -0.01 | 0.86 | 0.43 | 0.00 | 0.95 | 0.00 |

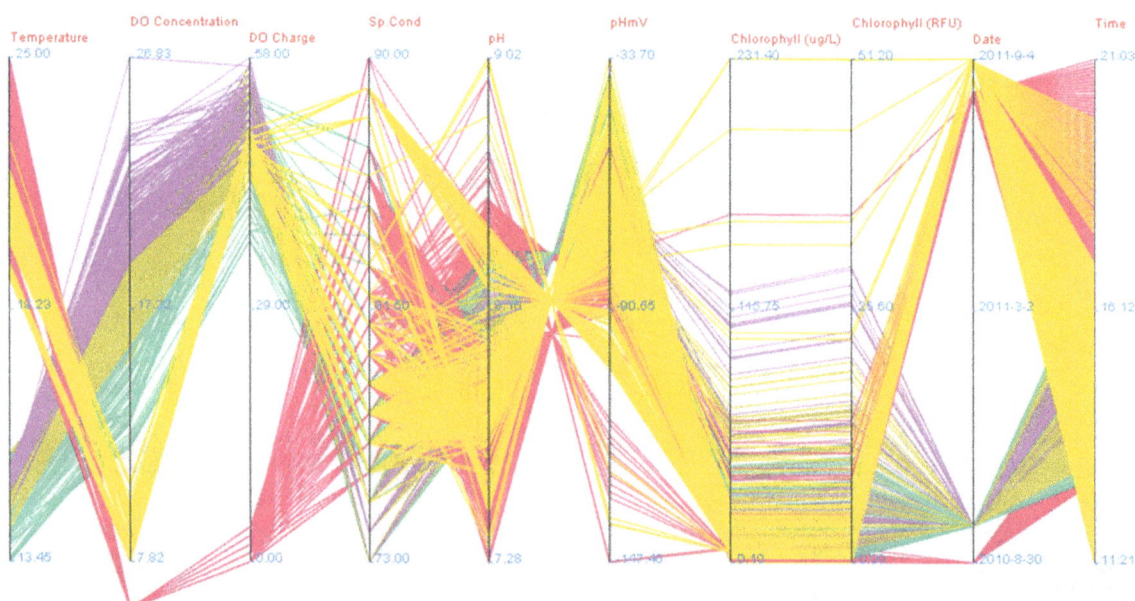

Figure 4. PCP of five clusters based on temperature and DO concentration. Invalid DO concentration readings are displayed below the DO axis

As expected, water temperature and dissolved oxygen increased and decreased, respectively, as the vessel approached shallower water near the Manitou Islands (10 km SW of North Bay) and North Bay waterfront. Spatial variation in these values showed that the sampling method was sensitive enough to detect even slight changes in temperature and dissolved oxygen.

### 3.2 Similarities across All Properties

Clustering enables many interesting features to be seen. For example, the PCP was clustered into three groups across eight (8) properties (all properties except date and time) to assess overall relationships (Figure 5). The 8D data were clearly clustered on date, with all data from transects in early August of 2011 (August 8 to August 11) categorized into the same cluster, all data from late August to early September of 2011 (August 22 to September 4) being placed into another cluster, and the data from September 26, 2010 comprising the third cluster. If the summers of 2010 and 2011 had similar conditions, it is expected that the readings from late August 2010 would be grouped with that from late August of 2011. However, the data from August 30, 2010 was placed in the same cluster as the 2011 early August readings, suggesting that there were substantial difference between the summers of 2010 and 2011, as is more clearly seen in the 3D view of the parallel coordinate plot (Figure 6).

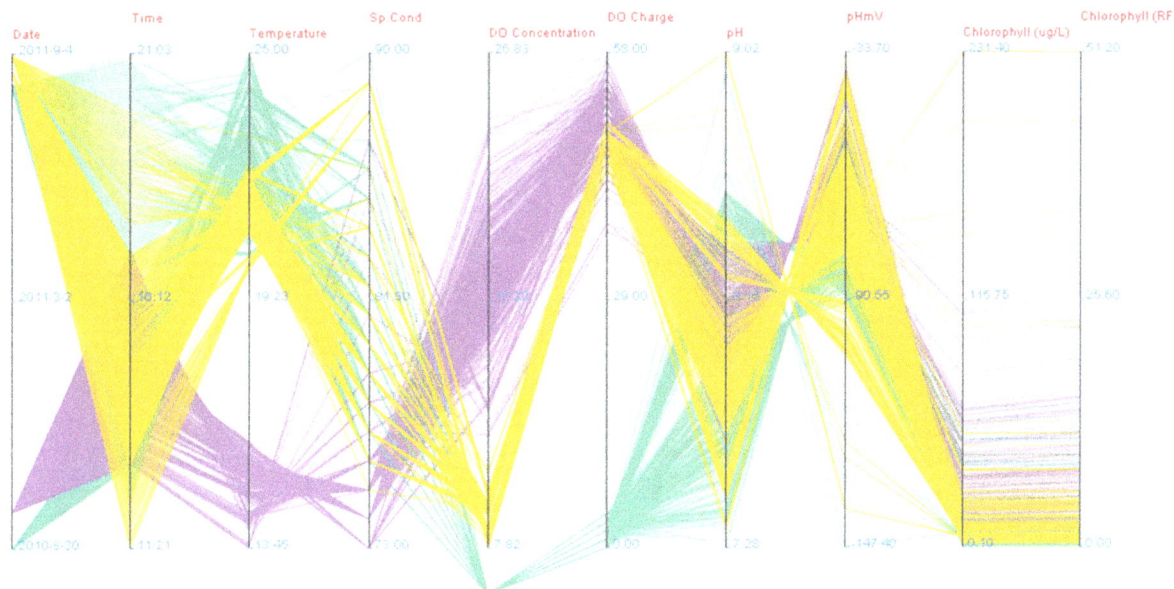

Figure 5. PCP of three clusters across all properties except date and time, indicating that the cruise from August 30, 2010 is in the same cluster as the cruises from early August 2011

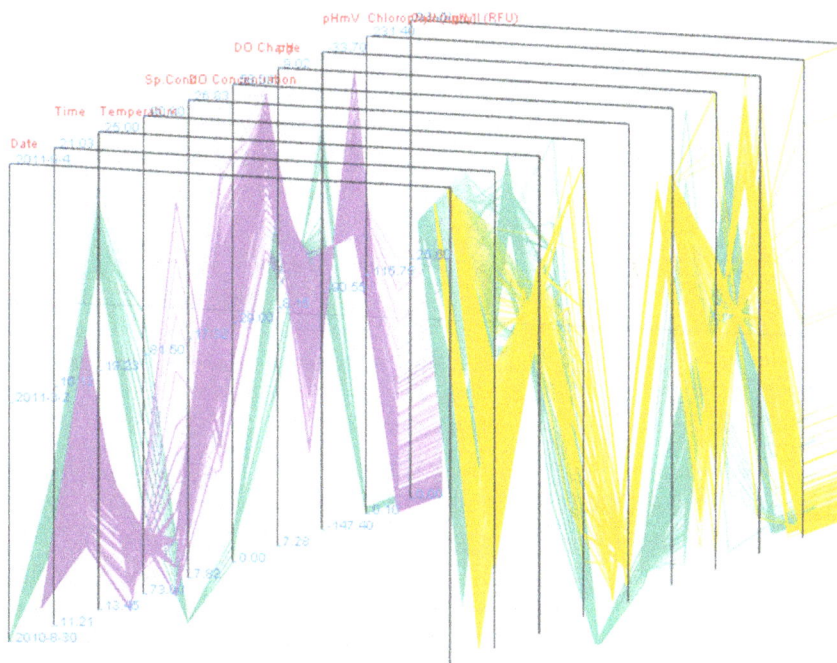

Figure 6. 3D representation of PCP for three clusters, showing the temporal separation of the clusters

### 3.3 Total Algae

Algae analysis is another example of how the visualization can be used. Densities of algae (as estimated by fluorescence) were recorded for four sampling transects (September 2 [two trips], September 3 and September 4, 2011). For most of the entire dataset, algae readings stayed below 10.00 Relative Fluorescence Units (RFUs). RFU values "spiked" at several points of the September 4, 2011 sampling transect. However, results explaining these fluctuations have been inconclusive. On the one hand, chlorophyll readings rose at the same time (Figure

7), suggesting that there were potentially high levels of primary production. However, the high values were only seen on one leg of the transect, despite the fact that the return leg sampled water that was sometimes 0.25 km away or less. Table 1 shows the detailed correlation analysis of chlorophyll and total algae, and suggests a strong positive relationship in regions close to the shore (Callander Bay and the French River), and not in more interior regions of the lake (Manitou Islands). Furthermore, pH and DO did not change significantly with high RFU readings, although such correlations are associated with post-bloom decomposition of algal material. High turbidity has been shown to also be correlated with high RFU values (Ahn et al., 2007). Unfortunately, turbidity was not recorded for these transects.

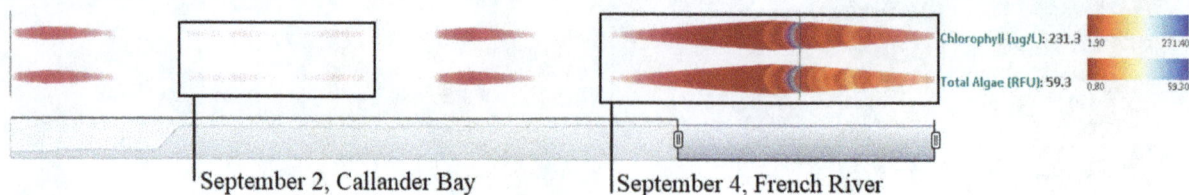

Figure 7. Cloud line plots of total algae Relative Fluorescence Units and chlorophyll readings for September 2 and September 4, 2011

## 4. Discussion

From the results presented above, the interactivity of the linked visualizations is expected to encourage the community to examine different visualizations to draw conclusions and to formulate hypotheses. It is again noted that the focus of the current study is to demonstrate the potential for visualizing and interacting with limnological data and the transparent presentation of these data. The specific data discussed here should not be treated as representative. However, although the data used in this pilot study may lead to incongruous results, the visualizations provide new impetus for demonstrating outliers instead of eliminating them, as well as for querying sensors (see the discussion of the temperature/DO relationship in Section 3.1).

As shown in Section 3.1, a combination of Hovmöller plots, cloud lines, and PCP were used to assess a relationship that, with a cursory look, could be misinterpreted (e.g. by viewing only the space-time plots). The ability to identify outliers or invalid sensor readings through PCP, and the time selection/zooming property of the cloud lines allows observers to see both positive and negative correlations between the properties, which can then be subsequently statistically analyzed (e.g. Table 1).

The clustering example in Section 3.2 demonstrates spatial and temporal separation, where the $k = 3$ clusters ($k$ was chosen empirically) indicates a division between sampling years. The choice of $k$, as well as which and how many properties on which to cluster, is dependent upon whether the investigator is researching specific phenomena, or exploring the data for unforeseen relationships. In the latter case, clustering parameter selections are generally empirical. In both cases, an intuitive interface and easy human-computer interaction allows efficient exploration of multiple aspects of the situation in the lake.

Finally, the cloud line visualizations suggest spatially and temporally varying relationships between chlorophyll and total algae, which were subsequently investigated with statistical methods.

Even the more complex visualizations, such as PCP, are supplied with features that allow the general public to make sense of high dimensional data. The interactive 3D aspect of the PCP, while not common for scientific investigations (Johansson et al., 2014), can aid discovery and insights that can be tested with other investigations. For instance, in the example above, PCP was used to further investigate the insights obtained through space-time plots and cloud lines. Brushing dense areas in the PCP, which are linked to the interactive map showing those cruise routes, identified specific geographic regions of interest.

The visualizations draw attention to patterns in the data (e.g. a negative correlation in the case of temperature and dissolved oxygen) that may be subsequently explored with more rigorous comparative statistical methods (e.g. computing the correlation coefficient).

Such web-based visualizations allow the community and other stakeholders to perform in-depth explorations of high-dimensional spatial-temporal data, or of single properties of interest.

## 5. Conclusion

The data collected for the present study are an important first step in gathering baseline data about Lake Nipissing's environment. Baseline data are especially crucial given the variability of the lake ecosystem.

Further investigation with the visualizations also shows that this variability was particularly apparent in 2010, when lake levels were several meters lower than average. As such, it was expected that the recorded temperature and oxygen values also depart from the norm. However, historical data with which to compare them is not available.

This pilot project indicates that remote monitoring of aquatic parameters in Lake Nipissing is feasible. The method will be adapted to allow the YSI 6600 V2 sonde readings and spatial data to be integrated on a data logger, which will simplify the data collection procedure and make detection of anomalous readings easier and quicker.

The community-based research pilot project presented in this paper is part of a larger initiative for monitoring Lake Nipissing, which presently includes collecting data from sensors at different depths on buoys placed at various locations around the lake. It is also anticipated that the visualizations presented in this web service will be of interest to other community, government, and academic organizations that focus on lake monitoring, such as the Global Lake Ecological Observatory Network (www.gleon.org).

As the applications for real-time aquatic monitoring expand, the potential of using multi-parameter data to better understand and anticipate adverse events in Lake Nipissing – such as anoxia or blue-green algae blooms – is also seen. For example, recent work has demonstrated how evolutionary algorithms can be used to model and forecast algal blooms in a dam reservoir in China (Ye, Cai, Zhang, & Tan, 2014). The visualization of these analyses could potentially aid in identifying correlations and increasing model accuracy.

With the rapid collection and analysis of data, and the implementation of a publicly-available web service, it is hoped that the flow of information will be two-way, where interested parties can comment on the data, as well as contribute their own, including location-tagged images, thereby fulfilling the mandate of citizen science and community-based participatory research.

### Acknowledgments

This research was made possible by funding from Nipissing University. M.P.W. is supported by NSERC Grant #386586-2011. We also thank Capt. R. Stirvins and the crew of the *Chief Commanda II* for their cooperation and assistance. Jacquelyne Forgette, Christopher Boasman, Kyle Jaynes, and Bryan Sarlo provided valuable technical assistance. Marlena Pearson helped to proofread and improve the text. Finally, we thank the anonymous reviewers for their constructive suggestions and criticisms.

### References

Abbott, J. G., & Campbell, L. M. (2009). Environmental histories and emerging fisheries management of the upper Zambezi river floodplains. *Conservation and Society, 7*(2), 83.

Ahn, C., Joung, S., Yoon, S., & Oh, H. (2007). Alternative alert system for cyanobacterial bloom, using phycocyanin as a level determinant. *Journal of Microbiology, 45*(2), 98.

Blaas, J., Botha, C. P., & Post, F. H. (2008). Extensions of parallel coordinates for interactive exploration of large multi-timepoint data sets. *IEEE Transactions on Visualization and Computer Graphics, 14*(6), 1436-1451. http://dx.doi.org/10.1109/TVCG.2008.131

Bonney, R., Cooper, C. B., Dickinson, J., Kelling, S., Phillips, T., Rosenberg, K. V., & Shirk, J. (2009). Citizen science: A developing tool for expanding science knowledge and scientific literacy; improving and integrating data on invasive species collected by citizen scientists. *Biological Invasions, 59; 12*(11; 10), 977; 3419-984; 3428. http://dx.doi.org/10.1525/bio.2009.59.11.9

Buzzelli, C. P., Ramus, J., & Paerl, H. W. (2003). Ferry-based monitoring of surface water quality in North Carolina estuaries. *Estuaries, 26*(4), 975-984. http://dx.doi.org/10.1007/BF02803356

Conrad, C. C., & Hilchey, K. G. (2011). A review of citizen science and community-based environmental monitoring: Issues and opportunities. *Environmental Monitoring and Assessment, 176*(1-4), 273-291. http://dx.doi.org/10.1007/s10661-010-1582-5

Crall, A. W., Newman, G. J., Jarnevich, C. S., Stohlgren, T. J., Waller, D. M., & Graham, J. (2010). Improving and integrating data on invasive species collected by citizen scientists. *Biological Invasions, 12*(10), 3419-3428. http://dx.doi.org/10.1007/s10530-010-9740-9

Danielsen, F., Burgess, N. D., Balmford, A., Donald, P. F., Funder, M., Jones, J. P. G., . . . Yonten, D. (2009). Local participation in natural resource monitoring: A characterization of approaches; participación local en el monitoreo de recursos naturales: Una caracterización de métodos. *Conservation Biology, 23*(1), 31-42. http://dx.doi.org/10.1111/j.1523-1739.2008.01063.x

Dykes, J., MacEachren, A. M., & Kraak, M. (2005). Exploring geovisualization. In J. Dykes, A. M. MacEachren & M. Kraak (Eds.), *Exploring geovisualization* (pp. 3-19). Elsevier.

Edsall, R. M. (2003). The parallel coordinate plot in action: Design and use for geographic visualization. *Computational Statistics and Data Analysis, 43*, 605-619.

Eghbalnia, C., Sharkey, K., Garland-Porter, D., Alam, M., Crumpton, M., Jones, C., & Ryan, P. H. (2013). A community-based participatory research partnership to reduce vehicle idling near public schools. *Journal of Environmental Health, 75*(9), 14-19. Retrieved from http://europepmc.org/abstract/MED/23734527

Filion, J. (2011). *Summer 2010 collapse of the Lake Nipissing zooplankton community subsequent to the introduction of the invasive zooplankter bythotrephes longimanus.* (Environmental report). North Bay, ON: Lake Nipissing Partners in Conservation.

Gold, C. M., & Condal, A. R. (1995). A spatial data structure integrating GIS and simulation in a marine environment. *Marine Geodesy, 18*, 213-228.

Goodchild, M. F. (2007). Citizens as sensors: Web 2.0 and the volunteering of geographic information. *GeoFocus, 7*, 8-10.

Hand, E. (2010). Citizen science: People power. *Nature, 466*(7307), 685-687.

Hochachka, W. M., Fink, D., Hutchinson, R. A., Sheldon, D., Wong, W., & Kelling, S. (2012). Data-intensive science applied to broad-scale citizen science. *Trends in Ecology & Evolution, 27*(2), 130. http://dx.doi.org/10.1016/j.tree.2011.11.006

Hutchinson, N., Karst-Riddoch, T., & Köster, D. (2010). *Callander Bay subwatershed phosphorus budget.* Bracebridge, ON: Hutchinson Environmental Sciences Ltd.

Israel, B. A., Parker, E. A., Rowe, Z., Salvatore, A., Minkler, M., López, J., . . . Halstead, S. (2005). Community-based participatory research: Lessons learned from the centers for children's environmental health and disease prevention research. *Environmental Health Perspectives, 113*(10), pp. 1463-1471. Retrieved from http://www.jstor.org/stable/3436118

Israel, B. A., Schulz, A. J., Parker, E. A., & Becker, A. B. (1998). REVIEW OF COMMUNITY-BASED RESEARCH: Assessing partnership approaches to improve public health. *Annual Review of Public Health, 19*(1), 173-202. http://dx.doi.org/10.1146/annurev.publhealth.19.1.173

Johansson, J., Forsell, C., & Cooper, M. D. (2014). On the usability of three-dimensional display in parallel coordinates: Evaluating the efficiency of identifying two-dimensional relationships. *Information Visualization, 13*(1), 29-41.

Johansson, J., Ljung, P., Jern, M., & Cooper, M. (2005). Revealing structure within clustered parallel coordinates displays. Paper presented at the *Information Visualization, 2005. INFOVIS 2005. IEEE Symposium on Information Visualization,* 125-132. http://dx.doi.org/10.1109/INFVIS.2005.1532138

Kanjo, E., & Landshoff, P. (2007). Mobile phones to monitor pollution. *IEEE Distributed Systems Online, 8*(7).

Kéry, M., Royle, J. A., Schmid, H., Schaub, M., Volet, B., Haefliger, G., & Zbinden, N. (2010). Site-occupancy distribution modeling to correct population-trend estimates derived from opportunistic observations. *Conservation Biology, 24*(5), 1388-1397.

Krstajic, M., Bertini, E., & Keim, D. A. (2011). CloudLines: Compact display of event episodes in multiple time-series. *IEEE Transactions on Visualization and Computer Graphics, 17*(12), 2432-2439. http://dx.doi.org/10.1109/TVCG.2011.179

Miller, P., Waghiyi, V., Welfinger-Smith, G., Byrne, S. C., Kava, J., Gologergen, J., . . . Seguinot-Medina, S. (2013). Community-based participatory research projects and policy engagement to protect environmental health on St Lawrence Island, Alaska. *International Journal of Circumpolar Health, 72*(0). Retrieved from http://www.circumpolarhealthjournal.net/index.php/ijch/article/view/21656

Moreland, K. (2009). Diverging color maps for scientific visualization. Paper presented at the *Proceedings of the 5th International Symposium on Advances in Visual Computing: Part II,* Las Vegas, Nevada. 92-103. http://dx.doi.org/10.1007/978-3-642-10520-3_9

Morgan, G. E. (2013). *Lake Nipissing data review 1967 to 2011* (technical report). North Bay: Ontario Ministry of Natural Resources.

Newman, G., Wiggins, A., Crall, A., Graham, E., Newman, S., & Crowston, K. (2012). The future of citizen science: Emerging technologies and shifting paradigms. *Frontiers in Ecology and the Environment, 10*(6), 298-304. http://dx.doi.org/10.1890/110294

Newman, G., Zimmerman, D., Crall, A., Laituri, M., Graham, J., & Stapel, L. (2010). User-friendly web mapping: Lessons from a citizen science website. *International Journal of Geographical Information Science, 24*(12), 1851-1869. http://dx.doi.org/10.1080/13658816.2010.490532

Siirtola, H., & Räihäm, K. (2006). Interacting with parallel coordinates. *Interacting with Computers, 18*(6), 1278-1309.

St. Martin, K. (2001). Making space for community resource management in fisheries. *Annals of the Association of American Geographers, 91*(1), 122-142. http://dx.doi.org/10.1111/0004-5608.00236

Streit, M., Ecker, R. C., Österreicher, K., Steiner, G. E., Bischof, H., Bangert, C., . . . Radu, R. (2006). 3D parallel coordinate systems--A new data visualization method in the context of microscopy-based multicolor tissue cytometry. *Cytometry Part A, 69A*(7), 601-611. http://dx.doi.org/10.1002/cyto.a.20288

Sullivan, B. L., Wood, C. L., Iliff, M. J., Bonney, R. E., Fink, D., & Kelling, S. (2009). eBird: A citizen-based bird observation network in the biological sciences. *Biological Conservation, 142*(10), 2282. http://dx.doi.org/10.1016/j.biocon.2009.05.006

Theus, M. (2005). Statistical data exploration and geographical information visualization. In J. Dykes, A. M. MacEachren & M. Kraak (Eds.), *Exploring geovisualization* (pp. 127-142). Elsevier.

Tulloch, A. I. T., Possingham, H. P., Joseph, L. N., Szabo, J., & Martin, T. G. (2013). Realising the full potential of citizen science monitoring programs. *Biological Conservation, 165*(0), 128. http://dx.doi.org/10.1016/j.biocon.2013.05.025

Viegas, F. B., Wattenberg, M., Van Ham, F., Kriss, J., & McKeon, M. (2007). Manyeyes: A site for visualization at internet scale. *IEEE Transactions on Visualization and Computer Graphics, 13*(6), 1121-1128.

Voinov, A., & Gaddis, E. J. B. (2008). Lessons for successful participatory watershed modeling: A perspective from modeling practitioners; A review of citizen science and community-based environmental monitoring: Issues and opportunities. *Environmental Monitoring and Assessment, 216; 176*(2; 1-4), 197; 273-291. http://dx.doi.org/10.1016/j.ecolmodel.2008.03.010

Weckel, M. E., Mack, D., Nagy, C., Christie, R., & Wincorn, A. (2010). Using citizen science to map Human—Coyote interaction in suburban New York, USA. *The Journal of Wildlife Management, 74*(5), 1163-1171. http://dx.doi.org/10.2193/2008-512

Wegman, E. J., & Luo, Q. (1996). High dimensional clustering using parallel coordinates and the grand tour. *Computing Science and Statistics, 28*, 361-368.

Wright, D. J., & Goodchild, M. F. (1997). Data from the deep: Implications for the GIS community. *International Journal of Geographical Information Science, 11*(6), 523-528. http://dx.doi.org/10.1080/136588197242176

Ye, L., Cai, Q., Zhang, M., & Tan, L. (2014). Real-time observation, early warning and forecasting phytoplankton blooms by integrating in situ automated online sondes and hybrid evolutionary algorithms. *Ecological Informatics, 22*, 44-51.

# 11

# Fate and Potential Mobility of Arsenic (As) in the Soil of Mechanic Workshops

Femi Francis Oloye[1, 5], Isaac Ayodele Ololade[1], Oluyinka David Oluwole[1, 6], Marcus Oluyemi Bello[2], Oluwabunmi Peace Oluyede[3] & Oluwaranti Ololade[4]

[1] Department of Chemistry, Adekunle Ajasin University, Akungba Akoko, Nigeria

[2] Department of Microbiology, Adekunle Ajasin University, Akungba Akoko, Nigeria

[3] Department of Environmental Biology and Fisheries, Adekunle Ajasin University, Akungba Akoko, Nigeria

[4] Department of Chemistry, Federal University of Technology, Akure, Nigeria

[5] Department of Chemistry, University of Aberdeen, United Kingdom

[6] Department of Chemistry, Rhodes University, South Africa

Correspondence: Femi Francis Oloye, Department of Chemistry, Adekunle Ajasin University, Akungba Akoko, Nigeria. E-mail: pen2crown@gmail.com

**Abstract**

In order to determine Arsenic (As) content in soils, from the vicinity of an automobile mechanic's workshop and to evaluate the contamination levels, different soil layers (0 – 15cm, 15 – 30cm and 30 – 45cm depth) were collected and analyzed for As content using atomic absorption spectrophotometer (AAS). Soil texture, conductivity, pH, total organic content and cation exchange capacity were also measured. Sequential extraction was carried out to determine geochemical phases of As. In the investigated soil samples, the mean total As concentrations were 2.45-2.53, 2.14-2.26 and 2.69-2.79 mg/kg for 0-15 cm, 15-30 cm and 30-45 cm depth respectively. Generally the affinity of As with the soil fraction increases in the order F3 < F6 < F1 < F2 < F5 < F4 < F7. The levels found in this study exceeded the background concentrations and safe limits for agricultural and residential purposes. The reported results indicate that the enrichment factors (EF) of As was 3.55-3.66, 4.04-4.26 and 3.49-3.62 mg/kg for top, middle and inner soils respectively; while the geoaccumulation index (*Igeo*) values of the metals in the soils studied indicate that they are uncontaminated to slightly contaminated with As. The results indicate that waste from auto-mechanic workshops represent a potential source of heavy metal pollution to the surrounding environment.

**Keywords:** Arsenic, contamination, pollution

## 1. Introduction

Arsenic (As) is widely known for its adverse effects on human health, and this element has been found in the groundwater of at least 70 countries and could pose a significant health threat to more than 140 million people (Chemistry world). People living near factories, farms, or waste sites -where As or pesticides were once used – can be exposed to As through touching, breathing, eye contact, drinking and eating of As contaminated foods. The US Environmental Protection Agency (EPA) and other agencies are currently reevaluating the current maximum contamination level (MCL) (50 µg/l) based on the health risks associated with drinking water containing As (Cai *et al.,* 2002). According to Lee and Lee (2011), the permissible levels for As in agricultural, residential and industrial soil are shown in Table 1 below.

Table 1. permitted As level in selected country

| Country | As / mg/kg |
| --- | --- |
| Germany | 50 |
| Canada | 30 |
| Taiwan | 60 |
| US | 50 |

As reacts with other elements to form organic and inorganic compounds. Inorganic As has been shown to be deadly. Unlike some toxic elements, As can cause negative outcomes as it directly effects systems in the human body (Chemistry world). The increased risk to human health from As is seen as a driver for increased research of As biogenical cycling in the environment (Kim and Nriagu, 2000).

As can enter terrestrial and aquatic environments through both geogenic and anthropogenic activities. As pools naturally in surface soils, and such can arise from the net effects of geological, hydrological, soil forming biogeochemical processes and fossil fuels. Under typical soil forming conditions, the nature of As in the soil is controlled by the lithology of the parent rock materials, volcanic activity and precipitation (Cullen and Reimer, 1989). The primary anthropogenic contributions of As in soils arise from the combustion of municipal solid waste, application of arsenical pesticides, solid waste, sewage sludge, mining and smelting of Arsenic. The combustion of fossil fuels, industrial waste, wood preservation and metal refining also add to the overall cumulative effects of As contamination (Ning, 2002).

In many countries, the current standards for metal evaluation/contamination are based on the total concentrations of metals obtained using strong acid digestion solutions such as nitric acid or aqua regia (Gupta *et al.*, 1996; Ololade *et al.*, 2014). However, mobility of metals in soils, and their toxicity to the biosphere, is related to their association with various soil constituents rather than total concentrations (Sadiq, 1997) Important factors affecting As chemistry, and its mobility in soils, are soil solution chemistry (pH and redox conditions), solid composition, As bearing phases of adsorption and desorption, biological transformation, volatilization and cycling of As in soil (Baroni *et al.*, 2004). In addition, metal sulfide and sulfide concentrations, temperature, salinity, distribution and composition of biota appear to be significant factors determining the fate and transport of As (Cai *et al.*, 2002). Lastly, organic matter improves both the physical and chemical properties of soil and hence affects the toxicity and fate of As in different types of soils.

Arsenate is a salt of As which is the basis of most As contamination in groundwater. Adsorption of arsenate unto soil particles depends on various parameters, such as Al and Fe oxides, clay content, pH and the redox condition of the soil (Baroni *et al.*, 2004). Its availability on the other hands depends on source, chemical speciation and soil parameters (pH, EC, organic matter and colloid contents, soil texture and drainage condition (Eisler, 1994) .This is important as many soils have different physiochemical forms which are associated with soil constituents. It is the chemical forms of As associated with various soil phases, rather than their total concentrations that affects its mobility, bioavailability and toxicity to the biosphere (Cullen and Reimer, 1989).

Even though a lot of work had been reported on the chemical reactivity of As in soil, very little work has been carried out on As mobility in soil systems, and even less on determining the fate and mobility of As soils surrounding auto- mechanic workshops.

## 2. Material and Methods

### 2.1 Study Area

This investigation was carried out in auto-mechanic workshops in Ondo State, South-Western part of Nigeria (see Fig 1).

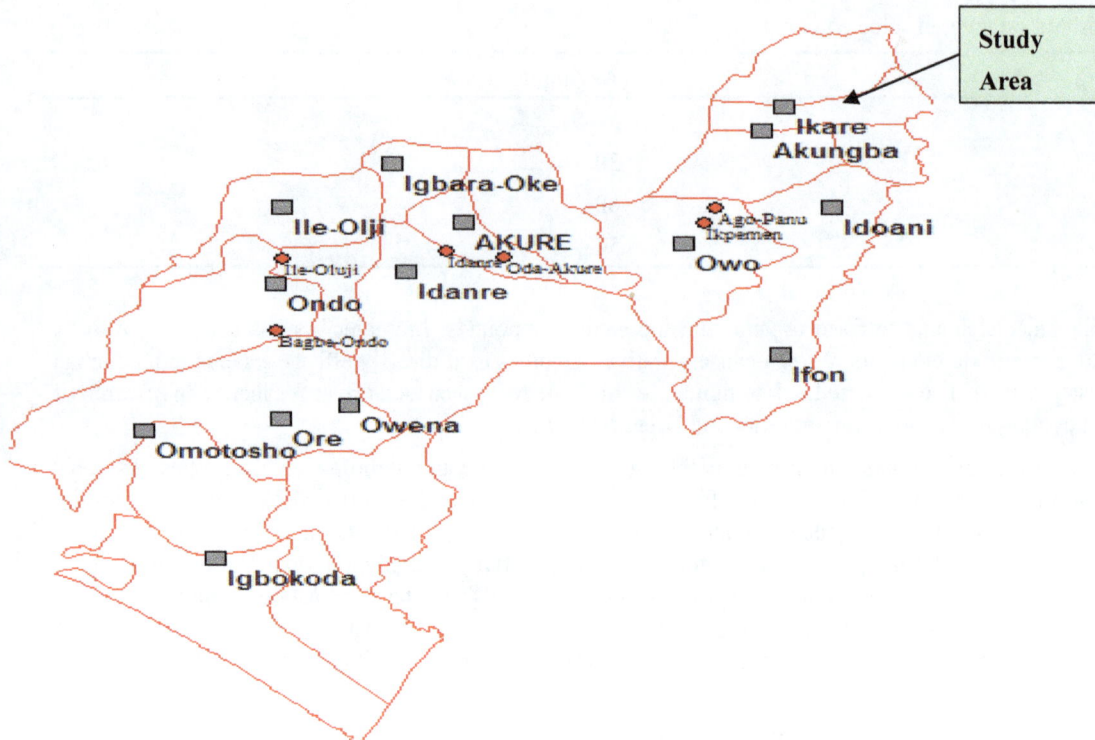

Figure 1. Map of Ondo state showing the area of study where the auto mechanic workshops are located

## 2.2 Sampling

The soil samples were obtained from three separate auto mechanic workshops in Akungba, Ikare and Akoko areas of Ondo State, Nigeria. At each of the workshops, there were three designated sites for sampling. At each site, samples were taken from different sampling depths: top soil surfaces (0 – 15 cm), middle soil (15 – 30 cm) and bottom soil (30 – 45 cm). Sampling was restricted to this depth because it provides the bulk of plant nutrients. During the sampling, the selected sampling sites were subdivided into grids of 20m x 20m, and samples were taken from the centre of each grid. To obtain composite samples of each site, a bulking method was employed to harmonize the samples. A conning and quartering method was applied repeatedly to reduce the sample volume for each site (Ololade *et al.*, 2014; Jackson, 1958). Each representative site sample was prepared by mixing one sample with the other replicates from different grids of the same site to overcome spatial variability. The soil samples were air – dried, ground, and sieved mechanically using a 2 mm sieve. The control (background) samples were obtained from a remote location within the industrial zone at the Adekunle Ajasin University Campus and from Akungba-Akoko which is far removed from the influence of any industrial activities.

## 2.3 Experimental

Soil physical and chemical properties were measured using standard operating procedures. Soil pH was measured in 1:2 (soil: 0.01 M $CaCl_2$) using a digital pH meter, and the particle size distribution was then determined by the hydrometer method (Bouyoucos, 1962). The temperature and relative humidity of the soil samples were taken in the field using appropriate instruments. Soil organic carbon and organic matter were determined using the Walkley-Black Method, and exchangeable acidity determination was done using a titration method, after extraction with 1N KCl. Exchangeable Ca, Mg, K, Na and effective cation exchange capacity (CEC) were also determined for pseudo total metal content analysis.

Quantification of total metallic content of digested soil samples and blanks was carried out with an atomic absorption spectrometer (AAS) (AA6300, Shimadzu, Japan), which was pre-calibrated using standard methods. To ensure that the atomic absorption spectrometer remained calibrated during the experiments, standards were analysed after every 10 runs. Soil samples were digested in accordance with the procedure used by Francis (2004). One gram of finely ground dried soil samples were mixed with 20 mL (1:1) $HCl/HNO_3$ acid mixture, and the contents were heated until dry. The residue was then extracted using a 2 M HCl solution and mixed with 50 mL of distilled water. The solution was then directly aspirated into an air-acetylene flame of AAS to obtain a metal profile of the soils. To validate the digestion protocol of the soil samples, the quality assurance method of Uzairu *et al.,* 2009, was utilized through a spiking experiment, which involves spike recoveries using a multi-element standard solution (MESS). The amounts of spiked metals recovered, after digestion of the spiked samples, were used to calculate the percentage of metals recovered. Finally, a triplicate digestion and analysis of the soil samples and controls were carried out.

One of the latent problems with any sequential extraction is the lack of specificity of extractants toward particular elemental forms of association Shiowatana *et. al.,* 2001. In order to overcome this problem, and to gain insight into fate and mobility of As in the tested soil samples, the approach described by Zein and Brummer, Wenzel et al., and adopted by Ololade et al., was adopted in accordance with Table 2; and levels of Arsenic in the fractions were analysed using the same AAS as described elsewhere (Ololade *et al.,* 2014).

All the chemicals (> 95 % purity) used in this work were purchased from Sigma – Aldrich (Steinheim, Germany). They were used without further purification.

Table 2. Sequential extraction scheme for As extractions

| Fractions | Sample | Extractant | |
|---|---|---|---|
| 1 | Soil sample | $1 \text{ mol.L}^{-1} \text{ NH}_4\text{NO}_3$ | Mobile (water-soluble and exchangeable bonds and easily soluble metal- organic complexes |
| 2 | Residual | $+ 1 \text{ mol.L}^{-1} \text{ NH}_4\text{OAc (pH 6.0)}$ | Easily available bonds |
| 3 | Residual | $+ 1 \text{ mol.L}^{-1} \text{ NH}_2\text{-OH.HCL} + 1 \text{ molL}^{-1} \text{ NH}_4\text{OAc (pH 6.0)}$ | Bonds with Mn oxides |
| 4 | Residual | $+ 0.025 \text{ mol.L}^{-1} \text{ NH}_4\text{-EDTA (pH 4.6)}$ | Organic bonds |
| 5 | Residual | $+ 0.02 \text{ mol.L}^{-1} \text{ NH}_4\text{-Oxalate (pH 3.25)}$ | Bonds with amorphous Fe oxides |
| 6 | Residual | $+ 0.02 \text{ mol.L}^{-1} \text{ NH}_4\text{-Oxalate} + \text{ascorbic acid (pH 3.25)}$ | Bonds with crystalline Fe bonds |
| 7 | Residual | $+ \text{HF}$ | Residual |

## 3. Results and Discussion

### 3.1 Method Validation

The mean percentage recovery of As was used to spike the soil sample represented in Table 3. The mean percentage recovery from the spiked soil sample for As was 97.6, which is similar to the reported values in a similar matrix by Awofolu *et al.,* (2005). Therefore, the reproducibility of the methods and the precision and accuracy of the AAS machine was adjudged reliable since the acceptable recovery percentage was obtained and the standard error was less than 5%.

Table 3. Means % recoveries (± SD) of metal standard added to pre-digested soil sample

| Metal | Spiked conc / $\text{mgL}^{-1}$ | % recovery |
|---|---|---|
| As | 3.0 | $97.6 \pm 2.9$ |

Values are mean of triplicate analyses, SD is the standard deviation

### 3.2 Physico-Chemical Properties

The results for the soil physicochemical characterization are given in Table 4, the pH across the entire study area ranged in a narrow interval. Top soils across the study area suggested slightly acidic to neutral condition, while the middle and bottom layers suggest slightly acidic soil conditions. The relative humidity (RH) ranged from 56.5-61.4 %, which could be accounted for by the climatic variation of the study area. Sand is the major soil component in the size distribution (Cai *et al.,* 2002) and gives a decreasing order of sand > clay > silt in particle size analysis. The percentages of total organic matter content (TOM) in all the samples were low (< 10 %). The TOM of the control area is much higher when compared to samples from the mechanic workshops; this difference in the TOM could be due to reaction of soil colloids with toxic metals from the waste of mechanic workshop which are indiscriminately dispersed within the shop and also the TOM concentrations in top soil > middle soil > bottom soil.

The cation exchange capacity (CEC) is highest in all the top soils, followed by the middle soils and the bottom soils; however, there is a variation in the soil samples where bottom soil has a higher CEC than the middle soil. In a similar manner to TOM, the soil CEC in the control is high when compared with that from the mechanic shops (Tlustors et al., 2000). This reduction in CEC and TOM could be accounted for as a reflection of nutrient depletion by waste from the activities of the mechanic workshops.

The total concentration of As in the soil samples ranged from 2.45-2.53, 2.14-2.26 and 2.69-2.79 mg/kg for top, middle and bottom soils respectively. It will be observed that the bottom soils are in the highest range followed by top soil; this could be as a result of leaching of arsenic from top soil to the bottom soil, although the levels found indicated a low concentration when compare with those in Table 1. It is important to determine the amount present in different phases since As is toxic in any concentration, and total content is less important in determining As lability. However, the results apparently suggest a pollution plume in the study areas. Hence, a methodology to separate the metal-bearing phases was also employed, since it is impossible to consider the presence of metals and their compounds in the environment and their potential release to the ecosystem without considering its geochemical forms (Adamu et al., 2013). In addition, distribution of a metallic contaminant amongst different phases profoundly affects its transport (Sager, 1992). It is the form in which As is associated with various phases rather than its total concentration that affects its mobility, bioavailability and toxicity. Thus, to assess the potential release of these As to localities in the vicinity, a sequential extraction protocol was employed, which could present characteristic bonding of the metals to the soils.

*3.3 Results*

Table 4. Chemical composition (Mean ± S.D) of soil samples

| Specifications | Soil Depth (cm) | Soil 1 | Soil 2 | Soil 3 | Control |
|---|---|---|---|---|---|
| pH | 0 – 15 | 7.10±0.21 | 6.90±0.11 | 6.51±0.22 | 6.71±0.11 |
| | 15 – 30 | 6.72±0.11 | 6.71±0.12 | 6.42±0.10 | 5.70±0.12 |
| | 30 – 45 | 5.61±0.22 | 6.60±0.21 | 6.22±0.21 | 5.71±0.10 |
| Relative Humidity (%) | | 61.41 | 56.51 | 58.30 | 59.00 |
| Organic Matter (%) | 0 – 15 | 3.04±0.11 | 2.67±0.09 | 2.12±0.12 | 7.21±0.10 |
| | 15 – 30 | 2.82±0.20 | 1.27±0.10 | 0.28±0.06 | 5.27±0.07 |
| | 30 – 45 | 1.31±0.13 | 0.69±0.13 | 0.21±0.01 | 4.51±0.09 |
| CEC (meq/100 g) | 0 – 15 | 10.74±1.31 | 9.69±1.85 | 9.85±1.89 | 5.17±0.15 |
| | 15 – 30 | 6.83±1.03 | 7.08±1.15 | 5.21±0.77 | 4.29±0.11 |
| | 30 – 45 | 8.07±1.34 | 6.58±1.22 | 3.26±0.79 | 3.52±0.09 |
| As (total, mg/kg) | 0 – 15 | 2.51±0.22 | 2.45±0.24 | 2.53±0.22 | 0.69±0.25 |
| | 15 – 30 | 2.14±0.34 | 2.23±0.21 | 2.26±0.25 | 0.53±0.21 |
| | 30 – 45 | 2.69±0.19 | 2.72±0.17 | 2.79±0.19 | 0.77±0.13 |
| Particle Size Distribution | Sand (%) | 67±3 | 63±3 | 71±4 | 55±2 |
| | Clay (%) | 24±3 | 26±2 | 22±3 | 34±2 |
| | Silt (%) | 09±2 | 11±3 | 09±3 | 11±2 |
| | T/C | SCL | SCL | SCL | SCL |

S.D: Standard deviation; T/C: Textural Class; SCL : Sand clay loam

Chemical fractionation of As

Table 5 gives the experimental results obtained for the seven fractions in sequential extraction. For exchangeable fractions, the fractions decreases in concentration down the soil profile, and the fractions in the control are low compared to the other samples. The fraction 2 represents easily available fractions and it decreased with soil depth. The sum of fraction 1 and 2 represent the bioavailability of As to plants (Shiowatana et al., 2001). The mean concentrations of fraction 1 is in the range of 0.04-0.06 mg/kg. The mean concentrations of fraction 2 ranges between 0.09 – 0.18 mg/ kg, which is greater than the amount presents in the fraction 1 and control samples. This is an indication that the As deposited by the activities of mechanic workshops are in mobile phase. The percentage that bonds to manganese is lowest in all fractions, while the fractions bonds to organic matters shows higher concentrations, significantly more than that of the control samples. Since this fraction relates to the portion released under strong oxidizing conditions, then it constitutes a source of potentially available As in soil (Ure and Davidson, 1995). The As contents were more in amorphous Fe than crystalline Fe (Shane Dever Whiteacre, 2009). The range of As in the soil is greater than that found in the control. The greatest part of As was associated with residual fractions. These fractions in their inert phase correspond to the part of the As which cannot be mobilized. The inner soil contains the highest proportion residual fraction.

Generally the affinity of As with the soil fraction increases in order: F3 < F6 < F1 < F2 < F5 < F4 < F7.

Table 5. Results of As binding fractions

| Soils | Soil Depth (cm) | FRACTIONS | | | | | | |
|---|---|---|---|---|---|---|---|---|
| | | $F_1$ | $F_2$ | $F_3$ | $F_4$ | $F_5$ | $F_6$ | $F_7$ |
| SOIL 1 | 0 – 15 | 0.06 | 0.09 | - | 1.13 | 0.10 | 0.03 | 1.75 |
| | 15 – 30 | 0.05 | 0.14 | 0.01 | 1.34 | 0.14 | 0.01 | 1.41 |
| | 30 – 45 | 0.04 | 0.18 | 0.02 | 1.46 | 0.20 | 0.02 | 1.81 |
| Mean | | 0.05 | 0.14 | 0.01 | 1.31 | 0.14 | 0.02 | 1.66 |
| S.D | | 0.01 | 0.05 | 0.01 | 0.17 | 0.05 | 0.01 | 0.22 |
| SOIL 2 | 0 – 15 | 0.05 | 0.07 | 0.01 | 1.24 | 0.13 | 0.01 | 1.50 |
| | 15 – 30 | 0.04 | 0.15 | 0.01 | 2.07 | 0.16 | 0.02 | 1.12 |
| | 30 – 45 | 0.04 | 0.16 | 0.01 | 1.40 | 0.23 | 0.02 | 1.78 |
| Mean | | 0.04 | 0.13 | 0.01 | 1.57 | 0.18 | 0.02 | 1.47 |
| S.D | | 0.01 | 0.05 | 0.00 | 0.44 | 0.05 | 0.01 | 0.33 |
| SOIL 3 | 0 – 15 | 0.07 | 0.09 | 0.01 | 0.73 | 0.16 | 0.01 | 1.57 |
| | 15 – 30 | 0.06 | 0.13 | 0.01 | 0.69 | 0.16 | 0.03 | 1.20 |
| | 30 – 45 | 0.04 | 0.15 | 0.02 | 0.80 | 0.19 | 0.02 | 1.83 |
| Mean | | 0.06 | 0.12 | 0.01 | 0.74 | 0.17 | 0.02 | 1.53 |
| S.D | | 0.02 | 0.03 | 0.06 | 0.06 | 0.02 | 0.01 | 0.32 |
| CONTROL | 0 – 15 | 0.02 | 0.02 | - | 0.12 | 0.03 | - | 0.34 |
| | 15 – 30 | 0.01 | 0.03 | - | 0.14 | 0.03 | - | 0.23 |
| | 30 – 45 | - | 0.03 | 0.01 | 0.16 | 0.05 | <0.01 | 0.42 |
| Mean | | 0.01 | 0.03 | 0.00 | 0.14 | 0.04 | - | 0.33 |
| S.D | | 0.01 | 0.01 | 0.01 | 0.02 | 0.01 | - | 0.10 |

The percent distribution of As across the various fractions are presented in Fig 2 below. Approximately 50 % of As concentration is in the residual fractions, followed by bonds to soil organic matters (i.e organic bonds). About 10 % is bound to amorphous Fe oxides, while about 8-9 % is bioavailable and mobile, bioavailable are the As content that are readily available for plant and mobile are those. About 1 % is associated with Fe oxides in the crystalline phase.

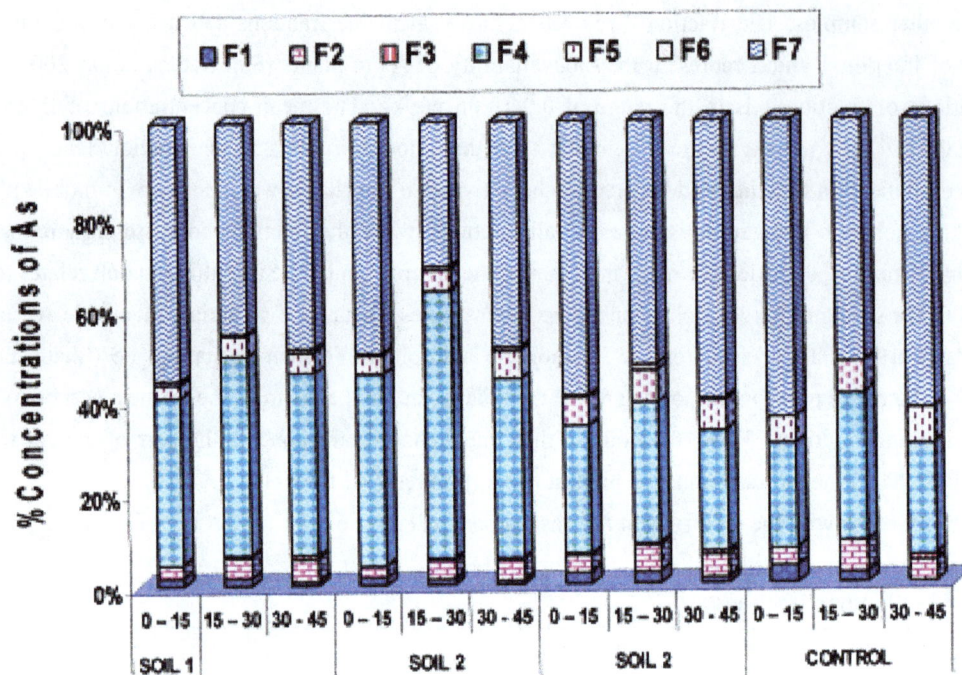

Figure 2. Percent distribution pattern of As across various fractions

In order to understand the extent of pollution, elemental contamination factors (CFs), soil enrichment factors (EFs) and geoaccumulation index (Igeo), results were determined using the formula described elsewhere (Likuku *et al.*, 2013; Loska *et al.*, 2003), The CFs of the entire study area ranges as follow 3.55-3.66, 4.04-4.26, 3.49-3.62 for top, middle and inner soil respectfully. The highest is found in the middle soil which negate what was observed for total concentration of As. The CFs from all the sample is higher than that of the control which ranges between 0.53-0.77 (Table 6).

Table 6. Elemental contamination factors (CFs) in different soil layers

| Element | Soil Layers | Soil 1 | Soil 2 | Soil 3 | Natural Background Concentration |
|---------|-------------|--------|--------|--------|----------------------------------|
|         | 0 – 15      | 3.64   | 3.55   | 3.66   | 0.69±0.25                        |
|         | 15 – 30     | 4.04   | 4.21   | 4.26   | 0.53±0.21                        |
|         | 30 – 45     | 3.49   | 3.52   | 3.62   | 0.77±0.13                        |

[a]Average natural background concentration (±SD, n=3)

The results for EFs and Igeo can be found in Table 7. In this study, iron was used as a conservative tracer to differentiate natural from anthropogenic components (Ergin *et al.*, 1991) The mean EFs of As in soils under study ranged between 1.77 and 2.40. This according to Zhang and Liu (2000), points to As being deposited from the influence of mechanic workshop waste. Zhang and Liu classified EFs between 0.5 and 1.5 as the natural metal content, while any amount greater than 1.5 was probably due to anthropogenic sources. In this study, the anthropogenic source referred to is mechanical wastes. Since the presence of heavy metals in the soil does not necessarily constitute a pollution risk, Igeo as proposed by Muller (1969) was used to access the pollution of As in the soil of the mechanic workshop. The Igeo of all the soil samples under study ranged from 0.37- 0.39. This is an indication that the soil samples were uncontaminated to moderately contaminated (Muller, 1969).

Table 7. Mean EF and *I*geo classes of As with respect to the natural background

| ELEMENT | Soil 1 | Soil 2 | Soil 3 |
|---|---|---|---|
| Enrichment Factor | | | |
| As | 1.77 | 2.40 | 2.18 |
| $I_{geo}$ | | | |
| As | 0.39 | 0.37 | 0.39 |

The normalizing element for calculation of the EF and Igeo is Fe with natural background value of 232.7 mg/kg. From both EF and Igeo, it is clear that the concentration of As in the soil is from the influence of auto-mechanic workshop waste, but it is not present in concentrations that pose a significant risk to the environment.

## 4. Conclusion

It is clear that the soils investigated were not contaminated with As, therefore plants and water in these areas are good for both residential and industrial purposes, although The CFs confirmed the influence of the auto-mechanic workshop's activities on the As mobility to the environment. The sequential extraction method adopted in this research provides dependable information on fractions of As in the soils of auto-mechanic workshops. The reagents were specific and selective for As in the phase examined. The fractions obtained from sequential extractions were small because the total As concentration in all soils were small. The results indicate that As is mainly associated with oxidizing and residual fractions, which allows us to predict their mobility. Hence periodic assessment of the mechanic workshops and waste sites is necessary to ensure the level is below the regulatory limit. This is of paramount importance since As from anthropogenic and geologic sources are considered one of the most toxic elements affecting millions of people around the world.

## Acknowledgement

The effort of Barr William McCallig in proof reading this work is highly appreciated.

## References

Adamu, H., Luter, L., Musa, M. L., & Umar, B. A. (2013). Chemical Speciation: A strategic pathway for insightful risk assessment and decision-making for remediation of toxic metal contamination. *Environment and Pollution, 2*(4), 92-99.

Awofolu, O. R., Mbolekwa, Z., Mtshenla, V., & Fatoki, O. S. (2005). Levels of trace metals in water and sediment from Tyume river and its effects on an irrigated farmland. *Water SA., 31*(1), 87-94. http://dx.doi.org/10.4314/wsa.v31i1.5124

Baroni, F., Boscagli, A., Di Lella, L. A., Protano, G., & Riccobono, F. (2004). Arsenic in soil and vegetation of contaminated areas in southern Tuscany (Italy). *Journal of Geochemical Exploration, 81*(1), 1-14. http://dx.doi.org/10.1016/S0375-6742(03)00208-5

Bouyoucos, G. J. (1962). Hydrometer method improved for making particle size analyses of soils. *Agronomy Journal, 5*(5), 464-465. http://dx.doi.org/10.2134/agronj1962.00021962005400050028x

Cai, Y., Cabrera, J. C., Georgiadis, M., & Jayachandran, K. (2002). Assessment of arsenic mobility in the soils of some golf courses in South Florida. *The science of the total Environment, 291*, 123-134. http://dx.doi.org/10.1016/S0048-9697(01)01081-6

Chemistry world. (2014). *Royal society of chemistry bulletin, 11*(4), 55-57. Retrieved from http://www.chemistryworld.org

Cullen, W. R., & Reimer, K. J. (1989). Arsenic speciation in the environment. *Chemi. Rev., 89*, 713-764. http://dx.doi.org/10.1021/cr00094a002

Eisler, R. (1994). A review of arsenic hazards to plants and animals with emphasis on fishery and wildlife resources.

Ergin, M., Saydam, C., Basturk, O., Erdem, E., & Yoruk, R. (1991). Heavy metal concentrations in surface sediments from the two coastal inlets (golden Horn Estuary and Izmit bay) of the northeastern sea of Marmara. *Chem., Geo., 91*, 269-285. http://dx.doi.org/10.1016/0009-2541(91)90004-B

Francis, O. (2004). ATrace metals contamination of soils and vegetation in the vicinity of livestock in Nigeria. *EJEAFche, 4*(2), 863-870.

Gupta, S. K., Vollmer, M. K., & Krebs, R. (1996). The importance of mobile, mobolizable and pseudo total heavy metal fractions in soil for three-level risk assessment and risk management. *Science total environment, 178*, 11-20. http://dx.doi.org/10.1016/0048-9697(95)04792-1

Jackson, M. L. (1958). *Soil chemistry analysis prentice Hall.* London, U.K, pp. 54-59.

Kim, M. J., & Nriagu, J. (2000). Oxidation of arsenic in groundwater using ozone and oxygen. *Science total environment, 247*, 71-79. http://dx.doi.org/10.1016/S0048-9697(99)00470-2

Lee, D. Y., & Lee, C. (2011). Regulatory standard of heavy metal pollutant in soil and groundwater in Taiwan.

Likuku, A. S., Gaboutloeloe, G. K., & Mmolawa, K. B. (2013). Determination and source apportionment of selected heavy metals in aerosol samples collected from Sebele. *American Journal of Environmental Science, 9*(2), 188-200. http://dx.doi.org/10.3844/ajessp.2013.188.200

Loska, K., Wiechula, D., Barska, B., Cebula, E., & Chojnecka, A. (2003). Assessment of Arsenic Enrichment of Cultivated Soils in Southern Poland. *Polish Journal of Environmental Studies, 12*(2), 187-192.

Muller, G. (1969). Index of geoaccumulation in sediments of Rhine river. *Geol. Journal, 2* 109-118.

Ning, R. Y. (2002). Arsenic removal by reverse osmosis. *Desalinasation, 143*(3), 239-241.

Ololade, I. A., Oloye, F. F., Adamu, H., Oluwole, O. D., Oluyede, O. P., Alomaja, F., & Ololade, O. (2014). *Distribution and Potential Mobility Assessment of Toxic Metals in Soils of Mechanic Workshops in Akoko, Ondo State Nigeria.* In press.

Sadiq, M. (1997). Arsenic chemistry in soils: an overview of thermodynamic predictions and field observations. *Water, air, and soil pollution, 93*(1-4), 117-136. http://dx.doi.org/10.1007/BF02404751

Sager, M. (1992). Hazardous metals in the environment: Techniques and instruments in analytical chemistry. *Amsterdam: Elsevier.,* 133-175. http://dx.doi.org/10.1016/S0167-9244(08)70106-9

Shane Dever Whiteacre, B.S. (2009). Soil control on Arsenic bio accessibility, Arsenic fractions and soil properties, MSc thesis, Ohio state university.

Shiowatana, J., Mclaren, R. G., Chanmekha, N., & Samphao, A. (2001). *Journal of environmental quality, 30*, 1940-1949. http://dx.doi.org/10.2134/jeq2001.1940

Tlustors, P., Szakova, J., Starkova, A., & Pavlikova, D. A. (2000). A comparison of sequential extraction procedures for fractionation of arsenic, cadmium, lead, and zinc in soil. *Central European journal of chemistry, 3*(4), 830-851. http://dx.doi.org/10.2478/BF02475207

Ure, A. M., & Davidson, C. M. (1995). Chemical speciation in the environment. Blackie Glasgow.

Uzairu, A., Harrison, G. F. S., Balarabe, M. L., & Nnaji, J. C. (2009). Concentration levels of trace metals in fish and sedment from Kubanni river, Northern Nigeria. *Bull. Chem. Soc. Ethiop., 23*(1), 9-17. http://dx.doi.org/10.4314/bcse.v23i1.21293

Wenzel, W. W., Kirchbaumer, N., Prohaska, T., Stingeder, G., Lombi, E., & Adriano, D. C. (2001). Arsenic fractionation in soils using an improved sequential extraction procedure. *Analytica chimica acta, 436*, 309-323. http://dx.doi.org/10.1016/S0003-2670(01)00924-2

Zein, H., & Brummer, G. W. (1989). Mittiling. Dtsch. Bodenkundl, *Gesllsch, 59*, 505.

Zhang, J., & Liu, C. L. (2000). Riverine composition and estuaine geochemistry of particulate metals in China-weathering features, anthropogenic impact and chemical fluxed. *Estuarine coastal shelfs, 54*, 1051-1070. http://dx.doi.org/10.1006/ecss.2001.0879

# 12

# Spatial and Seasonal Variation of Dissolved Nitrous Oxide in Wetland Groundwater

Xing Li[1], Changyuan Tang[1], Zhiwei Han[1], Piao Jingqiu[1], Cao Yingjie[1] & Zhang Chipeng[1]

[1] Graduate School of Horticulture, Chiba University, Matsudo, Japan

Correspondence: Changyuan Tang, Graduate School of Horticulture, Chiba University, 648 Matsudo, Matsudo-shi, Japan. E-mail: tangchangyuan@gmail.com

**Abstract**

Understanding the spatial and temporal pattern of dissolved nitrous oxide ($N_2O$) in groundwater is essential to estimate the $N_2O$ emissions from groundwater to the unsaturated zone and to the atmosphere. In order to study the spatial distribution and seasonal change of dissolved $N_2O$ in wetland, a headwater wetland in Ichikawa, Chiba Prefecture, Japan, was chosen. Variations of nitrate ($NO_3^-$), dissolved $N_2O$ and $\delta^{15}N\text{-}NO_3^-$ indicated that the dissolved $N_2O$ in the groundwater of study wetland consists of two parts, one from denitrification within the wetland, and another from nitrification at upland. Principal component analysis (PCA) was used to assess the shallow groundwater parameters in the wetland. And t-test was conducted to find statistically significant differences of the variables between the ASW and NS, warm season and cool season. The concentrations of dissolved $N_2O$ increased from the upland to the zone of adjacent area between slope and wetland (ASW) and then decreased at the zone near the stream (NS). In sight of dissolved $N_2O$ associated nitrogen migration, groundwater in the study area can be divided into three stages: upland as the stage 1, ASW as the stage 2, and NS as the stage 3. Higher temperature results in higher denitrification rate, lower dissolved oxygen (DO) and oxidation-redox potential (ORP), yielding higher concentration of $N_2O$ in the warm season. Therefore, the seasonal change of dissolved $N_2O$ in study wetland can be mainly interpreted by the variation of temperatures of groundwater.

**Keywords:** dissolved $N_2O$, spatial distribution, seasonal change, denitrification, stage

## 1. Introduction

Over the last few decades, much interest has been focused on specific natural systems, such as wetland (or riparian zone) which are vulnerable to improve water quality by physical, chemical and biological process that remove N from groundwater (García-García, Gómez, Vidal-Abarca, & Suárez, 2009; Groffman, Gold, & Simmons, 1992; Sabater et al., 2003). Wetlands offer an abundant organic C supply and dominated by inherently wet surface soil create anaerobic environment to consume nitrate via denitrification that is considered the most important reaction for nitrate removal in aquifer (Bastviken, Olsson, & Tranvik, 2003; Burgin & Hamilton, 2007; Whitmire & Hamilton, 2005). Especially in the shallow ground water of riparian areas, redox conditions are often favorable for intense denitrification processes (Ross, 1995).

The trace gas $N_2O$ is an obligate intermediate product of biological denitrification. And it is known to contribute to global warming and the destruction of stratospheric ozone. A significant amount of $N_2O$ emissions originates denitrification (Mathieu et al., 2006). Emissions from aquifers are most likely to occur from shallow aquifers, where $N_2O$ can be quickly transferred through the unsaturated zone to the atmosphere by diffusion (Rice & Rogers, 1993). $N_2O$ emission from wetland system has been estimated by numerous studies (Dhondt, Boeckx, Hofman, & Van Cleemput, 2004; Groffman, Gold, & Addy, 2000; Verhoeven, Arheimer, Yin, & Hefting, 2006). Understanding the spatial and seasonal pattern of dissolved $N_2O$ is essential to assess the indirect emission of $N_2O$ from groundwater (Geistlinger, Jia, Eisermann, & Florian Stange, 2010). Level of dissolved $N_2O$ in groundwater has been paid lots of attentions. For example, $N_2O$ concentration in groundwater was reported to exceed greatly those of atmospheric equilibration (with a mean value of 28.98 $\mu g\ L^{-1}$) under aerobic condtion in Kanto district, Japan (Ueda, Ogura, & Yoshinari, 1993), and the maximum up to 30000 times of that in the ambient air (Heincke & Kaupenjohann, 1999). However, few studies estimated level of dissolved $N_2O$ in wetland groundwater.

According literature review, study of spatial pattern of dissolved $N_2O$ has been focused on surface water, such as river, lake and ocean (Butler, Elkins, Thompson, & Egan, 1989; Ferrón, Ortega, & Forja, 2010; Hinshaw & Dahlgren, 2013; Wang et al., 2009; Zhang, Zhang, Liu, Ren, & Zhao, 2010). The pattern of seasonal and spatial of dissolved $N_2O$ is related to denitrification or nitrification depending on the environment in watershed. For example, the highest concentrations of dissolved $N_2O$ were observed in the riparian zone in May (warm season), when the nitrate ($NO_3^-$) and temperature were conducive for denitrification (Dividson, Stark, & Firestone, 1990). However, Kim, Isenhart, Parkin, Schultz, and Loynachan (2009) found that dissolved $N_2O$ concentrations were with the highest value in cool season and the lowest value in warm season. Thus, the pattern of seasonal change of dissolved $N_2O$ in wetland is unclear. In addition, $N_2O$ also could product from nitrification (fertilizer and manure ammonium-nitrogen is oxidized to nitrate-nitrogen) in unsaturated zone. $N_2O$ could leach to groundwater at upland and discharge to wetland through the groundwater flow system (Mühlherr & Hiscock, 1998; Spalding & Parrott, 1994). However, few studies estimated the contribution of the $N_2O$ from nitrification at upland to dissolved $N_2O$ in wetland.

Therefore, the objectives of this study were 1) to identify the source of dissolved $N_2O$ and its evolution stages based on $\delta^{15}N\text{-}NO_3^-$, $NO_3^-$ and dissolved $N_2O$; and 2) to understand comprehensively the spatial distribution and seasonal change of dissolved $N_2O$ concentration in shallow groundwater of headwater wetland. As a matter of convenience, we define the groundwater are at stage 1 in upland where the dissolved $N_2O$ is produced from nitrification, stage 2 where more dissolved $N_2O$ is produced than consumed in denitrification in wetland, and stage 3 where the net of dissolved $N_2O$ decreases resulting from little available nitrate and its reduction to $N_2$ as a proceed stage of stage 2.

## 2. Sites Description and Method

### 2.1 Site Description

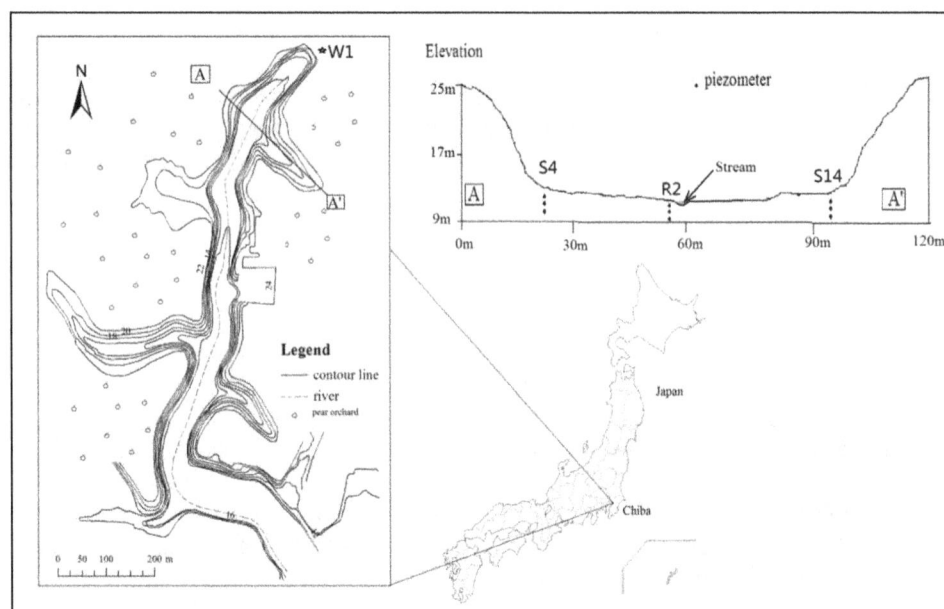

Figure 1. Study sites in Ichikawa, Japan. S4, R2 and S14 constitute transect A within the wetland and W1 is located in the upland site

The study area is a headwater wetland, located at Ichikawa City (35.76 °N, 139.97 °E), Chiba Prefecture, Japan (Figure 1). The wetland valley is U-shaped with an elevation of about 12m above sea level. The wetland receives discharge (both groundwater and overland flow) from an adjacent upland (elevation 25-28 m) area mostly pear orchard vegetation. A stream flowing through the wetland valley is recharged by spring water and groundwater in the wetland. Previously, this wetland used to be paddy field and had been redeveloped to a wetland park. Average yearly flow of the stream from the study weland is about $6.70\times10^5$ $m^3year^{-1}$. Dominating vegetation in the wetland are Houttuynia, Calamus and Japanese pampas grass. The slope is close to the orchard edge and is decreasing towards the wetland. It acts as a transitional zone linking the upland and the experimental wetland. The slope is covered by Acer, Pinophyta and Bambuseae. The annual average precipitation is 1,316 mm, with a maximum monthly precipitation of 226.5 mm month$^{-1}$ in May of study area. The annual average temperature is 15.6 °C with a highest temperature of 36.9 °C in August and a lowest temperature of -3.4 °C in January. The nitrogen load in pear orchard of the upland is estimated about 400 kg ha$^{-1}$ year$^{-1}$ (Agriculture and Forestry Research Center of Chiba Prefecture, 2003). The area of pear orchard in Ichikawa city is 272 ha which account for 45% of agricultural land of Ichikawa city. The upland is covered by Kanto Loam about 4 m in thickness. It is underlain in a sequence by Joso clay layer and narita sand. That is a fine sand layer which is the major aquifer and the water table was about 17 m above sea level at upland all around the year. Within the wetland, the aquifer is a fine sand layer overlaid by cohesive soil and sandy clay with the water table depth > 11.6 m above sea level.

*2.2 Sampling Procedures and Measurements*

Field surveys were conducted in May, July, September and November in 2011, March and June in 2012. The study area is characterized by a temperate climate with warm season from June to September and cool season from November to May. Water samples were taken from well W1 at the upland and piezometers which were installed at S4, S14 and R2 with depths of 1 m, 2 m and 3 m in the wetland, for a total of 9 piezometers (Figure 1). S4 and S4 and S14 were at the sides of wetland and R2 was placed approximately 0.3 m on the west side of the stream. In order to get the fresh groundwater, we withdrew water from the piezometers and waited the fresh groundwater flowing in. In order to avoid the loss of dissolved gas during the sampling, a new sampler has been developed (Figure 2a). The sampler was inserted into the bottom of piezometer slowly with the outlet opened and the inlet closed. The inlet was opened by drawing the rope stopper to let the fresh groundwater flow in gently, make the vial (35 ml) full and push out the air inside through tube with the three-way stopcock. After closing the three-way stopcock, the sampler was taken out from the piezometer and the vial was sealed with a rubber cap under the water in the sampler as soon as possible. 1 ml sterilant (hibitane) was injected into the vial through the cap after water collection. Dissolved $N_2O$-N concentration was determined by headspace method. 10 ml pure $N_2$ gas was injected into the vial to push out an equal volume of water from the vial. Vials were shaken for 1minute and stored at 40 °C for 24 h to equilibrate. The gas samples were analyzed for target gas ($N_2O$) by a gas chromatography (GC14B, Shimazu) equipped with an electron capture detector operated at 280 °C, injector at 100 °C and column at 700 °C.

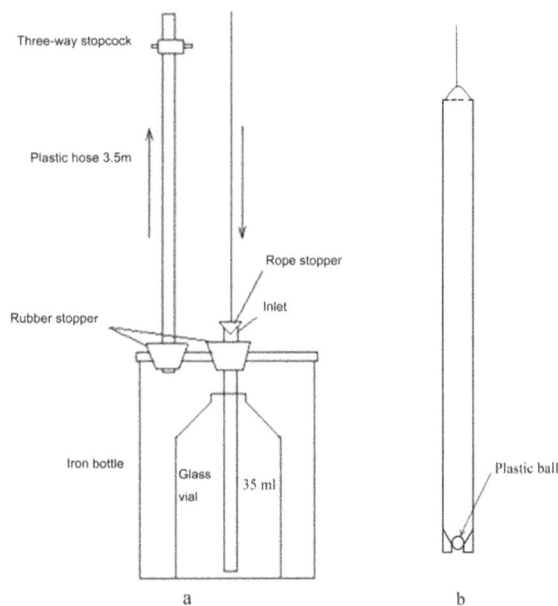

Figure 2. The schematic diagram of sampler for dissolved $N_2O$ (a) and ion, parameters and $\delta^{15}$N-NO$_3^-$ (b)

After sample for gas analysis was taken, water sample for ion, parameters and $\delta^{15}$N-NO$_3^-$ analysis were collected by a pipe sampler (Figure 2b). When the pipe is put into water, a plastic ball at the bottom of the pipe can go up due to floatage that let the water into the pipe. The ball go back to the bottom when the pipe out of water that seal the pipe. Then, water samples were taken from the piezometers separately. Samples were brought back to the laboratory and stored at 4 °C before laboratory analysis. DO, pH, ORP, and temperature of groundwater were measured in situ with sensors (HIROBA). All water samples were filtered (0.45 µm) before analysis for major ions by ion chromatography (Shimadzu CDD-6A and CDD-10Avp). 2 L water for each sample was collected for $\delta^{15}N$-$NO_3^-$ analysis in March 2012. $NO_3^-$ was collected by passing the water through pre-filled, disposable, anion exchanging resin columns in the field and then was eluted by 3 M HCl from the column. The nitrate-bearing acid eluant was neutralized with $Ag_2O$, filtered to remove the AgCl precipitate, then freeze dried to obtain solid $AgNO_3$, which was then combusted to $N_2$ in sealed quartz tubes for analysis by Integra CN mass spectrometer (Pdz Europa LTD) at Chiba University, Japan (Yingjie Cao, Tang, Song, Liu, & Zhang, 2012). All the samples were measured twice and the result showed the difference between the two measurements was less than ± 5%. Then the mean of two measurements was used as the value of $\delta^{15}N$-$NO_3^-$ in this study.

*2.3 Statistical Analysis*

Variables were tested using student t-test and principal component analysis (PCA), with SPSS 8.0 for Windows (SPSS, 1997, IL, USA). T-test was used to determine if two sets of data are significantly different from each other. The PCA is a data transformation technique that attempts to reveal a simple understanding structure that is assumed to exist within a multivariate dataset (Davis, 1986).

# 3. Results

*3.1 Basic Parameters and Dissolved $N_2O$ in Upland Shallow Groundwater*

Samples were taken from W1 in July and November 2012, respectively. DO and ORP concentrations were higher in July (Table 1). pH values of groundwater were lower than 7 both in July and November. Groundwater temperature was little higher in July than that in November. $NO_3^-$-N and $N_2O$-N concentrations were both higher in July than that in November.

Table 1. Basic parameters and dissolved $N_2O$ of upland shallow groundwater in July and November 2012

|       | $N_2O$-N ($\mu g\ L^{-1}$) | DO ($mg\ L^{-1}$) | ORP (mv) | $NO_3^-$-N($mgL^{-1}$) | T (°C) | pH   |
|-------|---------------------------|-------------------|----------|------------------------|--------|------|
| Jul   | 14.73                     | 9.29              | 325      | 34.74                  | 18.3   | 6.94 |
| Nov   | 8.11                      | 5.70              | 295      | 17.30                  | 16.9   | 6.05 |

*3.2 Basic Parameters and Dissolved $N_2O$ in Wetland Shallow Groundwater*

Groundwater temperatures in the wetland ranged from 14.2 to 24.8 °C during the study period (Table 2). pH values of groundwater ranged from 6.53 to 7.97, indicating that the groundwater was alkaline except S14 which pH was lower than 7 during the warm season. DO concentrations ranged from 0.07 to 11.50 mg $L^{-1}$. It was lower than 4 mg $L^{-1}$, and as low as 0.07 mg $L^{-1}$ in June at R2. At S4 and S14, the DO concentrations were lower than 5 mg $L^{-1}$ in the warm season, but up to 11.5 mg $L^{-1}$ in the cool season (S14-3 m in November). ORP values ranged from -244 to 303 mV. At 1 m and 2m depth of R2, ORP values were below about 0 mV in the warm season with the lowest value of -189 mV. However, ORP was up to 175 mV in March. At 3 m depth of R2, ORP was above 0 mV except in September (-244 mV). The $NO_3^-$-N concentration changed from 0 to 114.0 mg $L^{-1}$ in study sites. At S4 and S14, most $NO_3^-$-N concentrations are clearly above the standard of the drinking water (10 mg $L^{-1}$) set by United States Environment Protection Agency (Figure 3), whereas $NO_3^-$-N concentration was extremely low for detection at R2. $NH_4^+$ and $NO_2^-$ were also measured with other major ions, and found below the detectable limit.

Table 2. T (°C) pH DO (mg L$^{-1}$), and ORP (mV) of the groundwater at S4, S14 and R2 in 1m, 2 m and 3 m depth

| Site | May/11 | | Jul/11 | | Sep/11 | | Nov/11 | | Mar/12 | | Jun/12 | |
|------|-----|-----|-----|-----|-----|-----|-----|-----|-----|-----|-----|-----|
| | T | pH | T | pH | T | pH | T | pH | T | pH | T | pH |
| S4-1 | 14.7 | 7.70 | 19.7 | 6.99 | 24.8 | 7.56 | 14.7 | 7.48 | 15.4 | 7.38 | 19.4 | 7.15 |
| S4-2 | 15.2 | 7.62 | 19.1 | 7.05 | 24.8 | 7.50 | 15.2 | 7.41 | 14.8 | 7.36 | 19.1 | 7.23 |
| S4-2 | 14.9 | 7.42 | 19.9 | 7.18 | 18.4 | 7.53 | 14.9 | 7.40 | 14.5 | 7.54 | 18.5 | 7.52 |
| R2-1 | 16.2 | 7.48 | 23.2 | 6.90 | 24.7 | 7.15 | 16.2 | 7.63 | 19 | 7.49 | 22.5 | 6.94 |
| R2-2 | 16.4 | 7.72 | 21.0 | 7.17 | 21.4 | 7.47 | 16.4 | 7.59 | 15.3 | 7.59 | 20.3 | 7.07 |
| R2-3 | 15.5 | 7.97 | 20.7 | 7.30 | 20.8 | 7.72 | 15.5 | 7.65 | 17.7 | 7.63 | 20.3 | 7.17 |
| S14-1 | 15.4 | 7.84 | 20.5 | 6.68 | 20.9 | 7.17 | 15.4 | 6.90 | 17.2 | 7.66 | 19.5 | 6.87 |
| S14-2 | 15.0 | 7.63 | 19.4 | 6.58 | 18.3 | 7.03 | 15.0 | 6.71 | 17.1 | 7.79 | 18.9 | 6.84 |
| S14-3 | 14.2 | 7.41 | 20.1 | 6.53 | 20.4 | 6.84 | 14.2 | 6.61 | 15.4 | 7.66 | 18.2 | 6.95 |
| Site | May/11 | | Jul/11 | | Sep/11 | | Nov/11 | | Mar/12 | | Jun/12 | |
| | DO | ORP | DO | ORP | DO | ORP | DO | ORP | DO | ORP | DO | ORP |
| S4-1 | 7.31 | 287 | 2.90 | 185 | 3.26 | 284 | 6.82 | 287 | 7.41 | 267 | 3.16 | 269 |
| S4-2 | 7.05 | 278 | 2.03 | 178 | 3.06 | 303 | 6.00 | 278 | 5.74 | 258 | 4.00 | 269 |
| S4-2 | 8.09 | 260 | 2.46 | 191 | 1.80 | 287 | 8.26 | 260 | 8.26 | 272 | 2.73 | 280 |
| R2-1 | 3.23 | 18 | 0.21 | -77 | 0.14 | -26 | 3.20 | 18 | 2.8 | 128 | 1.84 | -27 |
| R2-2 | 3.95 | -173 | 1.55 | -177 | 0.13 | -189 | 2.80 | -173 | 2.59 | 175 | 0.13 | -150 |
| R2-3 | 1.41 | 232 | 2.19 | 50 | 0.68 | -244 | 3.78 | 232 | 2.65 | 146 | 0.07 | 151 |
| S14-1 | 7.2 | 280 | 2.35 | 147 | 4.59 | 197 | 9.35 | 280 | 4.75 | 287 | 3.13 | 126 |
| S14-2 | 7.2 | 311 | 1.53 | 169 | 2.10 | 230 | 11.00 | 311 | 7.18 | 279 | 3.70 | 168 |
| S14-3 | 6.69 | 292 | 2.44 | 177 | 2.36 | 229 | 11.50 | 292 | 6.61 | 279 | 4.03 | 164 |

Dissolved N$_2$O concentrations ranged from 0.09 to 100.62 µg L$^{-1}$ (Figure 3). At S4, dissolved N$_2$O ranged from 6.13 to 79.96 µg L$^{-1}$ with the highest concentrations in July and the lowest values in March. At S14, dissolved N$_2$O ranged from 7.8 to 100.62 µg L$^{-1}$ with the highest values in July and the lowest values in November. At R2, dissolved N$_2$O ranged from 0.09 to 2.29 µg L$^{-1}$ at 1 m in depth, and from 1.41 to 50.16 µg L$^{-1}$ at 2 m and 3 m in depth.

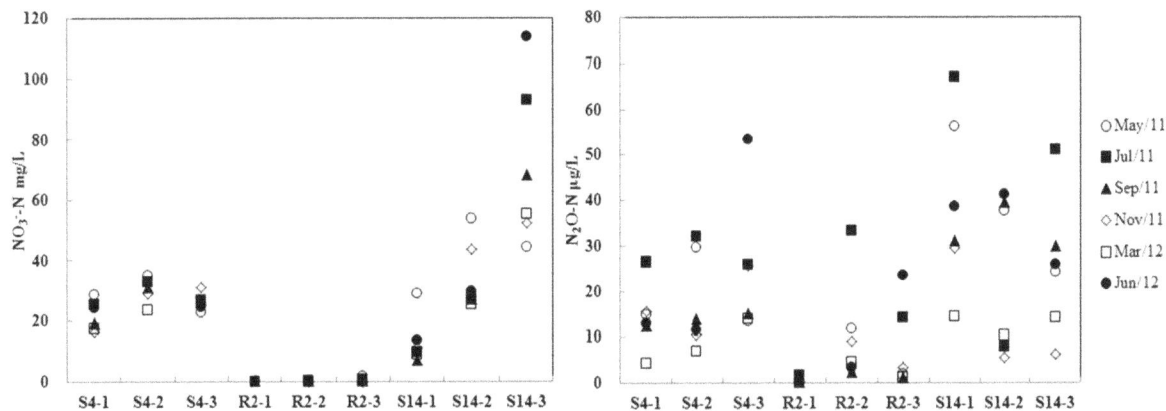

Figure 3. Variations of NO$_3^-$-N and N$_2$O-Nat S4, S14 and R2 in 1m, 2 m and 3 m depth (Open cycles: the cool season; Closed cycles: the warm season)

*3.3 Variations of $\delta^{15}N$-$NO_3^-$ in Shallow Groundwater*

$\delta^{15}N$-$NO_3^-$ in shallow groundwater of upland (W1) was 5.67‰ (Table 3). The $\delta^{15}N$-$NO_3^-$ in groundwater was 6.36‰ for 1m and 8.27‰ for 2m at S4, respectively. It was 8.7‰ for 1m, 9.81‰ for 2m and 7.67‰ for 3 m in depth at S14. The highest value was found in the groundwater 2m at S14 and the lowest value at W1. Comparing with groundwater in the upland, $\delta^{15}N$-$NO_3^-$ was enriched from 0.69‰ to 4.14 ‰ in the wetland. However, it was undetectable at R2 because little nitrate was available.

Table 3. $\delta^{15}N$-$NO_3^-$ in groundwater of wetland and upland

|  | W1 | S4-1m | S4-3m | S14-1 | S14-2m | S14-3m |
|---|---|---|---|---|---|---|
| $\delta^{15}N$-$NO_3^-$ | 5.67‰ | 6.36‰ | 8.27‰ | 8.7‰ | 9.81‰ | 7.67‰ |

*3.4 Statistical Analysis*

PCA was used to assess the shallow groundwater parameters in the wetland. To maximize the variance of the two principal axes, the varimax normalized rotation was applied. The load factors have been polarized after rotation of component matrix (Table 4). PCA results show two components with eigenvalues larger than 1, which explain 69.74% of the total variance. The first component explains about 39.894% of the observed variance. DO, T and ORP are correlated with the first component, representing the redox condition in the groundwater. The second component explains about 29.845% of the observed variation and is correlated with $NO_3^-$, $N_2O$ and pH which representing the reactants and products associated with denitrification process.

Table 4. Loadings for two principal components of groundwater variables in wetland

| Variable | Component 1 (Rotated) | Component 2 (Rotated) |
|---|---|---|
| $N_2O$-N | 0.089 | 0.735 |
| DO | 0.898 | 0.063 |
| T | -0.859 | 0.231 |
| ORP | 0.758 | 0.338 |
| $NO_3^-$-N | 0.426 | 0.720 |
| pH | 0.294 | -0.748 |
| Variance explained,% of total | 39.894 | 29.845 |

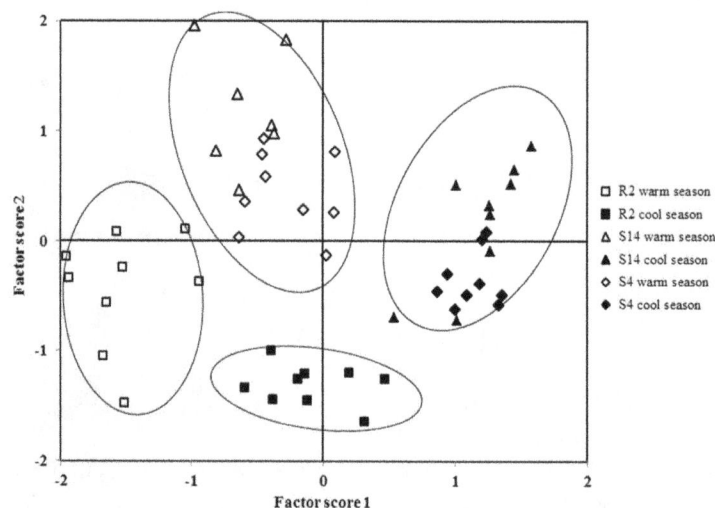

Figure 4. Bi-plot of the factor scores of the two principal components

The scores represent the influence of the component on the groundwater (Figure 4). It is possible to group the samples according to the axes of component 1 and 2. As a result, the samples are classified into four groups to showing seasonal and spatial patterns. The samples of NS are plotted at the down-left of the diagram for the warm season and the down-middle for the cool season. On the other hand, the samples of ASW are plotted at the upper-left of the diagram for the warm season and the upper-right for the cool season.

As the result of the groups from PCA, mean value, standard deviation and a t-test was conducted to find the difference and statistically significant differences of the variables between the ASW and NS, warm season and cool season (Table 5). $NO_3^-$-N, DO, ORP and $N_2O$-N in groundwater were significantly higher at ASW than those at NS, whereas there was no significantly difference of temperature and pH between ASW and NS. For $N_2O$-N, $NO_3^-$-N concentrations, and DO, variability was higher at ASW than those at NS on the basis of standard deviation. In contrast, variability of ORP was lower at ASW than it at NS. In addition, the mean $N_2O$-N concentration at ASW was high (36.14 $\mu g\ L^{-1}$), which was about 60 times of that in the ambient air. $N_2O$-N concentration and temperature in groundwater were significant higher in warm season than those in cool season, and DO, ORP and pH were significantly lower in warm season. There was no significant difference of $NO_3^-$-N concentrations between two seasons which seems to be the rule rather than the exception.

Table 5. Mean (m) and standard deviation (parentheses) of $N_2O$, DO, ORP, $NO_3^-$ and T in shallow groundwater of ASW (n=36) and NS (n=18)

| Zone | $N_2O$-N ($\mu g\ L^{-1}$) | DO (mg $L^{-1}$) | T (°C) | ORP (mV) | $NO_3^-$-N (mg $L^{-1}$) | pH |
|------|------|------|------|------|------|------|
| ASW | 36.14[*] | 5.22[*] | 17.60 n.s. | 237.00[*] | 33.00[*] | 7.24 n.s. |
| | (23.79) | (2.72) | (2.84) | (52.49) | (22.08) | (0.37) |
| NS | 9.27 | 1.85 | 19.10 | -5.00 | 0.01 | 7.42 |
| | (13.87) | (1.36) | (2.98) | (156.77) | (0.36) | (0.30) |
| Warm | 34.19** | 2.17* | 20.50* | 117.19** | 23.5 n.s. | 7.11* |
| | (27.01) | (1.30) | (1.93) | (162.66) | (28.34) | (0.30) |
| Cool | 20.19 | 6.03 | 15.6 | 209.67 | 20.6 | 7.49 |
| | (19.75) | (2.65) | (1.12) | (135.12) | (18.60) | (0.31) |

n.s., Not significant (p > 0.05)

*The difference between mean values is highly significant (p < 0.01)

** The difference between mean values is significant (p < 0.05)

## 4. Discussions

### 4.1 Source of Dissolved $N_2O$ of Shallow Groundwater

In order to estimate the concentration of $N_2O$ in groundwater, it is important to identify its source. Fertilizer and manure ammonium-nitrogen applied in the orchard are oxidized to nitrate-nitrogen and nitrous oxide in unsaturated zone of the upland. Nitrate leaches to the groundwater from unsaturated zone in the upland. $\delta^{15}N$-$NO_3^-$ value of W1 is coincided with range of $\delta^{15}N$-$NO_3^-$ (+4.5 ‰ to +8.5 ‰ ) in the area effected by mineral fertilizer (Cao, Sun, Xing, & Xu, 1991; Choi, Lee, & Ro, 2003; Choi, Han, Lee, Lee, & Yoon; Heaton, 1986; Singleton et al., 2007), indicating the dissolved $N_2O$ was produced via nitrification in the unsaturated zone of upland. DO concentrations were high at W1, indicating that denitrification could not occur. Nitrate and $N_2O$ transport from upland to wetland with groundwater consequently. $N_2O$ is difficult to denitrified to $N_2$ because the groundwater in upland is often assumed to have low biological activity due to low C content (Groffman, Gold, & Jacinthe, 1998). Geistlinger et al. (2010) found there will be a diffusive $N_2O$ flux from the deeper water to the capillary fringe. However the time scale of this process is very large i.e., for 10 cm travel distance, the $N_2O$ molecules need $\approx$ 230 d. Thus, diffusive loss to upward is considered to have little effect on $N_2O$ concentration in the groundwater during transporting from upland to wetland.

At the wetland, denitrification can enrich $^{15}N$ in the residual nitrate of groundwater (Cey, Rudolph, Aravena, & Parkin, 1999; Lehmann, Reichert, Bernasconi, Barbieri, & McKenzie, 2003). $\delta^{15}N$-$NO_3^-$ in the residual nitrate enriched from 2.8 ‰ to 78.32 ‰ when the concentration of $NO_3^-$-N decreased from 35.68 mg $L^{-1}$ to 0.45 mg $L^{-1}$

in a sand aquifer (Böttcher, Strebel, Voerkelius, & Schmidt, 1990), and from 6.4 ‰ to 24.8 ‰ when the $NO_3^-$-N concentration decreased from 13.3 to 5.6 mg $L^{-1}$ in a riparian zone (Cey et al., 1999) . In this study wetland, the $\delta^{15}N$-$NO_3^-$ enriched by 9.81 ‰ or even higher when the $NO_3^-$-N was no longer detectable. Therefore, dissolved $N_2O$ in the shallow groundwater of wetland consists of two parts, one from denitrification within the wetland, and another from the upland where nitrification is dominant.

*4.2 Spatial and Seasonal Pattern of Dissolved $N_2O$ in Shallow Groundwater of the Wetland*

The previous section suggested that the source of dissolved $N_2O$ of groundwater in wetland comes from nitrification in upland and denitrification in wetland. In the study wetland, denitrification controls the behavior of dissolved $N_2O$. Because $N_2O$ is an intermediate product of denitrification that is producted when nitrate is reduced and is consumed by reduction to $N_2$. Denitrification is considered to be related to many factors (DO, ORP, T, pH and $NO_3^-$). For example, the highest concentrations of $N_2O$ were found in the aerobic section of a limestone aquifer with the DO concentration below 4.00 mg $L^{-1}$ and in a phreatic aerobic aquifers with the DO concentration below 3.15 mg $L^{-1}$ (Deurer et al., 2008; Ronen, Magaritz, & Almon, 1988). However, the optimal maximum DO concentration for nitrogen removal was determined to be around 2.0-2.5 mg $L^{-1}$ in the laboratory experiments (Yoo et al., 1999). According the early study, Nelson and Knowles (1978) reported that the startup of denitrification can be inhibited while the oxygen level is as low as 0.13 mg $L^{-1}$ in a dispersed-well sludge reactor. In the laboratory experiments, as the ORP drops below 0 mV, the nitrate begins to be converted to nitrite and nitrite accumulates continuously for ORP ranging from 0 to -225 mV. From -225 to -400 mV, the accumulated nitrite is converted to $N_2$. As the ORP below -400 mV, the nitrate is firstly converted first to nitrite then the nitrite is converted immediately to $N_2$ without accumulation (Lee et al., 2000). It also reported that ORP below about 200 to 300 mV were found to be conducive to denitrification, and the maximum $N_2O$ were found at a ORP value of 0 mV (Kralova, Masscheleyn, Lindau, & Patrick Jr, 1992). Therefore, the optimum value of DO and ORP for $N_2O$ accumulation is not consistent with the value of the optimum for denitrification due to the $N_2O$ is an intermediate product. For nitrate, DeSimone and Howes (1998) studied that kinetics of denitrification at nitrate concentrations >1 mg-N $L^{-1}$ is zero order and even small amount of nitrate (lower than 2 mg-N $kg^{-1}$) leached was sufficient to create a large amount of $N_2O$ in groundwater (Müller, Stevens, Laughlin, & Jäger, 2004). Many studies suggested that high concentration of $NO_3^-$-N inhibits the $N_2O$ reductase yielding the higher concentration of $N_2O$ (Blackmer & Bremner, 1978; Deurer et al., 2008; Heisterkamp, Schramm, de Beer, & Stief, 2012). At ASW, the DO (m = 5.22 mg $L^{-1}$) and ORP (m = 237 mV) values were both higher than the optimum values respectively, as well as high concentrations of $NO_3^-$-N which were conducive to $N_2O$ accumulation (m = 36.14 μg $L^{-1}$ ) (Table 5). However, the mean value of DO concentrations (2.02 mg $L^{-1}$) and ORP were much lower (-5 mV) at NS. Additionally, $NO_3^-$-N is low or undetectable throughout the study. Under these conditions, the $N_2O$ is used as an electron acceptor instead of nitrate in denitrification process(Ishii, Ohno, Tsuboi, Otsuka, & Senoo, 2011), resulting in the lower concentration (m = 13.87 μg $L^{-1}$). Therefore, ASW and NS can be considered as in the stage 2 and stage 3, respectively. In addition, the average flux of $N_2O$ was found to be higher at ASW than it at NS (Li, Tang, Han, Cao, & Zhang, 2013) which is consistent with the trend of dissolved $N_2O$.

Seasonal changes of dissolved $N_2O$ are most associated with $NO_3^-$ concentration and water temperature (Bouwman, Boumans, & Batjes, 2002; Hinshaw & Dahlgren, 2013; Velthof, Oenema, Postma, & Van Beusichem, 1996). The T-test indicates that the concentrations of $NO_3^-$-N had no significant difference between the two seasons, which suggests $NO_3^-$-N is not the limited factor for denitrification rate in study wetland (Table 5). Temperature affected the dissolved $N_2O$ directly by controlling the denitrification rate (Nowicki, 1994; Pfenning & McMahon, 1997; Saunders & Kalff, 2001). The threshold temperature for controlling the rate of denitrification was 20 °C (Halling-Sørensen & Jorgensen, 1993) or even below 17 °C (McCutchan & Lewis, 2008; Nowicki, 1994). A study in coarse sandy soils found that the denitrification activity was low at 10 °C and completely inhibited at 2 and 5 °C because lower temperature may regulate metabolic rates for denitrifying bacteria (Vinther & Søeberg, 1991). Temperature also influences the solubility of oxygen, the rates of aerobic respiration of bacteria and the ORP change in groundwater, all of which in turn limit dissolved $N_2O$ indirectly. For example, the oxygen solubility is 14.60 mg $L^{-1}$ at 0 °C , about double at 30 °C (7.54 mg $L^{-1}$)(Weiss, 1970). Oxygen consumption by aerobic respiration increases when the temperature increases (Thamdrup, Hansen, & Jørgensen, 1998). When the temperature increased from 15 °C to 25 °C, the average ORP decreased from +40 mV to -60 mV (Zhu, Ndegwa, & Luo, 2002). In warm season, denitrification rate supposed not to be inhibited by temperature (m = 20.5 °C). The lower DO and ORP of groundwater could be assumed as a response to the higher temperature in the warm season. The characteristics of these factors resulted in the higher $N_2O$ concentration in the warm season (m = 34.19 μg $L^{-1}$) than it in cool season (20.19 μg $L^{-1}$). In addition, the decrease of pH was interpreted as a sign of intense denitrification (Ilies & Mavinic, 2001). Mean value of pH is lower in the warm season (m = 7.11) than it in the cool season (m = 7.49), which also can explain the higher dissolved $N_2O$ concentrations in the warm season. The seasonal change of dissolved $N_2O$ coincides with $N_2O$ flux measured in the study wetland. In fact, the average monthly $N_2O$ flux ranged from 0.019 to 0.286 mg N $m^{-2}$ $h^{-1}$ with the highest value in the warm season and the lowest flux appeared in the cool season (Li et al., 2013).

## 5. Conclusions

$N_2O$ concentrations, denitrification related factors ($NO_3^-$, DO, ORP, pH and T) and $\delta^{15}N$-$NO_3^-$ values were investigated in a typical headwater wetland and watershed. The main findings and conclusions are as follows:

Spatially, $NO_3^-$, DO and ORP are main factors to control the dissolved $N_2O$ in groundwater of study area. DO, ORP and $NO_3^-$ decreased continuously from upland to the wetland. Along the groundwater flow, the dissolved $N_2O$ was produced through nitrification at the upland and denitrification in the wetland, which is supported by the variations of $\delta^{15}N$-$NO_3^-$ in the shallow groundwater. The mean value of dissolved $N_2O$-N increased from 11.42 µg $L^{-1}$ at upland to 36.14 µg $L^{-1}$ at the ASW and then decreased to 9.27 µg $L^{-1}$ at NS. The dissolved $N_2O$ in the ASW zone is expected to be composed of two parts. One is transported from the upland and the other is produced from denitrification in the wetland. As a result, the dissolved $N_2O$ in the groundwater can be classified into the stage 1 for the upland, the stage 2 for ASW and the stage 3 for NS in the study area. Seasonally, the $N_2O$ concentration was higher in the warm season (m = 34.19 µg $L^{-1}$) and lower in the cool season (m = 20.19 µg $L^{-1}$). Temperature and pH are main factors to control the dissolved $N_2O$ in groundwater of study area. Higher temperature results in higher denitrification rate by elevating metabolic rates for denitrifying bacteria directly, and creating the lower DO and ORP environment that affects the $N_2O$ concentration indirectly in the warm season. In addition, lower pH in the warm season also may explain the higher dissolved $N_2O$ concentrations because the decrease of pH is interpreted as a sign of intense denitrification.

This study put forward an understanding of spatial distributions of dissolved $N_2O$ from upland (agricultural area) which related the materials transformation to groundwater flow system. Temperature is considered as the main driver to seasonal change of dissolved $N_2O$ in wetland groundwater.

## Acknowledgements

We would like to thank the wetland park of Ichikawa city and the farmers in the pear orchard for the assistance of water sampling.

## References

Agriculture and Forestry Research Center of Chiba Prefecture. (2003). Soil improvement of pear orchards. Retrieved from http://www.pref.chiba.lg.jp/ninaite/fukyuushidou/kenkyuu-h15/documents/15chosa5.pdf

Bastviken, D., Olsson, M., & Tranvik, L. (2003). Simultaneous measurements of organic carbon mineralization and bacterial production in oxic and anoxic lake sediments. *Microbial ecology, 46*(1), 73-82. http://dx.doi.org/10.1007/s00248-002-1061-9

Blackmer, A., & Bremner, J. (1978). Inhibitory effect of nitrate on reduction of $N_2O$ to $N_2$ by soil microorganisms. *Soil Biology and Biochemistry, 10*(3), 187-191. http://dx.doi.org/10.1016/0038-0717(78)90095-0

Böttcher, J., Strebel, O., Voerkelius, S., & Schmidt, H.-L. (1990). Using isotope fractionation of nitrate-nitrogen and nitrate-oxygen for evaluation of microbial denitrification in a sandy aquifer. *Journal of Hydrology, 114*(3), 413-424. http://dx.doi.org/10.1016/0022-1694 (90)90068-9

Bouwman, A., Boumans, L., & Batjes, N. (2002). Modeling global annual $N_2O$ and NO emissions from fertilized fields. *Global Biogeochemical Cycles, 16*(4), 28-21-28-29. http://dx.doi.org/10.1029/2001GB001812

Burgin, A. J., & Hamilton, S. K. (2007). Have we overemphasized the role of denitrification in aquatic ecosystems? A review of nitrate removal pathways. *Frontiers in Ecology and the Environment, 5*(2), 89-96. http://dx.doi.org/10.1890/1540-9295(2007)5[89:HWOTRO]2.0.CO;2

Butler, J. H., Elkins, J. W., Thompson, T. M., & Egan, K. B. (1989). Tropospheric and dissolved $N_2O$ of the west Pacific and east Indian Oceans during the El Nino Southern Oscillation event of 1987. *Journal of Geophysical Research, 94*(D12), 14865-14814,14877. http://dx.doi.org/10.1029/JD094iD12p14865

Cao, Y., Sun, G., Xing, G., & Xu, H. (1991). Natural abundance of 15N in main N-containing chemical fertilizers of China. *Pedosphere, 1*(4), 377-382. Retrieved from http://pedosphere.issas.ac.cn/trqcn/ch/reader/view_abstract.aspx?file_no=19910410&flag=1

Cao, Y., Tang, C., Song, X., Liu, C., & Zhang, Y. (2012). Characteristics of nitrate in major rivers and aquifers of the Sanjiang Plain, China. *Journal of Environmental Monitoring, 14*(10), 2624-2633. http://dx.doi.org/10.1039/C2EM30032J

Cey, E. E., Rudolph, D. L., Aravena, R., & Parkin, G. (1999). Role of the riparian zone in controlling the distribution and fate of agricultural nitrogen near a small stream in southern Ontario. *Journal of Contaminant Hydrology, 37*(1), 45-67. http://dx.doi.org/10.1016/S0169-7722 (98)00162-4

Choi, W.-J., Han, G.-H., Lee, S.-M., Lee, G.-T., Yoon, K.-S., Choi, S.-M., & Ro, H.-M. (2007). Impact of land-use types on nitrate concentration and $\delta^{15}N$ in unconfined groundwater in rural areas of Korea. *Agriculture, Ecosystems and Environment, 120*, 259-268. http://dx.doi.org/10.1016/j.agee.2006.10.002

Choi, W.-J., Lee, S. M., & Ro, H. M. (2003). Evaluation of contamination sources of groundwater $NO_3^-$ using nitrogen isotope data: A review. *Geosciences Journal, 7*(1), 81-87. http://dx.doi.org/10.1016/S0038-0717(03)00199-8

Davis, J. C. (1986). *Statistical and data analysis in geology* (2nd ed.). New York, John Wiley and Sons.

DeSimone, L. A., & Howes, B. L. (1998). Nitrogen transport and transformations in a shallow aquifer receiving wastewater discharge: A mass balance approach. *Water Resources Research, 34*(2), 271-285. http://dx.doi.org/10.1029/97WR03040

Deurer, M., Von Der Heide, C., Böttcher, J., Duijnisveld, W., Weymann, D., & Well, R. (2008). The dynamics of $N_2O$ near the groundwater table and the transfer of $N_2O$ into the unsaturated zone: A case study from a sandy aquifer in Germany. *Catena, 72*(3), 362-373. http://dx.doi.org/10.1016/j.catena.2007.07.013

Dhondt, K., Boeckx, P., Hofman, G., & Van Cleemput, O. (2004). Temporal and spatial patterns of denitrification enzyme activity and nitrous oxide fluxes in three adjacent vegetated riparian buffer zones. *Biology and Fertility of Soils, 40*(4), 243-251. http://dx.doi.org/10.1007/s00374-004-0773-z

Dividson, E., Stark, J. M., & Firestone, M. (1990). Microbial production and consumption of nitrate in an annual grassland. *Ecology*, 1968-1975. http://dx.doi.org/10.2307/1937605

Ferrón, S., Ortega, T., & Forja, J. M. (2010). Nitrous oxide distribution in the north-eastern shelf of the Gulf of Cádiz (SW Iberian Peninsula). *Marine Chemistry, 119*(1), 22-32. http://dx.doi.org/10.1016/j.marchem.2009.12.003

García-García, V., Gómez, R., Vidal-Abarca, M., & Suárez, M. (2009). Nitrogen retention in natural Mediterranean wetland-streams affected by agricultural runoff. *Hydrology and Earth System Sciences, 13*(12), 2359-2371. http://dx.doi.org/10.5194/hess-13-2359-2009

Geistlinger, H., Jia, R., Eisermann, D., & Florian Stange, C. (2010). Spatial and temporal variability of dissolved nitrous oxide in near - surface groundwater and bubble - mediated mass transfer to the unsaturated zone. *Journal of Plant Nutrition and Soil Science, 173*(4), 601-609. http://dx.doi.org/10.1002/jpln.200800278

Groffman, P. M., Gold, A. J., & Addy, K. (2000). Nitrous oxide production in riparian zones and its importance to national emission inventories. *Chemosphere-Global change science, 2*(3), 291-299. http://dx.doi.org/10.1016/S1465-9972(00)00018-0

Groffman, P. M., Gold, A. J., & Jacinthe, P.-A. (1998). Nitrous oxide production in riparian zones and groundwater. *Nutrient Cycling in Agroecosystems, 52*(2-3). http://dx.doi.org/179-186.10.1023/A:1009719923861

Groffman, P. M., Gold, A. J., & Simmons, R. C. (1992). Nitrate dynamics in riparian forests: microbial studies. *Journal of Environmental Quality, 21*(4), 666-671. http://dx.doi.org/10.2134/jeq1992.00472425002100040022x

Halling-Sørensen, B., & Jorgensen, S. E. (1993). *The removal of nitrogen compounds from wastewater*. The Netherlands, Elsevier, Amsterdam.

Heaton, T. (1986). Isotopic studies of nitrogen pollution in the hydrosphere and atmosphere: a review. *Chemical Geology, 59*(1), 87-102. Retrieved from http://media.wix.com/ugd/adde68_965ca6eb9c72111a2c7fb3eee5f9a16f.pdf

Heincke, M., & Kaupenjohann, M. (1999). Effects of soil solution on the dynamics of $N_2O$ emissions: a review. *Nutrient Cycling in Agroecosystems, 55*(2), 133-157. http://dx.doi.org/10.1023/A:1009842011599

Heisterkamp, I. M., Schramm, A., de Beer, D., & Stief, P. (2012). Incomplete denitrification in the gut of the aquacultured shrimp Litopenaeus vannamei as source of nitrous oxide. *Microbial nitrous oxide production and nitrogen cycling associated with aquatic invertebrates*, 135. Retrieved from http://elib.suub.uni-bremen.de/edocs/00102789-1.pdf#page=135

Hinshaw, S. E., & Dahlgren, R. A. (2013). Dissolved Nitrous Oxide Concentrations and Fluxes from the Eutrophic San Joaquin River, California. *Environmental science & technology, 47*(3), 1313-1322. http://dx.doi.org/10.1021/es301373h

Ilies, P., & Mavinic, D. (2001). The effect of decreased ambient temperature on the biological nitrification and denitrification of a high ammonia landfill leachate. *Water Research, 35*(8), 2065-2072. http://dx.doi.org/10.1016/S0043-1354(00)00477-2

Ishii, S., Ohno, H., Tsuboi, M., Otsuka, S., & Senoo, K. (2011). Identification and isolation of active $N_2O$ reducers in rice paddy soil. *The ISME journal, 5*(12), 1936-1945. http://dx.doi.org/10.1038/ismej.2011.69

Kim, D.-G., Isenhart, T., Parkin, T., Schultz, R., & Loynachan, T. (2009). Nitrate and dissolved nitrous oxide in groundwater within cropped fields and riparian buffers. *Biogeosciences Discussions, 6*(1), 651-685. http://dx.doi.org/10.5194/bgd-6-651-2009

Kralova, M., Masscheleyn, P., Lindau, C., & Patrick Jr, W. (1992). Production of dinitrogen and nitrous oxide in soil suspensions as affected by redox potential. *Water, Air, and Soil Pollution, 61*(1-2), 37-45. http://dx.doi.org/10.1007/BF00478364

Lee, P. G., Lea, R., Dohmann, E., Prebilsky, W., Turk, P., Ying, H., & Whitson, J. (2000). Denitrification in aquaculture systems: an example of a fuzzy logic control problem. *Aquacultural Engineering, 23*(1), 37-59. http://dx.doi.org/10.1016/S0144-8609(00)00046-7

Lehmann, M. F., Reichert, P., Bernasconi, S. M., Barbieri, A., & McKenzie, J. A. (2003). Modelling nitrogen and oxygen isotope fractionation during denitrification in a lacustrine redox-transition zone. *Geochimica et Cosmochimica Acta, 67*(14), 2529-2542. http://dx.doi.org/10.1016/S0016-7037(03)00085-1

Li, X., Tang, C., Han, Z., Cao, Y., & Zhang, C. (2013). Relation between nitrous oxide production in wetland and groundwater: a case study in the headwater wetland. *Paddy and Water Environment, 11*(1-4), 521-529. http://dx.doi.org/10.1007/s10333-012-0345-z

Mathieu, O., Lévêque, J., Hénault, C., Milloux, M.-J., Bizouard, F., & Andreux, F. (2006). Emissions and spatial variability of $N_2O$, $N_2O$ and nitrous oxide mole fraction at the field scale, revealed with[15]N isotopic techniques. *Soil Biology and Biochemistry, 38*(5), 941-951. http://dx.doi.org/10.1016/j.soilbio.2005.08.010

McCutchan, J., & Lewis, W. (2008). Spatial and temporal patterns of denitrification in an effluent-dominated plains river. *Internationale Vereinigung für Theoretische und Angewandte Limnologie Verhandlungen, 30*(2), 323. Retrieved from http://cires.colorado.edu/limnology/pubs/pdfs/Pub192.pdf

Mühlherr, I. H., & Hiscock, K. M. (1998). Nitrous oxide production and consumption in British limestone aquifers. *Journal of Hydrology, 211*(1), 126-139. http://dx.doi.org/10.1016/S0022-1694 (98)00224-8

Müller, C., Stevens, R., Laughlin, R., & Jäger, H.-J. (2004). Microbial processes and the site of $N_2O$ production in a temperate grassland soil. *Soil Biology and Biochemistry, 36*(3), 453-461. http://dx.doi.org/10.1016/j.soilbio.2003.08.027

Nelson, L., & Knowles, R. (1978). Effect of oxygen and nitrate on nitrogen fixation and denitrification by Azospirillum brasilense grown in continuous culture. *Canadian journal of microbiology, 24*(11), 1395-1403. http://dx.doi.org/10.1139/m78-223

Nowicki, B. L. (1994). The effect of temperature, oxygen, salinity, and nutrient enrichment on estuarine denitrification rates measured with a modified nitrogen gas flux technique. *Estuarine, Coastal and Shelf Science, 38*(2), 137-156. http://dx.doi.org/10.1006/ecss.1994.1009

Pfenning, K., & McMahon, P. (1997). Effect of nitrate, organic carbon, and temperature on potential denitrification rates in nitrate-rich riverbed sediments. *Journal of Hydrology, 187*(3), 283-295. http://dx.doi.org/10.1016/S0022-1694 (96)03052-1

Rice, C. W., & Rogers, K. L. (1993). Denitrification in subsurface environments: potential source for atmospheric nitrous oxide. *Agricultural ecosystem effects on trace gases and global climate change* (agriculturaleco), 121-132. http://dx.doi.org/10.2134/asaspecpub55.c

Ronen, D., Magaritz, M., & Almon, E. (1988). Contaminated aquifers are a forgotten component of the global $N_2O$ budget. *Nature, 335*, 57-59 http://dx.doi.org/10.1038/335057a0

Ross, S. (1995). Overview of the hydrochemistry and solute processes in British wetlands. *Hydrology and hydrochemistry of British wetlands, 135*, 181. Chichester, UK, Wiley.

Sabater, S., Butturini, A., Clement, J.-C., Burt, T., Dowrick, D., Hefting, M., & Rzepecki, M. (2003). Nitrogen removal by riparian buffers along a European climatic gradient: patterns and factors of variation. *Ecosystems, 6*(1), 0020-0030. http://dx.doi.org/10.1007/s10021-002-0183-8

Saunders, D., & Kalff, J. (2001). Nitrogen retention in wetlands, lakes and rivers. *Hydrobiologia, 443*(1-3), 205-212. http://dx.doi.org/10.1016/S0022-1694 (96)03052-1

Singleton, M., Esser, B., Moran, J., Hudson, G., McNab, W., & Harter, T. (2007). Saturated zone denitrification: Potential for natural attenuation of nitrate contamination in shallow groundwater under dairy operations. *Environmental science & technology, 41*(3), 759-765. http://dx.doi.org/10.1021/es061253g

Spalding, R. F., & Parrott, J. D. (1994). Shallow groundwater denitrification. *Science of the total environment, 141*(1), 17-25. http://dx.doi.org/10.1016/0048-9697 (94)90014-0

Thamdrup, B., Hansen, J. W., & Jørgensen, B. B. (1998). Temperature dependence of aerobic respiration in a coastal sediment. *FEMS Microbiology Ecology, 25*(2), 189-200. http://dx.doi.org/10.1111/j.1574-6941.1998.tb00472.x

Ueda, S., Ogura, N., & Yoshinari, T. (1993). Accumulation of nitrous oxide in aerobic groundwaters. *Water Research, 27*(12), 1787-1792. http://dx.doi.org/10.1016/0043-1354(93)90118-2

Velthof, G., Oenema, O., Postma, R., & Van Beusichem, M. (1996). Effects of type and amount of applied nitrogen fertilizer on nitrous oxide fluxes from intensively managed grassland. *Nutrient Cycling in Agroecosystems, 46*(3), 257-267. http://dx.doi.org/10.1007/BF00420561

Verhoeven, J. T., Arheimer, B., Yin, C., & Hefting, M. M. (2006). Regional and global concerns over wetlands and water quality. *Trends in ecology & evolution, 21*(2), 96-103. http://dx.doi.org/10.1016/j.tree.2005.11.015

Vinther, A., & Søeberg, H. (1991). Mathematical model describing dispersion in free solution capillary electrophoresis under stacking conditions. *Journal of Chromatography A, 559*(1), 3-26. http://dx.doi.org/10.1016/0021-9673 (91)80055-L

Wang, S., Liu, C., Yeager, K. M., Wan, G., Li, J., Tao, F., & Fan, C. (2009). The spatial distribution and emission of nitrous oxide ($N_2O$) in a large eutrophic lake in eastern China: Anthropogenic effects. *Science of the Total Environment, 407*(10), 3330-3337. http://dx.doi.org/10.1016/j.scitotenv.2008.10.037

Weiss, R. (1970). *The solubility of nitrogen, oxygen and argon in water and seawater.* Paper presented at the Deep Sea Research and Oceanographic Abstracts. http://dx.doi.org/10.1016/0011-7471 (70)90037-9

Whitmire, S. L., & Hamilton, S. K. (2005). Rapid removal of nitrate and sulfate in freshwater wetland sediments. *Journal of Environmental Quality, 34*(6), 2062-2071. http://dx.doi.org/10.2134/jeq2004.0483

Yoo, H., Ahn, K.-H., Lee, H. J., Lee, K. H., Kwak, Y. J., & Song, K. G. (1999). Nitrogen removal from synthetic wastewater by simultaneous nitrification and denitrification (SND) via nitrite in an intermittently-aerated reactor. *Water Research, 33*(1), 145-154. http://dx.doi.org/10.1016/S0043-1354(98)00159-6

Zhang, G. L., Zhang, J., Liu, S. M., Ren, J. L., & Zhao, Y. C. (2010). Nitrous oxide in the Changjiang (Yangtze River) Estuary and its adjacent marine area: Riverine input, sediment release and atmospheric fluxes. *Biogeosciences, 7*(11), 3505-3516. http://dx.doi.org/10.5194/bg-7-3505-2010

Zhu, J., Ndegwa, P. M., & Luo, A. (2002). Bacterial responses to temperature during aeration of pig slurry. *Journal of Environmental Science and Health, Part B, 37*(3), 265-275. http://dx.doi.org/10.1081/PFC-120003104

# 13

# Assessment of Surface Water Quality in Hyderabad Lakes by Using Multivariate Statistical Techniques, Hyderabad-India

A. Sridhar Kumar[1], A. Madhava Reddy[2], L. Srinivas[3] & P. Manikya Reddy[4]

[1] Department of Environmental Science, Osmania University, Hyderabad, Telangana State, India

[2] Environmental Specialist, Hyderabad, Telangana State, India

[3] Research Scholar, Department of Botany, Osmania University of Hyderabad, India

[4] Professor, Department of Botany, Osmania University-Hyderabad, India

Correspondence: A. Sridhar Kumar, Department of Environmental Science, Osmania University, Hyderabad, Telangana State, India. E-mail: meetsreedhar1@gmail.com

## Abstract

Multivariate statistical techniques such as cluster analysis (CA), principle component analysis (PCA), factor analysis (FA) were applied for the evolution of temporal variations and the interpretation of large complex water quality data set of the Hyderabad city, generating during year 2013-14 monitoring of 16 parameters at 23 different sites of an average depth of 1m. Hierarchical clustering analysis (CA) is first applied to distinguish the three general water quality patterns among the stations. Data set thus obtained was treated using R-mode factor analysis (FA) and followed by principle component analysis (PCA). Factor analysis identified five factors responsible for data structure explaining 75% of total variance and allowed to group selected parameters according to common futures. WT, EC, TSS and Na were associated and controlled by mixed origin with similar contribution from natural and anthropogenic sources. Whereas $NO_3$, $PO_4$, $SO_4$, FC, TC, F⁻, K and B were derived from anthropogenic sources.

**Keywords:** cluster analysis, factor analysis, water quality, natural pollution, anthropogenic pollution

## 1. Introduction

The protection and restoration of urban lakes and wetlands, urban lakes are in extremely poor condition in Hyderabad, within last 12 years, Hyderabad has lost 3245 ha. area of its water in the form of lakes and ponds. There are endless examples in India that shows such devastating state of urban water bodies (Sridhar Kumar et al, 2014). Almost all urban water bodies in India are suffering because of pollution and are used for disposing untreated local sewage, industrial waste water and solid waste, and in many cases the water bodies have been ultimately turned into landfills. Point and non point sources of pollution degrade surface and ground water and impair their use for drinking, industrial, agricultural, recreation or other purposes (Carpenter et al., 1998; Howarth et al., 1996). A number of point and non point sources contaminate the water bodies by adding the excess nutrients and heavy metals. Over the years their capacities went on decreasing by rapid urbanization, encroachments into lake areas and increased sedimentation resulting from the high human interference in the catchment area (Ramachandraiah and Prasad, 2004). Urbanization increases in population density and the intensification of agricultural activities in the upstream areas is among the main causes of water pollution. Therefore, researchers have been paying more attention to the effects of natural and human activities on water quality, in particular, the key contributors of human activities to nutrients and heavy metals. The discharge of effluents and associated toxic compounds into aquatic ecosystem represents an ongoing environmental problem due to their possible impact on communities in the receiving aquatic water and a potential effect on human health (Abbas Alkarkhi et al., 2008). Further these materials enter the surface water resulting in pollution of irrigation and drinking water. Although, the government of India's (GOI, 1992) policy statement on abatement of pollution at source (GoI, 1992).

Many investigations have been conducted on anthropogenic contaminants of ecosystems. Because of the spatial and temporal variation in water quality conditions, a monitoring program which provides a representative and reliable estimation of the quality of surface waters is necessary (Dixon and Chiswell 1996). The monitoring results produce a large and complicated data matrix that is difficult to interpret to draw meaningful conclusions. Multivariate statistical techniques are powerful tools for analyzing large numbers of samples collected in surveys, classifying assemblages and assessing human impacts on water quality and ecosystem conditions.

The application of different multivariate statistical techniques, such as principal component analysis (PCA), factor analysis (FA), and cluster analysis (CA), assists in the interpretation of complex data matrices for a better understanding of water quality and ecological status of the studied system. These techniques provide the identification of possible factors/sources that affect water environmental systems and offer a valuable tool for reliable management of water resources as well as rapid solution for pollution issues (Palma et al., 2010; Morales et al., 1999). Multivariate statistical techniques have been widely adopted to analyze and evaluate surface and freshwater water quality, and are useful to verify temporal and spatial variations caused by natural and anthropogenic factors linked to seasonality (Wunderlin et al., 2001; Simenov et al., 2003).

The study area, Hyderabad consists of urban lakes situated on the Deccan Plateau at a height of 1788 feet above sea level, located at 17° 22' of northern latitude and 78° 29' of the eastern longitude with an area of 7,100 sq km. The city has been dotted with a number of lakes and almost all the lakes were artificially created, often some centuries back, by constructing bunds and dams in the downstream area of micro-catchments. From upstream of the reservoir to the downstream, these lakes form a cascading system with limited storage space. (Fig-1). Normal rainfall is 786.8 mm which increases from northwest to southeast. The mean maximum and minimum temperature vary from 40° to 14°C. The city is drained by river Musi and the drainage pattern is of dendritic and rectangular type.

## 2. Materials and Methods

In the present study, the data obtained during the year 2013-14, is subjected to different multivariate statistical techniques to extract about the similarities or dissimilarities between sampling stations, identification of water quality variables responsible for spatial and temporal variations in lakes water quality, the hidden factors explaining the structure of the data base and the influence of possible sources (natural and anthropogenic) on the water quality parameters of the lake basins.

The author has conducted a water quality survey during the year 2013-14 on few lakes & tanks because of either increase in the levels of critical parameters or on the point of conservation so as to improve the water quality and its management. These lakes have an average depth of 1 m and having the major human activities like cattle wading, boating, fishing, and the agriculture is melon farming, vegetables and paragrass. The major water quality issues are pathogenic (Bacteriological) pollution, oxygen depleting organic pollution, agricultural runoff, salinity and trace elements.

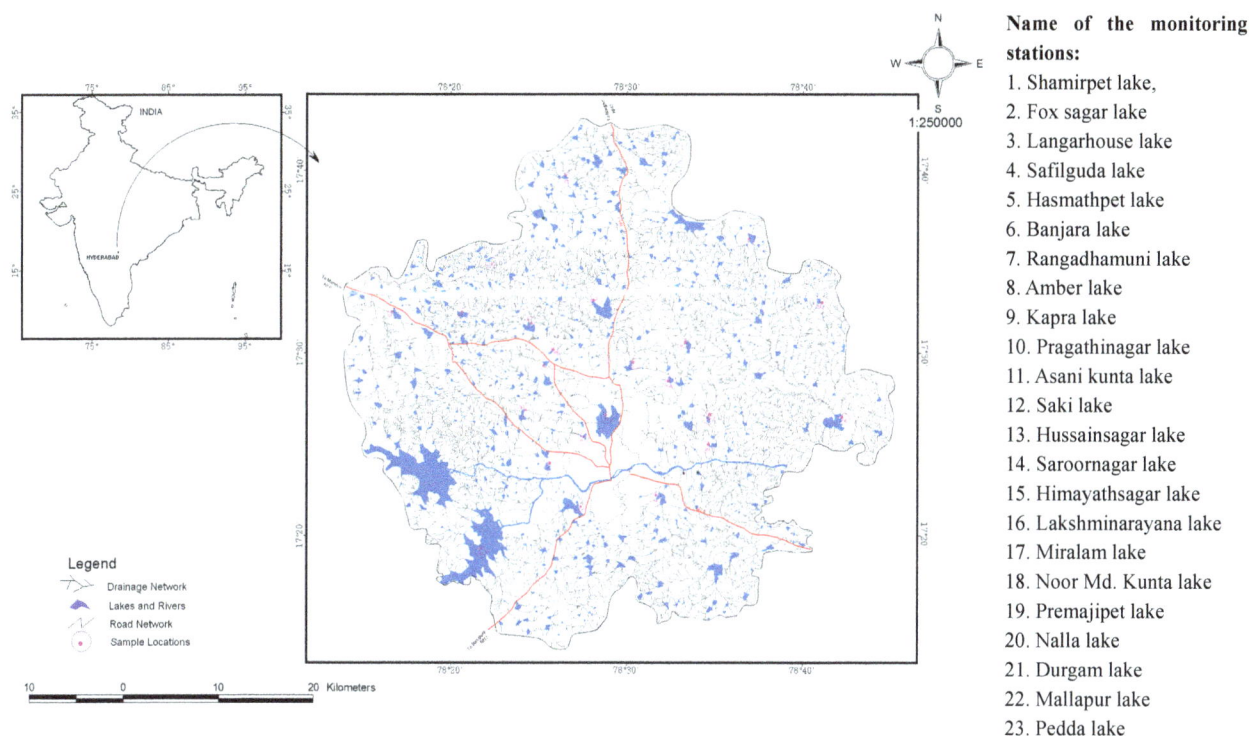

**Name of the monitoring stations:**
1. Shamirpet lake,
2. Fox sagar lake
3. Langarhouse lake
4. Safilguda lake
5. Hasmathpet lake
6. Banjara lake
7. Rangadhamuni lake
8. Amber lake
9. Kapra lake
10. Pragathinagar lake
11. Asani kunta lake
12. Saki lake
13. Hussainsagar lake
14. Saroornagar lake
15. Himayathsagar lake
16. Lakshminarayana lake
17. Miralam lake
18. Noor Md. Kunta lake
19. Premajipet lake
20. Nalla lake
21. Durgam lake
22. Mallapur lake
23. Pedda lake

Figure 1. Map of study area and water quality monitoring stations (listed 1-23) in Hyderabad basin

*2.1 Monitored Parameters and Analytical Methods*

The data generated about 23 water quality monitoring stations, comprising 16 water quality parameters monitored during the year (2013-14). The selected water quality parameters, their units and methods of analysis are summarized in Table 1.The author has sampled preserved and analyzed all the water quality parameters as per Indian inland surface water quality standards. All the samples were collected at center of the lake location. The depth of the sample is subsurface 0.5 m below the water surface. The basic statistics of the measured one year data set on Hyderabad lakes water quality are summarized in Table 2.

Table 1. The water quality parameters, their units and methods of analysis

| Parameters | Abbreviations | Units | Analytical methods |
|---|---|---|---|
| Temperature | WT | °C | Mercury thermometer |
| Total suspended solids | TSS | mg/l | Dried at 103 to 105 °C |
| Biochemical oxygen demand | BOD | mg/l | Winkler azide |
| Electrical conductivity | EC | μS/cm | Electrometric |
| Dissolved oxygen | DO | mg/l | Winkler azide |
| Fluoride | F- | mg/l | Ion selective electrode |
| Nitrate | $NO_3$ | mg/l | Nitrate electrode |
| Phosphate | $PO_4$ | mg/l | Stannous chloride |
| Sulphate | $SO_4$ | mg/l | Nephelometry |
| Boron | B | mg/l | Spectorphotometric |
| Sodium | Na | mg/l | Flame emission photometric |
| Sodium% | %Na | Meq/l | Atomic absorption spectrometry |
| Potassium | K | Mg/l | Flame emission photometric |
| Sodium absorption ratio | SAR | Meq/l | Atomic absorption spectrometry |
| Fecal coliform | FC | MPN/100 ml | Multiple tube dilution |
| Total coliform | TC | MPN/100 ml | Multiple tube dilution |

Table 2. Mean and S.D. of different lakes water quality parameters at various locations during the year 2013-14

| Parameters | | 1 | 2 | 3 | 4 | 5 | 6 | 7 | 8 | 9 | 10 | 11 | 12 | 13 | 14 | 15 | 16 | 17 | 18 | 19 | 20 | 21 | 22 | 23 |
|---|---|---|---|---|---|---|---|---|---|---|---|---|---|---|---|---|---|---|---|---|---|---|---|---|
| | | | | | | | | | | | | | | | | | | Stations (1 to 23) | | | | | | |
| WT | Mean | 26.00 | 27.30 | 27.70 | 29.70 | 27.00 | 25.30 | 28.70 | 28.30 | 26.70 | 28.00 | 27.30 | 27.70 | 26.30 | 25.00 | 25.30 | 29.00 | 24.30 | 26.70 | 26.70 | 25.00 | 25.70 | 29.70 | 30.00 |
| | S.D | 1.73 | 0.58 | 0.58 | 0.58 | 1.00 | 2.31 | 0.58 | 0.58 | 1.15 | 0.00 | 3.06 | 2.31 | 2.31 | 1.73 | 2.31 | 0.00 | 2.31 | 1.15 | 1.15 | 3.46 | 1.15 | 0.58 | 0.00 |
| TSS | Mean | 13.00 | 33.00 | 10.70 | 17.70 | 13.30 | 14.00 | 12.30 | 16.00 | 16.70 | 16.70 | 410.0 | 19.30 | 37.70 | 31.70 | 7.00 | 16.70 | 11.00 | 25.30 | 38.00 | 20.70 | 24.30 | 17.30 | 28.00 |
| | S.D | 5.57 | 29.31 | 4.16 | 7.77 | 1.53 | 5.29 | 8.50 | 3.00 | 5.13 | 2.31 | 190.0 | 16.17 | 39.37 | 13.87 | 3.61 | 6.03 | 1.73 | 11.02 | 10.58 | 10.26 | 7.09 | 1.15 | 10.39 |
| BOD | Mean | 37.70 | 59.00 | 54.30 | 47.30 | 41.70 | 53.00 | 50.30 | 29.30 | 25.30 | 33.30 | 74.30 | 6.70 | 43.00 | 62.30 | 13.70 | 28.30 | 34.00 | 114.7 | 117.7 | 74.70 | 28.00 | 51.30 | 43.30 |
| | S.D | 22.28 | 13.53 | 28.92 | 6.43 | 17.56 | 23.81 | 19.50 | 9.02 | 7.02 | 16.65 | 82.86 | 3.24 | 30.32 | 19.14 | 9.81 | 10.60 | 19.16 | 83.58 | 89.76 | 38.28 | 16.37 | 24.85 | 7.64 |
| EC | Mean | 626 | 1947 | 1117 | 1699 | 1443 | 918 | 1308 | 1386 | 1548 | 1374 | 9350 | 1948 | 1207 | 1542 | 420 | 1697 | 1291 | 1991 | 5862 | 1604 | 1393 | 1657 | 1965 |
| | S.D | 126 | 460 | 478 | 292 | 151 | 106 | 160 | 193 | 193 | 258 | 214 | 5534 | 504 | 321 | 132 | 70 | 323 | 303 | 165 | 3699 | 185 | 277 | 108 |
| DO | Mean | 4.60 | 4.00 | 0.60 | 1.90 | 4.10 | 1.30 | 3.50 | 5.40 | 3.00 | 5.40 | 0.30 | 6.00 | 0.60 | 1.30 | 5.10 | 5.90 | 0.60 | 2.00 | 0.60 | 1.10 | 0.20 | 1.80 | 1.30 |
| | S.D | 1.45 | 0.47 | 0.55 | 0.25 | 3.23 | 1.81 | 1.59 | 0.10 | 0.85 | 1.74 | 0.35 | 0.21 | 1.10 | 1.10 | 0.66 | 1.11 | 0.98 | 3.18 | 1.10 | 1.22 | 0.29 | 0.68 | 0.45 |
| F- | Mean | 1.00 | 1.60 | 1.20 | 1.90 | 1.20 | 1.60 | 0.80 | 1.20 | 1.60 | 1.00 | 0.80 | 0.50 | 1.20 | 0.90 | 0.30 | 1.10 | 1.20 | 2.30 | 1.10 | 1.40 | 1.10 | 1.50 | 1.70 |
| | S.D | 0.36 | 0.85 | 0.78 | 0.95 | 0.46 | 1.00 | 0.31 | 0.21 | 1.12 | 0.40 | 0.04 | 0.02 | 0.75 | 0.61 | 0.32 | 0.56 | 0.60 | 0.44 | 0.72 | 0.32 | 0.72 | 0.64 | 0.61 |
| NO3 | Mean | 10.90 | 5.40 | 15.70 | 11.70 | 28.80 | 12.20 | 35.00 | 19.00 | 2.20 | 24.80 | 22.70 | 16.30 | 19.60 | 11.40 | 0.60 | 8.30 | 4.20 | 33.20 | 6.50 | 75.00 | 32.70 | 48.00 | 13.70 |
| | S.D | 5.52 | 4.37 | 8.92 | 7.59 | 6.88 | 5.65 | 2.19 | 13.23 | 1.04 | 5.25 | 11.02 | 8.08 | 17.89 | 3.99 | 0.35 | 1.57 | 3.36 | 33.46 | 3.77 | 39.74 | 11.20 | 23.12 | 1.79 |
| PO4 | Mean | 0.20 | 0.30 | 2.50 | 1.80 | 2.00 | 1.70 | 3.40 | 3.10 | 3.20 | 3.30 | 0.70 | 0.30 | 4.10 | 2.10 | 0.10 | 2.60 | 1.80 | 1.70 | 1.00 | 3.30 | 3.30 | 3.00 | 0.50 |
| | S.D | 0.06 | 0.17 | 1.22 | 1.36 | 1.65 | 1.46 | 1.47 | 1.18 | 0.09 | 1.58 | 0.58 | 1.00 | 0.29 | 2.07 | 0.05 | 1.75 | 1.15 | 1.01 | 0.26 | 0.58 | 1.61 | 1.53 | 0.46 |
| SO4 | Mean | 23.70 | 72.30 | 57.70 | 61.00 | 62.30 | 52.70 | 115.3 | 67.70 | 70.70 | 70.70 | 499.3 | 38.70 | 94.30 | 99.30 | 30.70 | 80.00 | 67.70 | 176.7 | 659.7 | 97.30 | 74.70 | 96.30 | 96.70 |
| | S.D | 16.62 | 11.15 | 19.35 | 7.94 | 10.50 | 17.56 | 22.94 | 3.79 | 28.01 | 11.02 | 99.90 | 9.02 | 37.58 | 28.36 | 24.54 | 31.19 | 23.12 | 91.36 | 936.7 | 3.79 | 9.29 | 17.21 | 46.23 |
| B | Mean | 0.70 | 1.80 | 0.50 | 0.40 | 0.60 | 0.60 | 0.80 | 0.80 | 1.20 | 0.60 | 0.30 | 0.00 | 0.70 | 0.60 | 1.00 | 1.10 | 0.40 | 0.30 | 0.50 | 0.70 | 0.70 | 0.90 | 0.30 |
| | S.D | 0.61 | 1.56 | 0.40 | 0.32 | 0.93 | 0.67 | 1.19 | 0.98 | 1.85 | 0.74 | 0.46 | 1.05 | 0.67 | 0.55 | 0.49 | 1.16 | 0.61 | 0.30 | 0.47 | 1.10 | 0.76 | 1.25 | 0.00 |
| Na | Mean | 70.70 | 220.0 | 136.7 | 170.0 | 170.0 | 82.30 | 140.0 | 233.3 | 183.3 | 176.7 | 931.7 | 256.7 | 139.0 | 163.3 | 43.30 | 216.7 | 115.0 | 240.0 | 293.3 | 193.3 | 160.00 | 190.0 | 190.0 |
| | S.D | 18.01 | 10.00 | 70.24 | 36.06 | 10.00 | 14.64 | 40.00 | 45.83 | 40.41 | 55.08 | 15.28 | 224.3 | 36.37 | 15.28 | 23.18 | 55.08 | 87.89 | 17.32 | 115.9 | 28.87 | 20.00 | 10.00 | 1120. |
| %Na | Mean | 41.60 | 54.50 | 49.10 | 44.00 | 47.60 | 46.60 | 46.90 | 53.40 | 49.00 | 50.90 | 61.50 | 62.40 | 49.40 | 46.70 | 41.50 | 53.40 | 32.60 | 50.00 | 26.50 | 51.10 | 48.60 | 47.00 | 41.70 |
| | S.D | 3.04 | 6.73 | 5.78 | 2.77 | 4.73 | 7.11 | 7.10 | 9.71 | 10.14 | 8.37 | 6.06 | 17.56 | 2.28 | 3.10 | 13.07 | 5.07 | 22.31 | 3.60 | 5.23 | 6.32 | 3.27 | 2.59 | 18.83 |
| K | Mean | 0.70 | 1.80 | 0.50 | 0.40 | 0.60 | 0.60 | 0.80 | 0.80 | 1.20 | 0.60 | 0.30 | 0.00 | 0.70 | 1.00 | 0.30 | 1.10 | 0.40 | 0.30 | 0.50 | 0.70 | 0.70 | 0.90 | 0.60 |
| | S.D | 0.61 | 1.56 | 0.40 | 0.32 | 0.93 | 0.67 | 1.19 | 0.98 | 1.85 | 0.74 | 1.05 | 0.00 | 0.67 | 0.55 | 0.49 | 1.16 | 0.61 | 0.30 | 0.47 | 1.10 | 0.76 | 1.25 | 0.46 |
| SAR | Mean | 1.70 | 3.60 | 2.60 | 2.50 | 2.80 | 3.10 | 2.50 | 3.40 | 3.00 | 3.10 | 11.20 | 8.10 | 2.60 | 2.60 | 1.30 | 3.40 | 1.80 | 3.40 | 2.10 | 3.20 | 2.70 | 2.90 | 1.70 |
| | S.D | 0.76 | 0.40 | 1.06 | 0.35 | 0.21 | 0.40 | 0.70 | 1.01 | 0.78 | 1.06 | 7.71 | 3.23 | 0.46 | 0.26 | 0.64 | 0.21 | 1.39 | 0.17 | 0.17 | 0.51 | 0.00 | 0.35 | 0.29 |
| FC | Mean | 290.0 | 600.0 | 866.7 | 800.0 | 766.7 | 666.7 | 400.0 | 340.0 | 450.0 | 283.3 | 100.0 | 37.70 | 966.7 | 283.3 | 115.3 | 483.3 | 873.3 | 566.7 | 1733. | 900.0 | 633.30 | 766.7 | 1156. |
| | S.D | 182.4 | 141.4 | 814.4 | 173.2 | 230.9 | 767.8 | 173.2 | 165.2 | 396.8 | 76.38 | 130.0 | 11.24 | 602.7 | 76.38 | 159.9 | 425.2 | 740.3 | 30.51 | 230.9 | 400.0 | 461.88 | 230.9 | 351.1 |
| TC | Mean | 950.0 | 1300. | 1666. | 1400. | 900.0 | 1193. | 1233. | 816.7 | 800.0 | 866.7 | 343.3 | 223.3 | 1300. | 950.0 | 240.0 | 1083. | 1133. | 1066. | 1900. | 1600. | 883.3 | 1300. | 1333. |
| | S.D | 626. | 424.2 | 700.0 | 346.4 | 461.8 | 704.3 | 635.0 | 682.5 | 700.0 | 635.0 | 309.2 | 37.86 | 519.6 | 626.5 | 233.0 | 894.8 | 808.2 | 472.5 | 519.6 | 0.00 | 725.14 | 519.6 | 115.4 |

From the table 2, it is observed that, stations 13, 19, 20 and 21 are receiving directly untreated waste water from the urbanized catchment and the parameters like Phosphates, Nitrates, Coil forms shows above the prescribed standards, whereas stations 2, 11, 12 and 18 are polluted due to agricultural runoff from the catchment area and the presence of parameters like Boron, Potassium, Conductivity, Sodium, SAR and Fluoride shows the above prescribed standards as per the Central Pollution Control Board (CPCB) standards.

## 3. Data Treatment and Multivariate Statistical Methods

The surface water quality data sets were subjected through three multivariate techniques: cluster analysis (CA), principle component analysis (PCA) and factor analysis (FA) (Singh et al., 2004; Dong et al., 2010 and Kim et al., 2009). Summary statistics of these data sets were first calculated to evaluate the distributions. FA was applied on standardized data through Z-scale transformation in order to avoid misclassification due to wide difference in data dimensionality (Liu et al., 2003; Kim et al., 2009), standardization tends to increase the influence of variables whose variance is small and vice versa. All the mathematical and statistical computations were made using Statistical Package for Social Sciences SPSS, 1995).

Cluster analysis is group of multivariate techniques whose purpose is to assemble objects based on the characteristic they possess. Hierarchical agglomerative clustering is the most common approach, which provides intuitive similarity relationships between any one sample and the entire data set, and is typically illustrated by a dendrogram (tree diagram) (McKenna, 2003). The Euclidean distance usually gives the similarity between analytical values from the samples (Otto, 1998). In this study hierarchical agglomerative CA was performed on the normalized data set by means of the Wards method, using squared Euclidean distances as a measure of similarity. The Wards method uses an analysis of variance approach to evaluate the distances between clusters in an attempt to minimize the sum of squares (SS) of any two clusters that can be formed at each step. The special variability of water quality in the city determined from CA, using the linkage distance, reported as $D_{link}/D_{max}$, which represent the quotient between the linkage distances for particular case divided by the maximal linkage distance. The quotient is then multiplied by the 100 as a way to standardize the linkage distance represented on the y-axis.

### 3.1 Factor Analysis/Principal Component Analysis (PCA)

Factor analysis technique extracts the eigen values and eigen vectors from co-variance matrix of original variables. The principle components (PC) are the uncorrelated (orthogonal) variables obtain by multiplying original correlated variables with eigen vector, which is a list of coefficients (loading or weightings). Thus principal components are weighted linear combinations of original variables. PC provides information on the most meaningful parameters, which describe whole data set affording data reduction with minimum loss of original information (Vega et al., 1998; Helena et al., 2000; Shrestha and Kazama 2007). It is a powerful technique for pattern recognition that attempts to explain the variance of large set of inter-correlated variables and transforming in to a smaller set of independent (uncorrelated) variables (principle component). Factor analysis further reduce the contribution of less significant variables obtained from PCA and the new group of variables known as varifactors, are extracted through rotating the axis defined by PCA. A varifactor can include unobservable, hypothetical, latent variables, while a PC is a linear combination of observable water quality variables (Panda et al., 2006; Davis,1986). PCA of the normalized variables was performed to extract significant PC's and to further reduce the contribution of variables with minor significance. These PC's were subjected to varimax rotation (raw) generating varifactors (Brumelis et al., 2000; Love et al., 2004; Abdul et al., 2005).

## 4. Results and Discussions

### 4.1 Spatial Similarity and Size Grouping

Cluster analysis was used to detect the similarity groups between the sampling sites. It yielded a dendrogram (Fig 2) grouping all 23 sampling sites of the city in to three statistically meaningful clusters at $(D_{link}/D_{max})$ x 100 < 60. Since we used hierarchical agglomerative cluster analysis, the number of clusters was also decided by practicality of the results as there is ample information (e.g. land use, location of industries etc.) available on the study sites. The results indicate that the CA technique is useful in offering reliable classification of surface water in the whole region and will make it possible to design a future spatial sampling strategy in an optimal manner, which can reduce the number of sampling stations and associated cost.

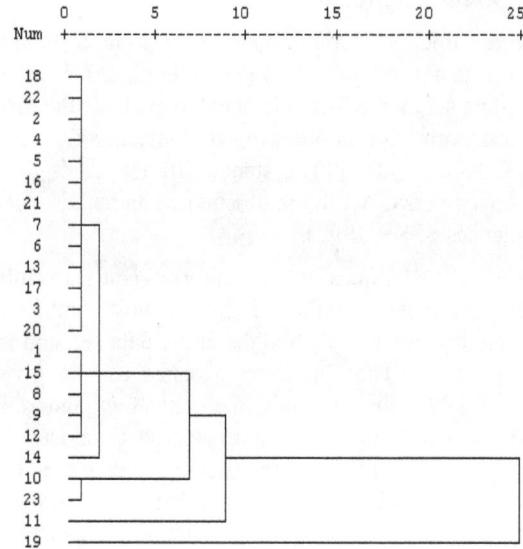

Figure 2. Dendrogram showing clustering of sampling sites according to water quality characteristics

The main descriptive statistics are shown in table 3. Statistical treatment of these data indicates their association and grouping with five factors in water bodies (Table 4). The presence of phosphate and nitrate in most of the sample stations were recorded high. Phosphate it varies from 0.00 to 4.85mg/l with an average of 1.99 mg/l. majority of the sample stations (16 out of 23) the phosphate recoded above the permissible limits of 1 or above 1.00 mg/l. Nitrate it varies from 0.20 to 112.00 mg/l with an average of 19.91 mg/l. sample stations 20, 21, 22 and 23 were recorded above the permissible limit of 45 mg/l. Phosphates are often considered a primary limiting element and Nitrates considered secondary limiting element in most of the lakes, and these concentrations are positively correlated in lakesIt was observed that the other high values of TSS, Conductivity, Na, SAR, F Coli, T Coli, $SO_4$, F-, K, Temperature and B due to point and non point sources which may be attributed to the industrial and agricultural activities.

Table 3. Descriptive statistical data of lakes water

| | WT | TSS | BOD | EC | DO | F- | $NO_3$ | $PO_4$ | $SO_4$ | B | Na | Sodium% | K | SAR | F. Coli | T. Coli |
|---|---|---|---|---|---|---|---|---|---|---|---|---|---|---|---|---|
| Mean | 27.10 | 36.97 | 48.84 | 1969.54 | 2.70 | 1.23 | 19.91 | 1.99 | 120.23 | 0.67 | 205.01 | 47.65 | 32.38 | 3.27 | 612.34 | 1061.03 |
| Median | 28.00 | 18.00 | 40.00 | 1511.00 | 2.15 | 1.10 | 15.00 | 1.72 | 72.00 | 0.30 | 180.00 | 49.00 | 30.00 | 2.70 | 500.00 | 1300.00 |
| SD | 2.09 | 87.52 | 39.76 | 2200.17 | 2.26 | 0.67 | 20.18 | 1.55 | 220.31 | 0.82 | 259.73 | 10.89 | 18.00 | 2.62 | 500.59 | 622.37 |
| Minimum | 23.00 | 4.00 | 3.00 | 361.00 | 0.00 | 0.10 | 0.20 | 0.00 | 6.00 | 0.00 | 0.00 | 6.80 | 6.00 | 0.20 | 20.00 | 50.00 |
| Maximum | 30.00 | 600.00 | 218.00 | 15740.00 | 7.10 | 2.90 | 112.00 | 4.85 | 1741.00 | 3.30 | 2175.00 | 74.00 | 99.00 | 20.00 | 2000.00 | 2500.00 |

Table 4. Factor analysis of lake water quality data

| Element | Communality | Eigen | Total variance | Cumulative total variance | Factor 1 | Factor 2 | Factor 3 | Factor 4 | Factor 5 |
|---------|-------------|-------|----------------|---------------------------|----------|----------|----------|----------|----------|
| WT | .661 | 4.411 | 27.569 | 27.569 | .172 | -.062 | .018 | .789 | -.064 |
| TSS | .810 | 2.568 | 16.052 | 43.621 | .889 | -.011 | -.088 | -.050 | .096 |
| BOD | .585 | 2.009 | 12.554 | 56.175 | .437 | .571 | -.053 | -.200 | .159 |
| EC | .951 | 1.722 | 10.760 | 66.935 | .918 | .322 | -.052 | -.029 | -.020 |
| DO | .653 | 1.398 | 8.739 | 75.674 | -.160 | -.460 | -.105 | .195 | -.606 |
| F- | .657 | .869 | | | -.056 | .237 | .725 | -.236 | -.129 |
| NO3 | .717 | .650 | | | .026 | -.009 | -.016 | .188 | .825 |
| PO4 | .690 | .626 | | | -.132 | -.039 | .778 | .092 | .240 |
| SO4 | .681 | .473 | | | .503 | .627 | -.139 | -.099 | -.075 |
| B | .682 | .434 | | | -.178 | .093 | -.282 | .713 | .232 |
| Na | .898 | .309 | | | .940 | .009 | .099 | .065 | .028 |
| Na% | .678 | .208 | | | .399 | -.564 | .072 | .342 | .281 |
| K | .868 | .153 | | | .563 | .032 | .717 | -.115 | -.149 |
| SAR | .913 | .101 | | | .906 | -.273 | .017 | .123 | .055 |
| F. Coli | .837 | .040 | | | -.162 | .801 | .303 | .233 | .154 |
| T. Coli | .825 | .030 | | | -.135 | .655 | .364 | .442 | .221 |

Extraction Method: Principal Component Analysis.

Rotation Method: Varimax with Kaiser Normalization.

### 4.2 Factor Analysis

By factor analysis complex linear correlation between metal concentrations was determined, which enabled interpretation of correlation of elements in the study area. Elements belonging to a given factor were defined by factor matrix after varimax rotation, with those having strong correlations grouped in to factors. Considering the influence they exerted in lakes by determining the distribution of parameters in the study area of Hyderabad, the said multiparameter factor was divided in to two groups: (i) factors with strong scattered anthropogenic influence and (ii) factors caused by predominantly natural processes or other anthropogenic influences. The identification of factors is based on dominant influence. The distribution manner of individual association of parameters in the lake waters was determined by principle component method (results are shown in table 4). Based on eigen values and varimax rotation five factors explained most of the variability (total variance explained was about 75.67%).

### Factor 1

Factor 1 exhibit 27% of the total variance of 75% with positive loading on TSS, Conductivity, Na and SAR. This factor can be attributed to the influence of agricultural activity in the study area. This factor indicates strong association (r=0.6-0.94) of TSS, Conductivity, Na and SAR. The high variability in the analytical data obtained is indicative of an external source for these parameters in water bodies. Total suspended solids levels were found to be high in few stations with concentration ranging from 4 to 600 mg/l with an average of 36.97 mg/l. The high TSS values reported at station 11. This may be due to direct discharge of untreated sewage from the nearby surroundings which was not having the proper diversion facilities and proves that source of TSS is anthropogenic addition.

Conductivity, it varies from 361 to 15740 μmhos/cm at 25$^0$C (average of 1969 μmhos/cm) and permissible limit is 2250 μmhos/cm, shows poor quality of water as per water class of irrigation guidelines in India. Na varying from 0.00 to 2175 mg/l with an average of 205 mg/l. and its tolerance limit is 60. Sample stations 2, 11 and 19 shows abnormal values greater than background mean distribution of 15740 μmhos/cm, and 238 mg/l is high in the area; high values of conductivity and Na which are near the vicinity of industrial area and found entry of industrial waste in to the water body. And agricultural runoff from the catchment causing its increase in water body as a point and non point sources of pollution. SAR values vary from 0.2 to 20 with an average of 3.5 and this value comes under water class Excellent to Good as per the irrigation guidelines in India.

*Factor 2*

Factor 2 exhibits 16% of the total variance with positive loading on fecal coli forms (FC), total coli forms (TC) and SO$_4$. Anthropogenic addition of FC in the water bodies ranging from (MPN) 20 to 2000/100ml (Most Probable Number) with an average of 612/100ml, the criteria as per CPCB surface water, fecal coli form (MPN) 500/100ml desirable and 2500/100mL maximum permissible limit. TC varies from 50 to 2500 MPN with an average of 1061 MPN were recorded at sample station 19 as against the tolerance limit of 50 MPN/100ml. Apart from the widespread nature in the environment, the presence TC may also found due to dumping of solid waste on lake shore area. SO$_4$, it varies from 6 to 1741 mg/l with average of 120 mg/l. The maximum value for SO$_4$ at sample station 19 (as against the tolerance limit of 400 mg/l) are due to entry of untreated industrial and domestic waste water into water body. Hence this factor can be attributed to origin of FC, TC and SO$_4$ in the area from anthropogenic source only.

*Factor 3*

It exhibits 12% of the total variance with positive loading on PO$_4$, F- and K. This factor can be attributed to the influence of industrial, municipal waste waters and agricultural runoff found on these parameters in the study area. PO$_4$ varies from 0.0-4.85mg/l (average=1.99 mg/l), sample stations 3,4,5,6,7,8,10,11,13,14,17,19,20,21,22, and 23 showing above the tolerance limit of 1mg/l. At these stations it was found that the entry of untreated industrial and domestic waste waters into water bodies as a point and non point source of pollution. F- Varies from 0.10-2.90mg/l (average=1.23mg/l), above the tolerance limit of 1.5mg/l found at stations 2,4,5,6,9,13,14,16,17,18,19,20,21,22 and 23, this may be the water bodies receiving high municipal sewage along with the solid waste. K varies from 6 to 99 mg/l (average = 32.38mg/l), the tolerance limit is <10 mg/l, except most of the sample stations were exceeded the limit and principle source of K may be due to entry of untreated industrial waste and municipal water into lakes.

*Factor 4*

Factor 4 exhibit 10.7% of the total variance and has positive loading on Temperature and Boron. Temperature varies from 23 to 30°C with an average of 27.12 °C, and B from 0 to 3.2 mg/l (average = 0.71mg/l). Most of the sample stations 1,2,5,7,8,9,10,13,14,16,17,21,22 and 23 shows more than the irrigation desirable limit of 1mg/l and the principle source of B are mainly from agriculture runoff and it is anthropogenic addition. The contamination due to B in water body and the values represent non point source pollution as an irrigation return flow from the catchment.

*Factor 5*

Factor 5 exhibit 8.73% of the total variance and has positive loading on NO$_3$. NO$_3$ concentration varies from 0.20 to 112 mg/l with an average of 19.91 mg/l. which exceeds the desirable limit of 20mg/l. This factor can be attributed to the influence of municipal waste waters and agricultural runoff found on these parameters in the study area. Sample stations 3,5,10,11,12,13,18,20,21,22 and 23 show comparatively higher concentration.

## 5. Conclusions

In this study, lakes getting polluted due to uncontrolled point and non point sources of pollution due to lack of proper sewage network. Results of factor analysis performed on 10 parameters and identified five factors controlling their variability in the study area. Multivariate statistical approaches show that the pathogenic (Bacteriological) pollution, organic pollution, salinity and Trace elements are highly polluting the lakes. The migration of pollutants in lakes in the form of untreated effluents in the catchment indicates the point source of pollution. The runoff from the agriculture fields also contributing the lake water pollution. The present study suggests that, the usefulness of multivariate statistical techniques for analysis and interpretation of complex data sets, water quality assessment and identification of pollution factors. Regular water quality monitoring for surface water should be undertaken for identification of pollution sources and understanding spatial variations in water quality for effective water quality management.

## 6. Recommendations

- Keeping in view of the urbanization and industrialization the organizations like municipal bodies need to conserve the water bodies around the Hyderabad.

- The untreated effluents emerging from the catchment must be diverted for maintaining the wholesomeness of the water bodies.

- The present study provides the baseline data for assessment of contaminations in the study area.

- The lake with sewage treatment plant (STP) was not giving much impact on water quality in lakes without first constructing the diversion sewers.

- Change of land use and construction activity of all types shall be prohibited in all water bodies. Construction should be avoided with in maximum water spread area.

## Acknowledgements

The authors are thankful to M.Satyanarayana, Managing Director, Clinicapro data research center, Hyderabad, for his continuous support and giving technical guidelines on statistical portion to this paper.

## References

Abbas Alkarkhi, F. M., Ismail, N., &Mat Easa, A. (2008). Assessment of arsenic and heavy metal contents in cockles (*Anadara granosa*) using multivariate statistical techniques. *Journal of Hazardous Materials, 150*, 783-789. http://dx.doi.org/10.1016/j.jhazmat.2007.05.035

Abdul-Wahab, S. A., Bakheit, C. S., & Al-Alawi, S. M. (2005). Principle component and multiple regression analysis in modeling of ground-level zone and factors affecting its concentrations. *Environmental Modeling &Software, 20*(10), 1263-1271. http://dx.doi.org/10.1016/j.envsoft.2004.09.001

Brumelis, G., Lapina, L., Nikodemus, O., & Tabors, G. (2000). Use of an artificial model of monitoring data to aid interpretation of principle component analysis. *Environmental Modeling & Software, 15*(8), 755-763. http://dx.doi.org/10.1016/S1364-8152(00)00060-8

Carpenter, S. R., Caraco, N. F., Correll, D. L., Howarth, R. W., Sharpley, A. N., & Smith, V. H. (1998). Nonpoint pollution of surface waters with phosphorus and nitrogen. *Ecological Applications, 8*, 559-568. http://dx.doi.org/10.1890/1051-0761(1998)008[0559:NPOSWW]2.0.CO;2

Davis, J. C. (1986). *Statistics and Data Analysis in Geology* (2nd ed.). Wiley, New York.

Dixon, W., & Chiswell, B. (1996). Review of aquatic monitoring program design. *Water Research, 30*, 1935-1948. http://dx.doi.org/10.1016/0043-1354(96)00087-5

Dong, J. D., Zhang, Y. Y., Zhang, S., Wang, Y. S., Yang, Z. H., & Wu, M. L. (2010). Identification of temporal and spatial variations of water quality in Sanya Bay, China by three-way principal component analysis. *Environ. Earth Sci., 60*, 1673-1682. http://dx.doi.org/10.1007/s12665-009-0301-4

GoI. (1992). *Policy Statement for Abatement of Pollution.* Delhi: Ministry of Environment and Forests, Government of India.

Helena, B., Pardo, R., Vega, M., Barrado, E., Fernandez, J. M., & Fernandez, I. (2000). Temporal evaluation of ground water composition in an alluvial aquifer (Pisuerga River, Spain) by principle component analysis. *Water Research, 34*, 807-816. http://dx.doi.org/10.1016/S0043-1354(99)00225-0

Howarth, R. W., Billen, G., Swaney, D., Townsend, A., Jaworski, N., Lajtha, K., ... Zhu, Z. L. (1996). Regional nitrogen budget and riverine N & P fluxes for the drainages to the North Atlantic Ocean: Natural and human influences. *Biogeochemistry, 35*, 75-139. http://dx.doi.org/10.1007/BF02179825

Kim, J. H., Choi, C. M., Kim, S. B., & Kwun, S. K. (2009). Water quality monitoring and multivariate statistical analysis for rural streams in South Korea. *Paddy Water Environ, 7*, 197-208. http://dx.doi.org/10.1007/s10333-009-0162-1

Liu, C. W., Lin, K. H., & Kuo, Y. M. (2003). Application of factor analysis in the assessment of groundwater quality in a Blackfoot disease area in Taiwan. *Science of the Total Environment, 313*, 77-89. http://dx.doi.org/10.1016/S0048-9697(02)00683-6

Love, D., Hallbauer, D., Amos, A., & Haranova, R. (2004). Factor analysis as a tool in ground water quality management: two southern African case studies. *Physics and Chemistry of the Earth, 29*, 1135-1143. http://dx.doi.org/10.1016/j.pce.2004.09.027

McKenna, Jr., J. E. (2003). An enhanced cluster analysis programme with bootstrap significance testing for ecological community analysis. *Environmental Modeling & Software, 18*(3), 205-220. http://dx.doi.org/10.1016/S1364-8152(02)00094-4

Morales, M. M., Marti, P., Lopis, A., Compos, L., & Sagrado, S. (1999). An Environmental study by factor analysis of surface seawaters in the gulf of Valenica (Western Medterranean). *Analytica Chimca Acta, 394*, 109-117. http://dx.doi.org/10.1016/S0003-2670(99)00198-1

Otto, M. (1998). Multivariate methods. In R. Kellner, J. M. Mermet, M. Otto & H. M. Widmer (Eds.),

*Analytical Chemistry*. Wiley-VCH, Weinheim.

Palma, P., Alvarenga, P., Palma, V. L., Fernandes, R. M., Soares, A. M. V. M., & Barbosa, I. R. (2010). Assessment of anthropogenic sources of water pollution using multivariate statistical techniques: A case study of Alqueva's reservoir, Portugal. *Environ. Monit. Assess, 165*, 539-552. http://dx.doi.org/10.1007/s10661-009-0965-y

Panda, U. C., Sundaray, S. K., Rath, P., Nayak, B. B., & Bhatta, D. (2006). Application of factor and cluster analysis for characterization of river and estuarine water systems - a case study: Mahanadhi River (India). *Journal of Hydrology, 331*(3-4), 434-445. http://dx.doi.org/10.1016/j.jhydrol.2006.05.029

Ramachandraiah, C., & Prasad, S. (2004). Impact of Urban Growth on Water Bodies. The Case of Hyderabad. *Working Paper* No. 60. pp no 16-22.

Shrestha, S., & Kazama, F. (2007). Assessment of surface water quality using multivariate statistical techniques: A case study of the Fuji river basin, Japan. *Environmental Modeling and Software, 22*(4), 464-475. http://dx.doi.org/10.1016/j.envsoft.2006.02.001

Simeonov, V., Stratis, J. A., Samara, C., Zachariadis, G., Vousta, D., Anthemidis, A., Sofoniou, M., & Kouimtzis, T. (2003). Assessment of the surface water quality in Northern Greece. *Water Research, 37*, 4119-4124. http://dx.doi.org/10.1016/S0043-1354(03)00398-1

Singh, K. P., Malik, A., & Sinha, S. (2005). Water quality assessment and apportionment of pollution sources of Gomti River (India) using multivariate statistical techniques: A case study. *Anal. Chim. Acta., 538*, 355-374. http://dx.doi.org/10.1016/j.aca.2005.02.006

Singh, K. P., Malik, A., Mohan, D., & Sinha, S. (2004). Multivariate statistical techniques for the evaluation of spatial    and temporal variations in water quality of Gomti River (India): A case study. *Water Research, 38*, 3980-3992. http://dx.doi.org/10.1016/j.watres.2004.06.011

SPSS® (Statistical Package for Social Sciences), version 6.1, USA. Professional Statistics 6.1, 385, Marija J. Norusis/SPSS Inc., Chicago, 1995.

Sridhar Kumar, A., Shnakaraiah, K., Rao., P. L. K. M., & Sathyanarayana, M. (2014). Assessment of water quality in Hussainsagar lake and its inlet channels using multivariate statistical techniques. *International Journal of Scientific & Engineering Research, 5*(9), 327-333. http://dx.doi.org/10.14299/ijser.2014.09.007

Vega, M., Pardo, R., Vega, M., Barrado, E., & Deban, L. (1998). Assessment of seasonal and polluting effects on the quality of river water by exploratory data analysis, *Water Research, 32*, 3581-3592. http://dx.doi.org/10.1016/S0043-1354(98)00138-9

Wunderlin, D. A., Diaz, M. P., Ame, M. V., Pesce, S. F., Hued, A. C., & Bistini, M. A. (2001). Pattern recognition techniques for the evaluation of spatial and temporal variations in water quality. A case study: Suquia River basin (Cordoba-Argentina). *Water Research, 35*, 2881-2894. http://dx.doi.org/10.1016/S0043-1354(00)00592-3

# 14

# Contamination of Roadside Soil and Bush Mint (*Hyptis suaveolens*) with Trace Metals along Major Roads of Abuja

Ahmed Usman[1] & Umar Ibrahim Gaya[1]

[1] Department of Pure and Industrial Chemistry, Bayero University, Kano, Nigeria

Correspondence: Umar Ibrahim Gaya, Department of Pure and Industrial Chemistry, Bayero University, P.M.B. 3011, Kano State, Nigeria. E-mail: uigaya.chm@buk.edu.ng

**Abstract**

There has been a growing concern over environmental pollution by trace metals from automobile source. Abuja, like most urban cities has a high road traffic density. The present study investigates the levels of trace metals in roadside plant and soils along some major roads in Abuja. Thirty samples, consisting of equal number of plants and soils from Airport, Kubwa and Nyanya road were analyzed for Pb, Fe, Cu, Zn, and Cr levels using atomic absorption spectroscopy. The findings reveal trace metal contamination gradient, with the maximum levels closer to the road. Copper is prevalent in the study area with concentrations standing at $76.66 \pm 12.02$ µg g$^{-1}$ and $300.00 \pm 50.00$ µg g$^{-1}$ in the plant and soil respectively. There is a significant correlation in the concentration of the metals studied regardless of sample class. The average distribution of the metals in the samples decreased in the order Cu > Zn > Fe > Pb > Cr with the exception of Nyanya soil and the plant samples from Kubwa road. Evidence for Pb transfer from soil to *Hyptis suaveolens* was established and accumulation of Pb, Cu and Fe has reached alarming levels. Chromium traces were as low as $11.91 \pm 1.38$ µg g$^{-1}$ in the plant but reached up to $39.68 \pm 6.87$ µg g$^{-1}$ in the soil. Concentration of the metals investigated in the soil except for Cu, are within the safety limit recommended by FAO/WHO.

**Keywords:** contamination, pollution, trace metals, bush mint, soil

## 1. Introduction

Soil and plant pollution by trace metals from automobile emissions is an important environmental issue. Metals are released in significant levels during different transport activities by different processes such as combustion, components wear, fluid leakage and corrosion of metal (Dolan et al., 2006). The United Nations (UN) estimated that over 600 million people worldwide are exposed to hazardous traffic generated pollutants (UNO, 1989). Mobile sources can contribute to the formation of toxic particulate matter leading to serious public health problems, including premature mortality, aggravation of respiratory and cardiovascular diseases, damage to lung tissues and structures, altered respiratory defense mechanisms, and chronic bronchitis (USEPA, 2007). Besides toxicity, trace metals are also persistent and pose serious danger to human and wildlife (Schwela, 2000).

Trace metals have varied toxicity and can act as biological hazards even at low levels. Toxicity studies have established that trace metals can directly damage the human body via impairment of mental and neurological function and alteration of numerous metabolic body processes (Greenwood & Earnshaw, 1986). For example, Pb diminishes Ca in the bones, precludes the synthesis of haemoglobin and affects the kidney and central nervous system (Essian, 1992; Bhata, 2002; Bridges & Zalups, 2005). In fact, this metal has no known importance in human biochemistry and physiology, and consumption even at very low concentration can have serious health implications (Bryan, 1976; Nolan, 2003). Copper is non-toxic, but its soluble forms are poisonous when present in large amounts (Scheinberg, 1991). Zinc is an essential micronutrient that can be found in all tissues of the body and is essential for cell growth, differentiation, healthy immune system and DNA synthesis (Sandstead, 1991; WHO, 1996). Zinc toxicity and gastric distress can occur from moderately high intakes of Zn greater than 150 mg day$^{-1}$ over long period of time (Samman, 2002). Although Cr$^{3+}$ is an essential dietary nutrient for normal glucose metabolism and to potentiate insulin (Cohen et al., 1993; Mertz, 1993), toxicity to lungs and gastrointestinal tract may occur when present in +4 oxidation state (Yu, 2008). Iron is the most abundant trace mineral in the body and is an essential element in most biological systems (Greentree, 1995; Goyer, 1996). It is likely that iron was essential for developing aerobic life on earth, but it is toxic to cells in excessive amount (Williams, 1990). Iron toxicity is largely based on Fenton and Haber Weiss chemistry, where catalytic amounts of iron are sufficient to yield hydroxyl radical ($OH^{\bullet}$) (Halliwel & Gutteridge, 1990). Free radicals formed in these reactions may promote oxidation of protein, peroxidation of membrane lipids and modification of nucleic acid.

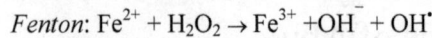

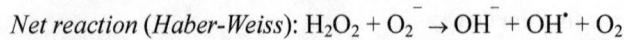

*Fenton*: $Fe^{2+} + H_2O_2 \rightarrow Fe^{3+} + OH^- + OH^{\bullet}$

$$Fe^{3+} + O_2^- \rightarrow Fe^{2+} + O_2$$

*Net reaction (Haber-Weiss)*: $H_2O_2 + O_2^- \rightarrow OH^- + OH^{\bullet} + O_2$

In order to prevent risk to natural life and public health, it is imperative to assess trace metal pollution in different components of the environment. Although trace metals are naturally present in soil, anthropogenically introduced trace metals from industrial processes, agricultural practices, combustion of fossil fuels and transport are more damaging to fauna and flora than the naturally occurring trace metals (EEA, 1995). Toxic elements such as aforementioned can accumulate in organic matter in soils and these may be uptaken by growing plants (Dara, 1993). Interestingly, high concentration of trace metals in soil is reflected by higher concentration of trace metals in plants and consequently in animals because of the food web (Farago, 1994). Plants having the ability to absorb and accumulate xenobiotics can therefore be used as indicators of environmental pollution (Farago, 1994). This study aimed at the determination of the level of trace metals (Pb, Fe, Cu, Zn, and Cr) as indicated in roadside bush mint plant (*Hyptis suaveolens*) and soil. *Hyptis suaveolens* (L.) (bush mint) is a popular medicinal plant which is found along roads of Abuja, the capital city of Nigeria. We preferred to examine Kubwa, Nyanya and Airport road which have traffic density of 28,000, 90,300 and 80,000 respectively (FRSC, 2011).

## 2. Materials and Methods

### 2.1 Sample and Sampling

The soil and plant materials of *Hyptis suaveolens* (L.) were collected from three different sampling areas along Kubwa (A1, A2, A3), Nyanya (B1, B2, B3) and Airport road (C1, C2, C3). The sampling areas were separated from one another by interval of 500 meters. There were 30 samples consisting of 15 plants and 15 soils collected; 5 sampling per sampling area. The sampling points were 2 m, 4 m, 6 m, 8 m & 10 m away from the edge of the road (Figure 1). Whole plant was cut excluding the root. Soil samples were taken down to 10 cm depth beneath each bush mint plant. All samples were collected in polythene bags and taken for further treatment.

### 2.2 Sample Preparation

Bush mint plant samples were prepared according to the procedures followed by Munson et al. (1990). Grass samples were washed, air dried, crushed to a powder and finally sieved using 25 μm sieve. One (1) g of each plant sample was ashed for 6 hr at 500 °C in a muffle furnace and kept in desiccators before use. The ash was moistened with water and 3 ml of nitric acid was added. The solution was heated on a hot plate to evaporate excess nitric acid. The solution was cooled, filtered into 50 ml volumetric flask using Whatman 40 filter paper. The filtrate was made up to mark with de-ionized water.

The procedures of Ayodele and Gaya (1998) were used for the pretreatment of soil samples. The soil samples were crushed in porcelain mortar to break the lumps, sieved through a 25 μm sieve and dried to constant weight at 100 °C. One (1) g portion of each soil sample was digested for 30 minutes with 30 ml of 6M $HNO_3$. The digest was filtered into 100 ml volumetric flask using Whatman 40 filter paper. The filtrate was made up to mark with de-ionized water.

Five (5) ml of aliquot of the resulting solutions from the foregoing steps was transferred into a 50 ml volumetric flask and was diluted to the mark with de-ionize water. Metal concentrations in these test samples were determined using a Buck Scientific model 210VGP Flame Atomic Absorption Spectrometer (FAAS) operated with a continuous source background correction.

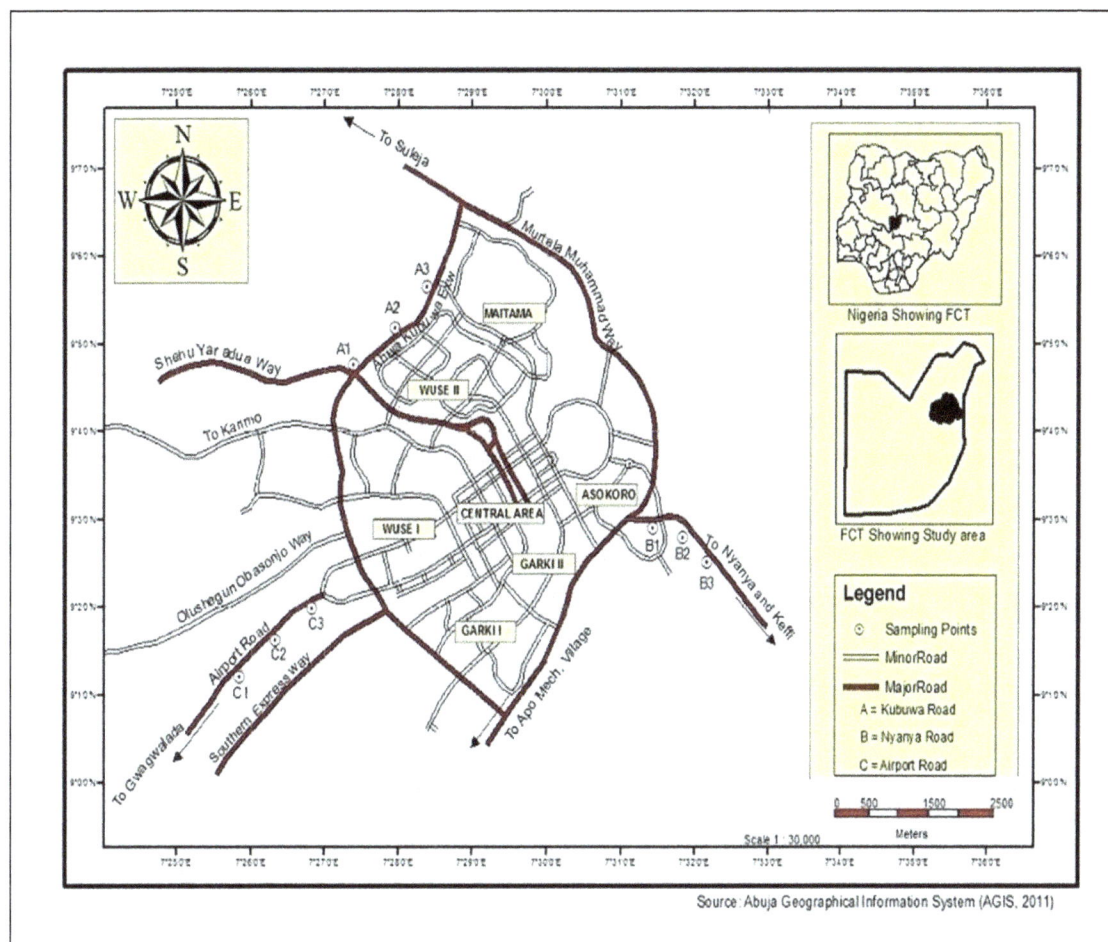

Figure 1. Map of Abuja showing sampling areas A, B and C along Kubwa, Nyanya and Airport road

The soil and plant samples were analysed in triplicate experiments. Analytical data were reported as mean ±SD. Statistical computations were done using Microsoft Excel. The correlation of trace metal in plant and soil samples were verified using Pearson correlation coefficient test, to determine significant difference between data sets of trace metal level in soil and plant, considering a level of significance of less than 5% ($p < 0.05$).

## 3. Results and Discussion

### 3.1 Levels of Trace Metals in Roadside Bush Mint

The average level of trace metals (Pb, Fe, Cu, Zn and Cr) in the three representative roads (Airport, Kubwa and Nyanya road) are presented in the following discussions. Figure 2 shows the level of Pb in *Hyptis suaveolens* on Airport road ranging between $32.59 \pm 7.14$ μg g$^{-1}$ and $9.63 \pm 3.57$ μg g$^{-1}$. There was a drastic decrease in the amount of Pb in *Hyptis suaveolens* between 2 m to 6 m which turns gradual between 8 m to 10 m. This may be due to the fact that lead particulate are only transported through a short distance before settling. Similar observation was made in the study earlier reported by Tjell et al. (1979). The Pb level in *Hyptis suaveolens* along Airport road was higher than that of Kubwa road ranging between $32.59 \pm 7.14$ μg g$^{-1}$ and $9.63 \pm 3.57$ μg g$^{-1}$). But the highest level of Pb in *Hyptis suaveolens* was recorded in the samples along Nyanya road ($35.92 \pm 7.14$ μg g$^{-1}$ to $11.11 \pm 5.09$ μg g$^{-1}$) which may be due to the fact that Nyanya road has the highest traffic density among the three roads under consideration. The levels Pb in *Hyptis suaveolens* and soil from the sampled areas were higher than that of control ($10.44 \pm 1.925$ μg g$^{-1}$ and $22.00 \pm 6.367$ μg g$^{-1}$ in plant and soil respectively). Concentration of Pb in *Hyptis suaveolens* is above the limit permitted by FAO/WHO (2011). In each case, lead was found in higher concentration in the samples closest to the road edge specifically 2 m away from the road. This suggests that, the volume of traffic affect significantly the extent to which road side plant can be contaminated with Pb. In fact, the correlation between the plant Pb level and the volume of traffic is significant at 1% (p<0.01) (Table 1), that of soil Pb level and traffic volume was significant at 5% (p<0.05) (Table 2).

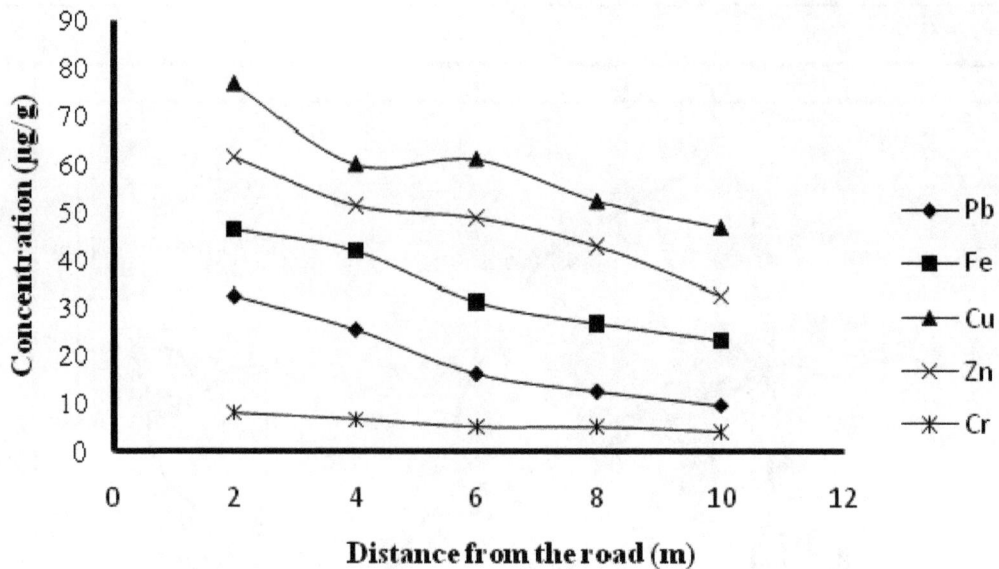

Figure 2. Average concentration of trace metals in *Hyptis suaveolens* away from Airport road

The amounts of Fe in plant tissues (*Hyptis suaveolens*) were of higher concentration on the sample along Nyanya road with maximum value of $50.00 \pm 14.45$ µg g$^{-1}$ compared to those along Airport road ($46.465 \pm 13.665$ µg g$^{-1}$) and Kubwa road ($35.354 \pm 6.123$ µg g$^{-1}$) as shown in Figure 2, 4 and 6 respectively. This suggests that plants along Nyanya road are more exposed to Fe contaminant owing to the heavy traffic along this road. Kubwa road with the least traffic flow show the least level of Fe in *Hyptis suaveolens*. The maximum level of Fe in *Hyptis suaveolens* from all the roads investigated occurred at 2 m from the road edge while the minimum was at 10 m. The content of Fe in *Hyptis Suaveolens* from all the roads under study was above that of control of 17.727 µg g$^{-1}$. There is significant correlation between Fe in the bush mint plant and traffic volume and between the soil Fe level and traffic volume ($p<0.05$) (Tables 1 & 2).

Highest level of Cu in *Hyptis suaveolens* were at Airport road with concentration ranging between $76.667 \pm 12.018$ µg g$^{-1}$ and $46.667 \pm 10.000$ µg g$^{-1}$ (Figure 2). The plant Cu level along this road is higher than that of Nyanya road (containing $58.89 \pm 5.09$ µg g$^{-1}$ to $35.56 \pm 5.09$ µg g$^{-1}$ Cu) which has more traffic volume. This suggested that, apart from traffic, the other anthropogenic activities may have lent to the high Cu. Kubwa road with the least traffic volume has the lowest content of Cu in the plant sample ($52.22 \pm 5.09$ µg g$^{-1}$ to $33.33 \pm 3.33$ µg g$^{-1}$) (Figure 4). The level of Cu in *Hyptis suaveolens* for the control was $22.667 \pm 3.054$ µg g$^{-1}$. Copper content in roadside plant were reported to be in the range of 10.7 to 45.0 µg g$^{-1}$ (Guan & Peart, 2006). Kabata-Pedias (1985) have reported that the normal content of Cu in most plants to be within 2 to 20 µg g$^{-1}$. Robson and Reuter (1981) explained that the critical level of Cu is 20 to 30 µg g$^{-1}$ for most plants. Result of this investigation shows that *Hyptis suaveolens* from the study areas is at risk of Cu pollution.

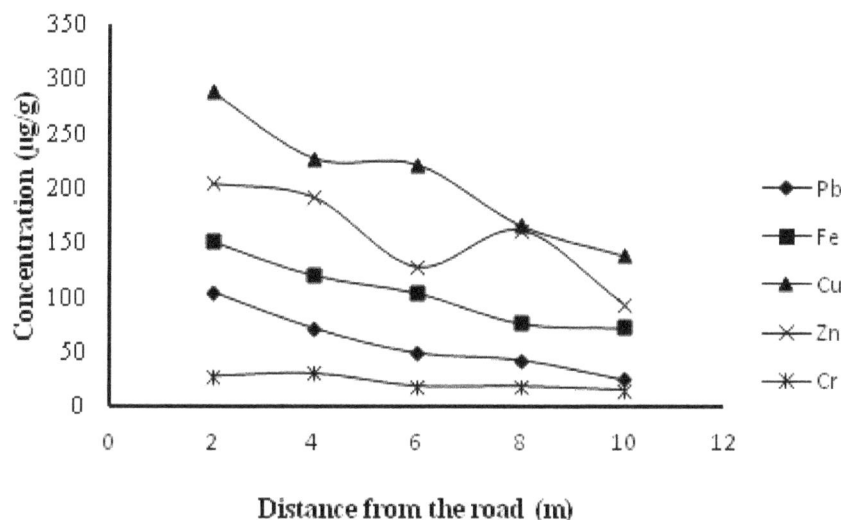

Figure 3. Average concentration of metals in soil away from Airport road

The concentration of Zn in *Hyptis suaveolens* along Airport road ranges from $61.54 \pm 6.35$ µg g$^{-1}$ to $32.48 \pm 3.92$ µgg$^{-1}$ (Figure 2). Zinc in *Hyptis suaveolens* along Kubwa road ranges from $64.54 \pm 14.29$ µg g$^{-1}$ and $41.15$ µg g$^{-1}$ $\pm 6.03$ µg g$^{-1}$ (Figure 4) while that of along Nyanya road ranges from $68.38 \pm 13.16$ µg g$^{-1}$ to $44.69 \pm 17.78$ µg g$^{-1}$ (Figure 6). The amount of Zn in the control was $25.641 \pm 4.263$ µg g$^{-1}$. The profile of Zn distribution in bush mint plant 2 to 10 m away from the three major roads of Abuja clearly indicates higher concentrations closer distance to the road (2 to 4 m). The critical toxic level of Zn for plant is 100 µg g$^{-1}$ (Allen et al., 1974). However, the safety limit of Zn in plant by WHO/FAO (2011) was 60.0 µg g$^{-1}$, hence, the level of Zn in *Hyptis suaveolens* within 2 m from all the road under investigation were little above the permissible limit, those at 6 to10 m were within the safety limit. On the other hand, the level of Cr in *Hyptis suaveolens* along Airport road ranges from $7.94 \pm 1.46$ µg g$^{-1}$ to $3.97 \pm 0.75$ µg g$^{-1}$ with the maximum in the sample at 2 m from away from the road. The distribution of Cr levels in *Hyptis suveolens* along Kubwa road was similar to that of Airport road, with concentrations ranging from $7.23 \pm 1.50$ µg g$^{-1}$ to $3.58 \pm 0.70$ µg g$^{-1}$. Nyanya road has the highest content of Cr in the plant sample which could be attributed to the density of traffic on this road. The Cr level in *Hyptis suaveolens* away from the road is in the range of $9.52 \pm 3.46$ µg g$^{-1}$ to $5.29 \pm 0.31$ µg g$^{-1}$. The level of Cr in the control sample was $2.381 \pm 0.873$ µg g$^{-1}$. Generally, the mean concentrations of 3 µg Cr g$^{-1}$ was obtained in the *Hyptis suaveolens* which indicates possible contamination or increased accumulation (Williams, 1988; Janus & Krajnc, 1989). This concentration is however lower than the critical toxic level (5 to 10 µg g$^{-1}$) (Cicek & Kopral, 2004). On the whole, the result of Cr level in bush mint plant from Nyanya road and Airport road in this investigation require some caution as the Cr concentrations in these areas are within this alarming range.

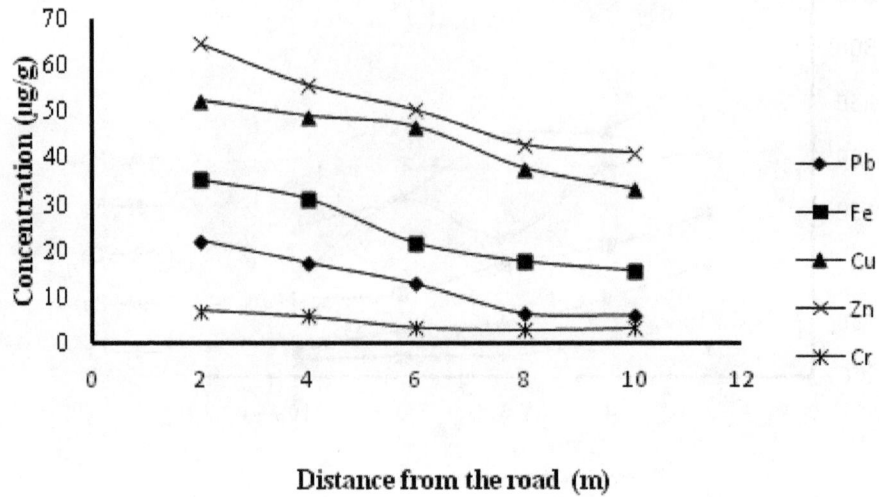

Figure 4. Average concentration of trace metals in *Hyptis suaveolens* along Kubwa road

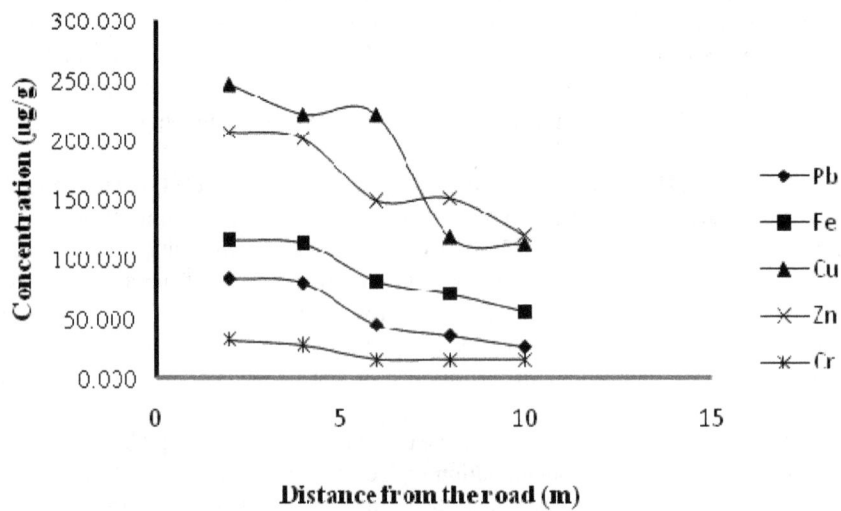

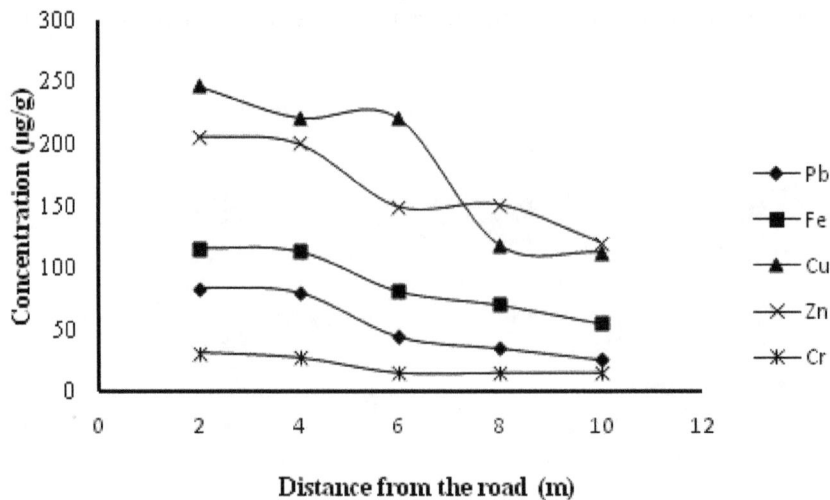

Figure 5. Distribution of trace metals in soil away from Kubwa road

## 3.2 Levels of Trace Metals in Roadside Soil

In the previous decades, vehicular traffic was the most widespread lead source (Lagerwerf & Specht, 1970) as a result of the use of tetramethyl and tetraethyl lead as anti-knock agent (Ewers & Schliphoter, 1991). Despite the fact that these chemicals are currently discontinued and replaced by environmentally friendly ones, traces of vehicular added lead need to be monitored. Nyanya road contains the highest soil Pb with value ranging from $118.00 \pm 7.00$ µg $g^{-1}$ to $26.93 \pm 4.34$ µg $g^{-1}$. The high density of traffic on this road may be responsible for this. The soil Pb level along Airport road which is next to Nyanya in terms of traffic volume ranges from $105.30 \pm 9.07$ µg $g^{-1}$ to $25.93 \pm 3.21$ µg $g^{-1}$ (Figure 3). The soil Pb level along Kubwa road was however the lowest. This is probably as a result of the relatively lower traffic volume along this road compared to that of Nyanya and Airport road. The soil Pb level was above that of control (containing $22.000 \pm 6.367$ µg $g^{-1}$ of Pb). The concentration of Pb in the soils was however within the safety limit specified by FAO/WHO (2011). Excessive concentrations of lead were observed closer to the road edge which may well be attributed to vehicular sources. Lead contamination of road side soil has previously been established (Jaradat et al., 1999; Abechi et al., 2010; Yahaya et al., 2010). The results obtained in this study agrees with previous study by Onder et al. (2007) to determine trace metal pollution in city green area, where Pb concentration was found to be higher at road edge and then decreases on increasing distance from the road.

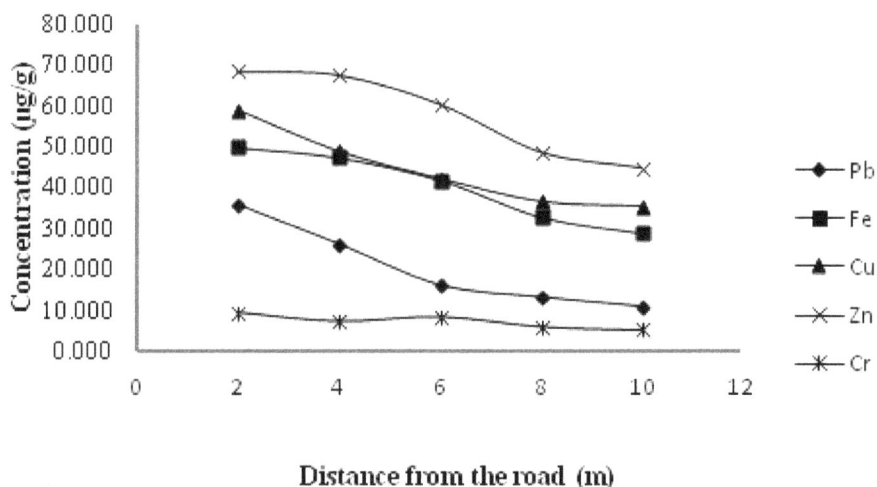

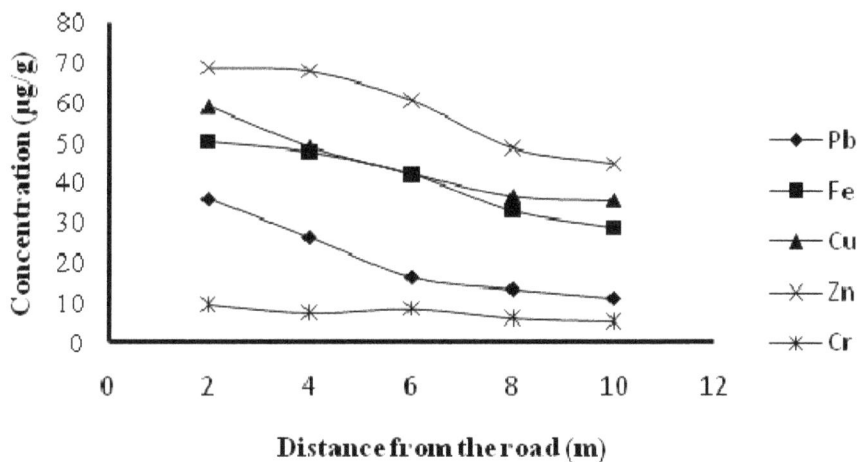

Figure 6. Profile of trace metals in *Hyptis suaveolens* away from Nyanya road

The variation in the level of Fe in the soil samples at various distances of the roadside follow a similar trend as observed in that of plant. The levels of Fe in the soil was however higher than that of the plant bush. The highest soil level of Fe was in the samples along Airport road with soil Fe level of 152.32 ± 16.38 µg g$^{-1}$ compared to Nyanya road (151.52 ± 13.12 µg g$^{-1}$) and that of Kubwa road of 116.16 ± 11.57 µg g$^{-1}$ (Figures 3, 5 & 7). There is no significant difference in the soil Fe content along Nyanya and Airport road, despite Nyanya road having a higher traffic volume. The soil along Airport road seems to be naturally richer in Fe compared to that of Nyanya. However, the decrease in the Fe level in the soil as the distance from the road increases is as a result of input from vehicular traffic. The soil level of Fe in the control was 68.182 ± 0.000 µg g$^{-1}$. Iron (Fe) is a major composition of soils, its distribution in decreasing trend as the distance from the road increase as observed in the figures above could be as a result of mechanical abrasion and component wear, from engine, from thrust bearings, bushing and bearing metals. The result obtained is in agreement with that of similar study of some roads in Jos (Abechi et al., 2010).

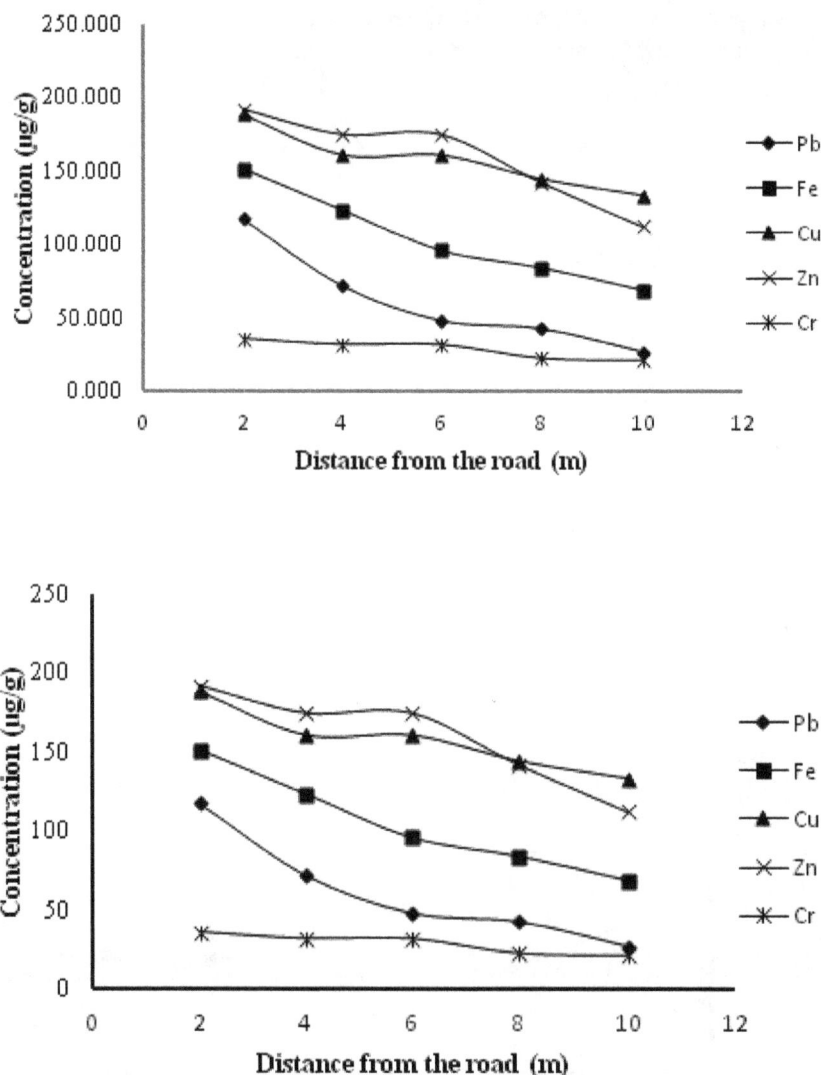

Figure 7. Trace metal concentrations in soils away from Nyanya road

The soil Cu levels in the three busy roads follow a similar pattern observed in the case of plant. Soil sample along Airport road has the highest Cu content (288.89 ± 19.245 µg g$^{-1}$ to 138.89 µg g$^{-1}$ ± 9.623 µg g$^{-1}$) (Figure 3) followed by Kubwa road (246 ± 16.67 µg g$^{-1}$ to 112.78 µg g$^{-1}$ ± 8.22 µg g$^{-1}$) (Figure 5). However, the soil samples along Nyanya road however have the lowest level of soil Cu with value (188.89 ± 25.46 µg g$^{-1}$ to

$133.33 \pm 16.67$ µg g$^{-1}$) (Figure 7), despite having the highest traffic volume. This observation may be attributed to slight change in the natural soil Cu content. The control sample contain $71.667 \pm 8.680$ µg g$^{-1}$ of Cu. Ordinarily, copper is usually present in soils in concentration range of 0 to 250 µg g$^{-1}$ (Alloway, 1995; Zheng et al., 2000). McGrath and Loveland (1992) reported the Cu content of 1.2 to 150.7 mg kg$^{-1}$ for soils of England and Wales. Relatively however, Cu content in urban and roadside soils, can reach 5 to 10 times higher than these normal concentrations (Nriagu, 1979; Baker & Senft, 1995).

The level of Zn in soils off Airport road ($205.13 \pm 12.82$ µg g$^{-1}$ to $94.02 \pm 7.40$ µg g$^{-1}$) (Figure 3) was lower than that of Kubwa road ($206.58 \pm 7.92$ µg g$^{-1}$ to $120.62 \pm 3.36$ µg g$^{-1}$) (Figure 5). Nyanya road with highest traffic volume has the least level of soil Zn ($192.31 \pm 46.23$ µg g$^{-1}$ to $112.74 \pm 5.91$ µg g$^{-1}$). Zinc levels in the study areas fall within the normal concentrations in soils (1 to 900 µg g$^{-1}$) (Alloway, 1995). The soil Zn level was higher than that of control ($88.205 \pm 6.206$ µg g$^{-1}$). There is positive correlation between the level of Zn in the plant sample and the traffic volume but this is less significant at 5% ($p > 0.05$) (Table 1). A decrease in the average level of zinc was observed as the distance from the road increases which may be as a result of roadside activities. Aksoy (1996) has reported higher zinc concentrations (410 µg g$^{-1}$) in urban roadside soils of Bradford.

Comparatively, soil Cr levels were lower than those of other trace metals investigated. The maximum level of Cr was in the soil samples along Nyanya road ($35.71 \pm 14.31$ µg g$^{-1}$). This decreased for Kubwa road ($746 \pm 3.97$ µg g$^{-1}$) and then for Airport road ($29.10 \pm 2.29$ µg g$^{-1}$) (Figures 3, 5 and 7). Since, the toxic level of Cr in soil is around 50 µg g$^{-1}$ (Bergmann, 1992), the study areas can be declared safe with respect to Cr toxicity.

*3.3 Relationship between Trace Metal Concentrations in Soil Vis-A-Vis Hyptis suaveolens*

The relationship between traffic volume and trace metal levels in the bush mint and soil is shown in Table 1 and Table 2 respectively. The correlation between the concentration of Pb, Fe, Zn, and Cu in soil and Plant was significant ($p < 0.05$). Positive correlation for metal content between *Hyptis suaveolens* and soil may be linked to the uptake of these elements from soil by the bush mint plant. However, there was insignificant correlation between Cr level and the plant. The concentration of Cr in the soil is higher than that of plant. Accordingly, many studies have demonstrated that chromium uptake from soils or nutrient solution and translocation to plant cells can be very low. However, concentrations of Cr in the edible portions of the plant may remain low, even when growing on chromium-contaminated soil (Dowdy & Ham, 1977; Lahouti & Peterson, 1979; Sykes et al., 1981; De Haan et al., 1985). Nyanya study area contained the highest Cu concentration in the soil. The changes in concentration of the trace metals in soil were always more drastic compared to that of the plant *Hyptis suaveolens*. Similarly, the bush mint plant contained more Cu than any other trace metal except along Kubwa roadside where Zn was present in exceeding concentrations. The variation in the Cu content in both the plant and soil samples from the road edge shows that the activities of Cu in the samples is as a result of road traffic. This is however independent of traffic as the correlation between the traffic volume, soil and plant Cu is less significant ($p > 0.05$) (Tables 1 and 2).

Table 1. Correlation analysis among traffic volume and metals in Bush mint

| Distance (m) | Pb | Fe | Cu | Zn | Cr |
|---|---|---|---|---|---|
| 2 | 0.994* | 0.994* | 0.616 | 0.197 | 0.817 |
| 4 | 0.999* | 0.976** | 0.386 | 0.375 | 0.900 |
| 6 | 0.992* | 0.911 | 0.168 | 0.491 | 0.798 |
| 8 | 0.874 | 0.961** | 0.326 | 0.588 | 0.973** |
| 10 | 0.884 | 0.953** | 0.525 | 0.110 | 0.765 |

*–Correlation is significant at 0.01; **–Correlation is significant at 0.05.

Chromium is a natural component of plant tissues, although concentrations vary considerably between different plant species, plant tissues, and soil types. Levels in shoots of plants grown on uncontaminated soil usually do not exceed 0.5 mg kg$^{-1}$. The concentration of Cr in plant and soil were highest in the samples close to the road edge and decreases as the distance from the road increases. Chromium concentration in plant and soil sample did not show any significant correlation. However, the plant and soil Cr levels are significantly correlated with the traffic volume ($p < 0.05$) (Tables 1 and 2).

Table 2. Correlation analysis among traffic volume and metals in soil

| Distance (m) | Pb | Fe | Cu | Zn | Cr |
|---|---|---|---|---|---|
| 2 | 0.970** | 0.989** | -0.215 | -0.837 | 0.240 |
| 4 | -0.992 | 0.993* | -0.531 | 0.756 | 0.992* |
| 6 | 0.896 | 0.866 | -0.173 | -0.277 | 0.779 |
| 8 | 0.969** | 0.921 | 0.935 | 0.177 | 0.961** |
| 10 | -0.338 | -0.106 | 0.984** | 0.991* | -0.091 |

*–Correlation is significant at 0.01; **–Correlation is significant at 0.05.

Lead is the most common environmental contaminant found in soils. Among trace metals, Pb has long been known as potential hazard to health (Rowchowdhury & Gautum, 1995; Nariagu et al., 1996; Shannon & Graef, 1996). Unlike most other trace metals, Pb has no biological role, and is potentially toxic to organisms (Sobolev & Begonia, 2008). Undesirable and unnatural concentrations of Pb are found in air, water, soil and vegetation, particularly near heavily ply automobile ways (Fuller, 1997; Habashi, 1992). Jaradat et al. (1999) reported that the increasing number of vehicle on the road during the last few years, in Jordan, mostly operated by leaded fuel, have lead to high levels of some trace metals and other pollutants in soil and plants near highways in both rural and urban areas. According to Fergusson (1990), Pb-containing particles in motor vehicle exhausts tend to be larger near motorways in urban areas. Roadsides soils and vegetations are prone to Pb particles and this could be as a result of wind action and vehicular emission (Naima et al., 2010). Plants near road ways have relative increase of Pb deposition due to vehicles using leaded petrol (Bu-Olayan & Thomas, 2002).

### 3.4 Transfer Factors from Soils to Plant

Transfer factor (TF) defined as the ratio of trace metal concentration in plant and trace metal concentration in soil were computed in order to surmise the extent of transfer of soil trace metals into the plant. The transfer factor ranges for Pb, Fe, Zn, Cu and Cr from the soil to the plant *Hyptis suaveolens* are displayed in Table 3. Notably, the TF values of lead from soil to plant varied from 0.304 to 0.413 by Nyanya road, 0.288 to 0.354 by airport road, 0.189 to 0.292 by Kubwa road. Generally, from the table it may be inferred that Cr is relatively transferred to the plant largely poorly while great uptake of Pb and Fe are observed. The accumulation factor of Cu in the plants did not show particular pattern along the roads.

Table 3. Transfer factor of trace metals from soil to plant 2-10 m away from Abuja roads

| Distance (2-10 m) | Pb | Fe | Cu | Zn | Cr |
|---|---|---|---|---|---|
| Nyanya road | 0.304-0.304 | 0.330-0.437 | 0.254-0.312 | 0.341-0.396 | 0.233-0.271 |
| Airport road | 0.288-0.371 | 0.298-0.346 | 0.263-0.336 | 0.263-0.380 | 0.208-0.273 |
| Kubwa road | 0.189-0.292 | 0.250-0.304 | 0.211-0.318 | 0.277-0.341 | 0.196-0.233 |

## 4. Conclusion

The result of this study revealed the distribution of trace metals (Pb, Fe, Zn, Cu, and Cr) in the roadside soils and bush mint plant along some major roads in the capital city of Nigeria. Trace metal profiles decreased with distance away from Abuja roads which indicate the impact of vehicular traffic on the environment. The concentrations of trace metals in both the soils and plants are in the order of Cu > Zn > Fe > Pb > Cr in decreasing order of concentration except in the soil sample from Nyanya and the plant sample from Kubwa road where Zn is present in the highest amount. The levels of the metals are high compared to those of the control samples which indicated that there is accumulation of these metals in the soil and subsequent transfer to plants growing along the highway. Transfer factors have shown relatively higher accumulation of Pb and Fe.

Generally, the maximum levels of Pb, Fe, Zn, and Cr in soil along each road were within the safety limit guidelines proposed by most regulatory bodies. The level of copper in soil however calls for concern. Similarly, the concentration of the metals Pb, Cu and Fe in the bush mint is alarming as this was found to exceed the safety limit guidelines set by most regulatory bodies. There is significant correlation between the Pb and Fe in plant tissues and traffic volume which indicates contributions from anthropogenic activities.

## Acknowledgements

The authors acknowledge the contribution of Mr Joshua Kubai of the Analytical laboratory at the Department of Pure and Industrial Chemistry, Bayero University, Kano.

## References

Abechi, E. S., Okunola, O. J., Zubairu, S. M. J., Usman, A. A., & Apene, E. (2010). Evaluation of heavy metals in roadside soils of major streets in Jos metropolis, Nigeria. *Journal of Environmental Chemistry and Ecotoxicology, 2*(6), 98-102.

Aksoy, A. (1996). *Autecology of Capsella bursa-pastoris (L) Medic* (Unpublished PhD thesis, University of Bradford, Bradford, 1996).

Allen, S. E., Grimshow, H. M., Parkinson, J. A., & Quarmby, C. (1974). *Chemical analysis of Ecological materials. Osney Mead.* Oxford, UK: Blackwell Scientific Publication.

Alloway, B. J. (1995). *Heavy Metals in Soils.* London: Chapman & Hall. http://dx.doi.org/10.1007/978-94-011-1344-1

Ayodele, J. T., & Gaya, U. M. (1998). Chromium Manganese and Zinc in Kano Municipality Street dust. *J. Chem. Soc. Nigeria, 23,* 24-34.

Baker, D. E., & Senft, J. P. (1995). Copper. In 'Heavy Metals in Soils. In B. J. Alloway (Ed.), *Heavy metals in soils* (2nd ed., pp. 179-205). London: Blackie Academic and Professional.

Bergmann, W. (1992). *Nutritional disorder of plants.* New York: Gustav Fischer.

Bhata, S. C. (2002). *Environmental Chemistry* (pp. 442). New Delhi: CBS Publishes and Distributors.

Bhatia, I., & Choudhri, G. N. (1991). Impact of automobile effusion on plant and soil. *International Journal of Ecology and Environmental Sciences, 17,* 121-127.

Bridges, C. C., & Zalups, R. K. (2005). Molecular and ionic mimicry and the transport of toxic metals. *Toxicol. and Appl. Pharmacol., 204*(3), 274-308. http://dx.doi.org/10.1016/j.taap.2004.09.007

Bryan, G. W. (1976). Heavy Metal contamination in sea. In R. Johnson (Ed.), *Marine pollution* (pp.185-302). Academic Press.

Bu-Olayan A. H., & Thomas, B. U. (2002). Biomonitoring studies on the Lead levels in Mesquite (Prosopis juliflora L.) in the arid ecosystem of Kuwait. *Kuwait J. Sci. Eng., 29*(1), 65-73.

Cicek, A., & Koparal, A. S. (2004). Accumulation of soil heavy metals in soil and trees leaves sampled from the surrounding of Tuncbilek Thermal Power Plant. *Chemosphere, 57,* 1031-1036. http://dx.doi.org/10.1016/j.chemosphere.2004.07.038

Cohen, M. D., Kargacin, B., Klein, C. B., & Costa, M. (1993). Mechanisms of chromium carcinogenicity and toxicity. *Critical Reviews in Toxicology, 23*(3), 255-81. http://dx.doi.org/10.3109/10408449309105012

Dara, S. S. (1993). *A text book of environmental chemistry and pollution control* (pp. 167-206). New Delhi: Rjendra Rarindra Printers (PVT) Ltd Ram Niger.

De Haan, S., Rethfeld, H., & van Driel, W. (1985). *Acceptable levels of heavy metals (Cr, Cu, Pb, Zn) in soils, depending on their clay and humus content and cation-exchange capacity.* Haren, The Netherlands: Instituut voor Boodenvruchtbaarheid.

Dolan, L. M. J., Van Bohemen, H., Whelan, P., Akbar, K. F., O'Malley, V., O'Leary, G., & Keizer, P. J. (2006). Towards the sustainable development of modern road ecosystem (pp. 275-331). In J. Davenport, & J. L. Davenport (Eds.), *The Ecology of Transportation: Managing Mobility for the Environment.* Netherlands: Springer. http://dx.doi.org/10.1007/1-4020-4504-2_13

Dowdy, R. H., & Ham, G. E. (1977). Soybean growth and elemental content as influenced by soil amendments of sewage sludge and heavy metals: Seedling studies. *Agron. J., 69,* 300-303. http://dx.doi.org/10.2134/agronj1977.00021962006900020024x

EEA. (1995). Soil Pollution by heavy metals. *Europe's environment the Dobris assessment* (p. 676). European Environmental Agency, Luxembourg: Office des Publications.

Essian, E. U. (1992). Differential accumulation of lead on selected edible vegetables associated with roadside gardening in Nig. Trop. *J. Applied Sciences, 1*(2), 83-86.

Ewers, U., & Schlipkoter, H. W. (1991). Chronic Toxicity of metals and metals compounds. In E. Merian (Ed.), *Metal and their compounds in the Environment* (pp. 591-603). VCH, Weinheim, New York: Verlag SegSell Schaft Mbh.

FAO/WHO. (2011). Joint FAO/WHO food standards Programme. *Report of the 5th session of the codex committee on contaminants in foods* (CCCF) (pp. 64-89), March 21-25, 2011, The Hague, The Netherlands.

Farago, M. E. (1994). Plants as indicators of mineralization and pollution, In M. E. Farago (Ed.), *Plants and chemical elements* (pp. 221). VCH, Wienheim.

Fergusson, J. E. (1990). *The heavy elements: Chemistry, environmental impacts and health effects.* Oxford: Pergamon press.

FRSC. (2011). *Road count details, Federal Road Safety Corps, Federal Capital Territory (FCT) sector command, Abuja.*

Fuller, E. C. (1997). *Chemistry and Man's Environment* (pp. 22-26). Boston: Houghton Mifflin Publishing Company.

Goyer, R. A. (1996). Toxic effects of metals. In C. D. Klaassen (Ed.), *Casarett & Doull's toxicology: the basic science of poisons* (5th ed., pp. 715-716). New York: McGraw-Hill.

Greentree, W. F., & Hall, J. O. (1995). Iron toxicosis. In J. D. Bonagura (Ed.), *Kirks current therapy XII. Small animal practice* (pp. 240-242). Philadelphia: W. B. Saunders Co.

Greenwood, N. N., & Earnshaw, A. (1986). *Chemistry of the elements.* Oxford, U.K.: Pergamon Press.

Guan, D., & Peart, M. R. (2006). Heavy metal concentration in plants at roadside locations and parks of urban guangzou, *Journal of Environmental Sciences, 18*(3), 495-502.

Habashi, F. (1992). *Environmental issues in the metallurgical Industry Progress and Problems, Environmental issues and Waste Management in Energy and Mineral Production* (pp. 1143-1153). Rotherdam: Balkama.

Halliwell, B., & Gutteridge, J. M. C. (1990). The role of free radicals and catalytic metal ions in human disease: an overview. *Methods Enzymol. 186*, 1-85. http://dx.doi.org/10.1016/0076-6879(90)86093-B

Janus, J. A., & Krajnc, E. I. (1989). *Integrated criteria document chromium: Effects* (Appendix to report no. 758701001). National Institute of Public Health and Environmental Protection, Bilthoven, Netherlands.

Jaradat, Q. M., & Moman, K. A. (1999). Contamination of Roadside Soil, Plants, and Air With Heavy Metals in Jordan, A Comparative Study. *Turk J. Chem, 23*, 209-220.

Kabata-Pedias, A. (2000). *Trace elements in soil and plants* (pp. 57-59). CRC Press Book.

Lagerwerf, J. V., & Specht, A. W. (1970). Contamination of roadside soil and vegetation with cadmium, nickel, lead and zinc. *Environmental Science and Technology, 4*, 583-586. http://dx.doi.org/10.1021/es60042a001

Lahouti, M., & Peterson, P. J. (1979). Chromium accumulation and distribution in crop plants. *J. Sci. Food Agri., 30*, 136-142. http://dx.doi.org/10.1002/jsfa.2740300207

Mcgrath, S. P., & Loveland, P. J. (1992). *The Soil Geochemical Atlas of England and Wales.* London: Blackie Academic & Professional.

Mertz, W. (1993). Chromium occurrence and function in biological systems. *Physiol. Rev. 49*, 163-239.

Munson, R. D., & Nelson, W. L. (1990). Principle and Practices in Plant Analysis. In R. L.Westerman. (Ed.), *Soil testing and Plant analysis* (pp.359-389). Madison, USA.

Naima, H. N., Aima, I. B., Fayyaz, U., & Uzma, H. (2010). Leaves of roadside plants as bioindicator of traffic related lead pollution during different seasons in Sargodha, Pakistan. *African Journal of Environmental Science and Technology, 4*(11), 770-774.

Nariagu, J. O., Blankson, M. L., & Ocran, K. (1996). Childhood lead poisoning in Africa: A growing public health problem. *Sci. Total Environ, 181*(2), 93. http://dx.doi.org/10.1016/0048-9697(95)04954-1

Nolan, K. (2003). Copper toxicity syndrome. *J. Orthomol. Psychiatry, 12*(4), 270-282.

Nriagu, J. O. (1979). Copper in the atmosphere and precipitation. In J. O. Nriagu (Ed.), *Copper in the Environment. Part 1* (pp. 43-75). Chichester: John Wiley.

Onder, S., Dursun, Gezgin, .S., & Demirbas, A. (2007). Determination of Heavy Metal Pollution in Grass and Soil of City Centre Green Area. *Polish Journal of environ. Stud., 16*(1), 145-154.

Robson, A. D., & Reuter, D. J. (1981). Diagnosis of Copper deficiency and toxicity. In J. F. Loneragen, A. D. Robson, & R. B. Graham (Eds.), *Copper in soils and plants* (pp. 287-312). London: Academic press.

Rowchowdhury, A., & Gautum, A. K. (1995). Alteration of human sperms and other seminal constituents after lead exposure. *Ind. J. Physiol. Allied Sci., 49*(2), 68.

Samman, S. (2002). Trace Elements. In J. Mann, & A. S. Truswell (Eds.), *Essentials of human nutrition* (2nd ed.). New York: Oxford University Press.

Sandstead, H. H. (1991). Zinc deficiency. A public health problem? *Am. J. Dis. Child, 145*(8), 853-9.

Scheinberg, I. H. (1991). Copper. In E. Meria (Ed.), *Metals and their compounds in the environment* (pp. 893-908). Weinheim, New York: VCH. Verlags segsellschaft Mbh.

Schwela, D. (2000). Air pollution and health in urban areas. *Reviews on Environmental Health, 15*(1-2), 13-42. http://dx.doi.org/10.1515/REVEH.2000.15.1-2.13

Shannon, M., & Graef, J. W. (1996). Lead intoxication in children with pervasive developmental disorders. *J. Toxicol. Clin. Toxicol., 34*(2), 177. http://dx.doi.org/10.3109/15563659609013767

Sobolev, D., & Begonia, M. F. T. (2008). Effects of Heavy Metal Contamination upon Soil Microbes: Lead-induced Changes in General and Denitrifying Microbial Communities as Evidenced by Molecular Markers. *Int. J. Environ. Res. Public Health, 5*(5), 451. http://dx.doi.org/10.3390/ijerph5050450

Sykes, R. L., Corning, D. R., & Earl, J. (1981). The effect of soil-chromium III on the growth and chromium absorption of various plants. *J. Am. Leather Chem. Assoc, 76*, 102-126.

Tjell, J. C., Hovmand, M. F., & Mosback, H. (1979). Atmospheric pollution of grass grown in a Background area in Denmark. *Nature, 2080*, 425-426. http://dx.doi.org/10.1038/280425a0

UNO. (1988). Prospects of world urbanization. *United Nations Organisation, Population study no. 112*. New York.

USEPA. (2007). *Mobile source air toxics-regulations. Control of hazardous air pollutants from mobile sources*. Office of Transportation and Air Quality, United State Environmental Protection Agency. Retrieved from http://www.epa.gov/otaq/toxics-regs.htm#02262007

Williams, J. H. (1988). *Chromium in sewage sludge applied to agricultural land*. Office of Official Publications for the Commission of the European Communities, Brussels.

Yahaya, M. I., Ezeh, M. I., Musa, Y. F., & Mohammad, S. Y. (2010). Analysis of heavy metals concentration in road sides soil in Yauri, Nigeria. *African Journal of Pure and Applied Chemistry, 4*(3), 22-30.

Yu, D. (2008). Chromium toxicity. In: *Case studies in Environmental Medicine*. Agency for toxic substances and disease registry (ATSDR). Retrieved from http://www.atsdr.cdc.gov/csem/chromium/docs/chromium.pdf

Zheng, C. R., Tu, C., & Shen, Z. G. (2000). Chemical method and phytoremediation of soil contaminated with heavy metals. *Chemosphere, 41*, 229-234. http://dx.doi.org/10.1016/S0045-6535(99)00415-4

# 15

# Assessment of Drinking Water Quality of Groundwaters in Bunpkurugu-Yunyo District of Ghana

Maxwell Anim-Gyampo[1], Musah Saeed Zango[1] & Boateng Ampadu[1]

[1] Department of Earth and Environmental Science, University for Development Studies, Navrongo, Ghana

Correspondence: Maxwell Anim-Gyampo, Department of Earth and Environmental Science, University for Development Studies, P. O. Box 24, Navrongo, Ghana. E-mail: gyampom@gmail.com

**Abstract**

Water quality assessment of nineteen (19) boreholes sampled during the two climatic regimes (raining and dry seasons) in Bunkpurugu-Yunyo District of Ghana has been carried out using standard methods. Analysis of results showed that all the parameters fell within World Health Organisation (WHO, 2008) acceptable limits with exception of turbidity, nitrates, lead and cadmium. The order of major cations and anions in water samples obtained during the raining and dry seasons were Na>Ca>Mg>K and $HCO_3$>$NO_3$>Cl>$SO_4$>F>$PO_4$, and Na>Ca>K>Mg and $HCO_3$>Cl>$SO_4$>F>$PO_4$>$NO_3$ respectively. Groundwaters were fresh (TDS< 1000mg/l) and generally of temporal hardness. High mean values of Pb and Cd above acceptable limits implied potential health hazards to inhabitants over a long period of water consumption. Agro-chemicals could be the major sources of Pb and Cd contamination to the groundwaters while the potential of contamination from natural sources may be a possibility. Assessment of water quality index (WQI) showed remarkable variation of water quality with respect to climatic conditions, with 94.7% samples falling within "Excellent" and "Good" categories in the raining season while conversely, about 89.5% fell within "Poor" and "Unsuitable" categories during the dry season.

**Keywords:** Bunkpurugu-Yunyo, Ghana, groundwater, lead, water quality index

## 1. Introduction

Groundwater plays a very pivotal role in the domestic water delivery system in Ghana as it is officially, the considered source of potable water for rural water supply delivery in Ghana (CWSA, 2009). The preference stems from the fact that groundwater is generally of better quality, less polluted and requires little or no bacteriological treatment prior to consumption. Furthermore, groundwater sources are more reliable for utilization in rural areas than surface water sources such as rivers, streams and dams in water-stressed semi-arid regions like northern Ghana. Bunkpurugu-Yunyo district in the north-eastern part of Ghana has about 90% of the population living in rural areas. The area falls within a semi-arid region with very little surface water resources, and where available, it is mostly unreliable throughout the greater part of the year; highly polluted by grazing cattle and other livestock thus rendering it very unsuitable for human consumption. Groundwater sources are therefore the most reliable and sometimes, the only source of potable drinking water for the inhabitants as well as livestock. Despite the general acceptance of groundwater as the preferred source of potable drinking water for rural communities in Ghana (MWWH-NWP, 2007), the qualities of some sources had been found to be unacceptable for domestic use with attendant adverse health implications due to either natural and/or anthropogenic factors that have the potential to adversely affect the health of consumers. Natural factors may include the dissolution of several anions and cations into groundwaters in an aquifer system due to groundwater-rock interactions and weathering of rocks. For instance, groundwaters in parts of northern Ghana including the Bongo, Tongo, Talensi and Bolgatanga districts in upper east regions (Apambire et al., 1997; Anongura, 1995) and Gushiegu, Karaga, Saboba, Yendi and Chereponi districts of northern region of Ghana Anongura et al. (2003) and Anim-Gyampo et al. (2012) have been found to contain excessive fluoride up to a level of over 4.5mg/l. Climatic variations (rainy and dry seasons) may have impact on groundwater quality either positively or negatively making it suitable or otherwise for human consumption. Anthropogenic activities such as the use of agro-chemicals (i.e. chemical fertilisers, weedicides, and pesticides), release of toxic heavy metal and non-metals during mining activities, industrial and domestic waste disposals can in the long term contribute to the deterioration in the quality of groundwater resources (Todd, 1980). The Bunkpurugu-Yunyo District is predominantly a rural district with crop farming being the main stay of the inhabitants. Cultivation of crops is achieved by the intense use of agro-chemicals such as weedicides, chemical fertilisers and pesticides, which contain some heavy metals such as Pb, Cd, As, Co, Hg etc as traces could infiltrate and contaminate the groundwaters, which is the only source of potable drinking water for the inhabitants in the area. This study therefore assesses the water quality of groundwater in aquifers in the Bunkpurugu district with the aim of ascertaining its suitability for human consumption throughout the year.

### 1.1 The Study Area

Bunkpurugu-Yunyo District in northeastern Ghana (Figure 1), carved out from the East Mamprusi District, was created in 2004 with the Bunkpurugu town as the capital. It covers an area of approximately 3079.7 km$^2$ and located between latitude 10° 14' 44.252 N to 10° 42' 34.087 N and longitude 0° 5' 15.485 W to 0° 18' 10.49 W. It is bounded in the north by Garu-Timpani, east with Republic of Togo, west by West Mamprusi and south by Gushiegu and Chereponi districts. It is predominantly a rural community and farming is the main occupation of the inhabitants with groundwater serving as the main source of potable drinking water, especially during the long dry season where almost all surface water sources dry-up. The position of the district as a border to north-western Togo affords it the opportunity as a potential business hub of the eastern corridor of Ghana and therefore attracts lots of mobile population. The study area falls within the tropical continental climatic region of Ghana and is influenced by the movement of the Inter-Tropical Convergence Zone (ITCZ) and it is among the areas with lowest rainfall values in Ghana with mean annual rainfall of about 100-115 cm. Rainfall is very erratic and intermittent droughts and floods within the season are quite common. The study area is characterised by generally high temperature and is among the driest places in Ghana. The highest mean monthly and daily temperatures of 33 and 42 °C, respectively are recorded in March-April; whilst the lowest mean monthly value of 26.5 °C is registered during the peak Harmattan season in December and January each year. Relative humidities of 70-90% are recorded during the rainy season and in the dry season; the lowest value of about 20% could be observed (Dickson & Benneh, 1985).

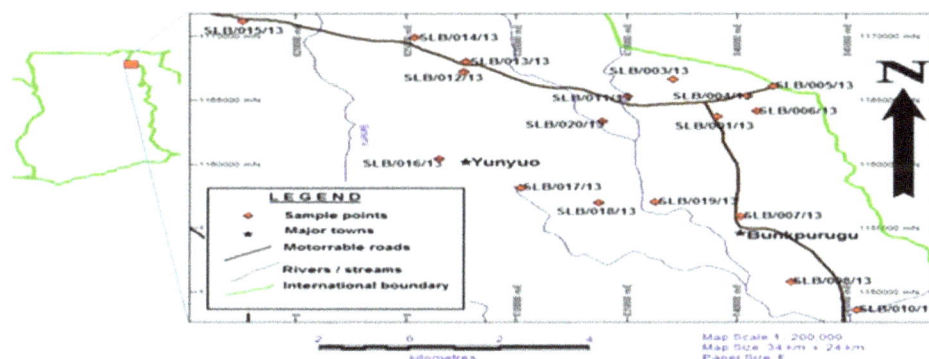

Figure 1. Location of the study area showing sampling sites

The major river draining the study area is the White Volta, which enters the region in the northeast after it has been joined by the much smaller and shorter minor rivers (Nawonga and Moba) draining the southwestern part. All the rivers in the study area dry up during greater part of the year (approximately five months) except the White Volta, which is perennial. The area is semi-arid and experiences very short wet (raining) seasons to a maximum of about four months, followed by a prolonged dry season of approximately seven to eight months. The prolonged dry season renders many people in the area seasonally unemployed because farming activities are basically rainfall dependent. The total number of wet days in a year has been estimated to about 44 (Anim-Gyampo et al., 2013), which are woefully inadequate to support plant life, resulting in low agricultural productivity and consequential high poverty levels. The vegetation type is the Interior Savannah Woodland, which is characterised by tall grass interspersed with drought-resistant trees such as Neem, Shea, Dawadawa, Baobab, Acacia and Mahogany. Grasses grow in tussocks and can reach a height of approximately three meters or more. There is a marked change in vegetation depending on the prevailing climatic conditions (dry and wet). The vegetation is largely affected by indiscriminate and uncontrolled bush fires, indiscriminate felling of trees for housing, charcoal production for fuel wood especially during the dry periods (Anongura et al., 2003). The topography is generally gently rolling with the Nakpanduri (formerly Gambaga) escarpment marking the northern limits of the paleozoicvoltaian sedimentary basin. Apart from the mountainous areas bordering the escarpment, there are generally little runoffs when it rains and the greater part of rainwater seeps into the ground to recharge aquifers, which had resulted in the development of relatively good groundwater aquifers and therefore forming part of the areas with high groundwater potential in northern part of Ghana. According to Obeng (2000), two main types of soils are found in the study area namely, the Savannah Onchrosols and the Groundwater Laterites. The Savannah Ochrosols which covers almost the entire district is moderately well drained and is developed mainly within the quartzites. The texture of the surface soil is loamy sand with good water retention. Savannah Ochrosols has high potential for wide range of crops. Some areas do not appear to be fully utilized though some lands are under considerable pressure in the district. The Groundwater Laterites type of soils covers a relatively smaller portion of the study area, and is found mainly in the north- eastern parts. These are concretionary soils developed mainly in the shale, mudstone and argillaceous sandstone materials. The concretionary soils are associated with frequent exposures of iron pan and boulders. The soils are perfectly drained during the wet season with perched water tables developing in some areas, and become extremely dry during the dry season. Exposure enhances the formation of ironstone (fericate); which can result in soil degradation by capping potential arable soils. Geologically, the area is underlain by the neo-proterozoic Voltaian sedimentary formation of Ghana (Figure 2). The area is predominantly (about 60% of the study area) underlained by the Panabako group, which consists of medium-grained quartzites with distinctive structural feature of cross-bedding. South of the Panabako group is a relatively thin belt of the Kodjari formation, which consists of silliceous tuff and fine-grained laminated arkosicsanstones. South of the Kodjari formation is the Oti-Penjari group, which consists of mudstones and siltstones, weakly micaceous thin-bedded arkosic and lithic sandstones (Griffiths et al., 2002).

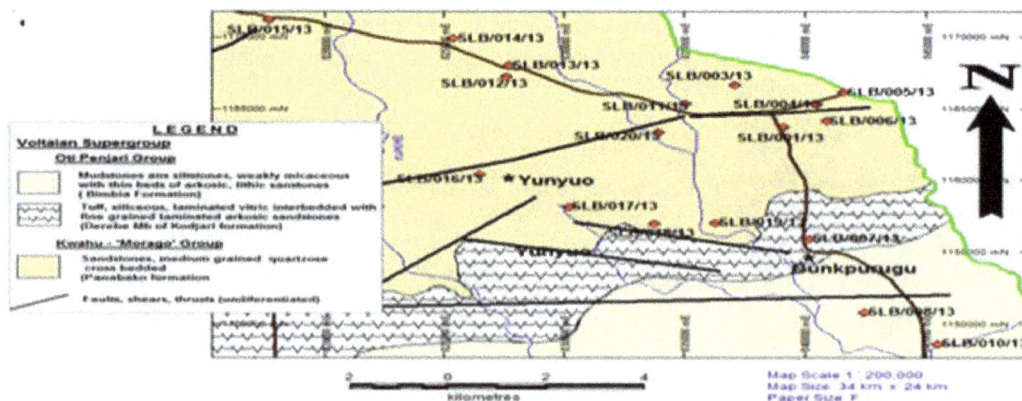

Figure 2. Geological map of the study area showing sampling sites

## 2. Materials and Methods

### 2.1 Water Samples Collection

A total of nineteen (19) groundwater samples (Figure 1) were collected from identified boreholes in 0.5 litre polythene bottles and their respective geographical locations (i.e. elevation, longitude and latitude) were measured. Sampling was carried out in accordance with protocols described by (Claasen, 1982) and (Barcelona et al., 1985). Sampling bottles were initially conditioned by washing with detergent, then with ten per cent (10%) nitric acid, and finally rinsing several times with distilled water. This was carried out to ensure that the sample bottles were free from contamination, which could affect the concentrations of various ions in the groundwater samples. Boreholes were pumped for at least five minutes to purge the aquifer of stagnant water so as to acquire fresh samples for analysis. Hand-held syringes fitted with a filter head with 0.45 $\mu$m cellulose filter membrane were used to filter the water samples in the field. Two samples were collected at each site; one was acidified by adding 2% of concentrated nitric acid ($HNO_3$). The acidified water samples were used for metal analysis while the non-acidified water samples were used for physico-chemical analysis. The sampled waters were tightly capped and preserved in an ice-chest at a temperature of 4 °C and transported to the laboratory of the Ghana Atomic Energy Commission at Kwabenya within the shortest time for analysis.

*2.2 Sample Analysis*

2.2.1 Physico- Chemical Parameters

Unstable hydrochemical parameters such as electrical conductivity (EC), pH and alkalinity were measured in situ (in the field) immediately after collection of samples, using a WTW field conductivity meter model LFT 91, WTW field pH meter model pH 95 and a HACH digital titrator respectively, that had been calibrated before use. Major ions such as Sodium ($Na^+$) and Potassium ($K^+$) were analyzed in the laboratory using the flame photometer. Calcium ($Ca^{2+}$) and Magnesium ($Mg^{2+}$) were analyzed using the AA240FS Fast Sequential Atomic Absorption Spectrometer. The ICS-90 Ion Chromatograph (DIONEX ICS -90) was employed in the analysis of Chloride ($Cl^-$), Fluoride ($F^-$), Nitrate ($NO_3^-$), and Sulphate ($SO_4^{2-}$). Phosphate ($PO_4^{3-}$) was determined by the ascorbic acid method using the ultraviolet spectrophotometer (UV-1201). A multipurpose electronic DR/890 Colorimeter was used to measure the color, turbidity, total dissolved solids and a HACH SEN 523 pH meter was used to measure the pH and temperature. An electronic HACH SEN ION 5 conductimeter was used to measure the conductivity, salinity and total dissolved solids of all the samples.

2.2.2 Heavy Metal Analysis

5 ml of each acidified water sample was measured and 6 ml of nitric acid, 3 ml of HCl and 5 drops of hydrogen peroxide ($H_2O_2$) were added for acid digestion and placed in a milestone microwave lab station ETHOS 900. The digestate was then assayed for the presence of Zinc (Zn), lead (Pb), Copper (Cu), Chromium (Cr) and Cobalt (Co) using VARIAN AAS240FS Atomic Absorption Spectrum in an acetylene-air flame. Arsenic (Ar) and Mercury (Hg) were determined using argon-air flame.

## 3. Results

Table 1. Statistical summary of physico-chemical parameters from sampled wells

| Parameter | Wet Season | | | Dry Season | | | WHO, 2008 |
|---|---|---|---|---|---|---|---|
| | Min | Max | Mean | Min | Max | Mean | |
| pH | 5.77 | 7.36 | 6.76 | 4.96 | 8.3 | 6.97 | 6.5-8.5 |
| Temperature | 23.5 | 25.9 | 24.49 | 23.5 | 25.9 | 24.49 | n.a |
| Conductivity | 57.9 | 830 | 413.46 | 47.6 | 709 | 356.86 | 250 |
| TDS | 34.8 | 502 | 248.83 | 23.9 | 355 | 178.51 | 1000 |
| TSS | 0 | 75 | 13.57 | 0 | 75 | 13.57 | n.a |
| Turbidity | 1 | 73 | 8.81 | 0 | 96 | 13.24 | 5 |
| Colour | 0 | 30 | 13.09 | 0 | 30 | 3.29 | 15 |
| Alkalinity | 16 | 302 | 163.21 | 24 | 352 | 183.14 | 1000 |
| Hardness | 20 | 214 | 97.33 | 10 | 218 | 87.43 | 500 |
| Calcium | 2.4 | 49.7 | 18.69 | 3.2 | 77 | 26.4 | 200 |
| Magnesium | 2.4 | 36.9 | 12.96 | 0 | 1.3 | 0.4 | 150 |
| Sodium | 3.9 | 150 | 51.44 | 2.1 | 78.1 | 30.9 | 200 |
| Potassium | 1.3 | 8.3 | 4.1 | 2.9 | 20.5 | 9.7 | 30 |
| Chloride | 6 | 67.5 | 19.65 | 3.2 | 48.1 | 15.6 | 250 |
| Bicarbonate | 19.5 | 368 | 199.62 | 29.3 | 429.2 | 212.6 | n.a |
| Sulfate | 1.6 | 16.7 | 6.46 | 0.4 | 32.9 | 8.4 | 400 |
| Nitrate | 0.01 | 118.7 | 20.69 | 0 | 0.2 | 0.1 | 10 |
| Phosphate | 0.01 | 0.14 | 0.05 | 0 | 0.5 | 0.1 | 30 |
| Fluoride | 0.1 | 0.13 | 0.34 | 0.1 | 0.6 | 0.3 | 0.5-1.5 |

As shown in Table 1, the statistical summary of the results of the physico-chemical analysis of water samples from nineteen (19) boreholes obtained from the study area during the two climatic regimes (wet and dry seasons), show that the pH of the groundwater samples ranged from 5.77 to 7. 36 with a mean of 6.76, and 4.96 to 8.3 with a mean of 6.96, EC ranged between 57.9-830 with a mean of 413.46 and 47.6-709 with a mean of 356.86µS/cm, Turbidity ranged from 1-73 with a mean of 8.81 and 0-96 with a mean of 13.24NTU for wet and dry seasons, respectively. All anions and cations were within permissible levels for human consumption except nitrate. Nitrate concentration in the wet season varied from 0.01 to 118.7 mg/l with a mean of 20.69 mg/l.

Table 2. Statistical summary of trace metals of sampled wells in the study area

| Parameter | Wet Season | | | Dry Season | | | WHO, 2008 |
| --- | --- | --- | --- | --- | --- | --- | --- |
| | Min | Max | Mean | Min | Max | Mean | |
| Zinc | 0.007 | 0.077 | 0.028 | 0.008 | 0.092 | 0.046 | 5 |
| Copper | 0.013 | 0.114 | 0.045 | 0.044 | 0.044 | 0.044 | 1 |
| Cobbolt | 0.048 | 0.144 | 0.096 | 0.048 | 0.140 | 0.096 | n.a |
| Arsenic* | | | | | | | 0.01 |
| Chromium | 0.028 | 0.064 | 0.04 | 0.028 | 0.064 | 0.04 | n.a |
| Cadmium | 0.001 | 0.01 | 0.003 | **0.064** | **0.12** | **0.092** | 0.003 |
| Mercury* | | | | | | | 0.05 |
| Lead | | | | **0.16** | **0.78** | **0.34** | 0.001 |

*measured concentrations were below detection limit.

As shown in Table 2, the heavy metals detected during the analysis were zinc, copper, cobbolt, chromium and lead. Arsenic and mercury were not detected at all while lead was detected only during the dry season. The concentrations of lead in sampled waters in the dry season ranged from 0.16-0.78 mg/l with a mean value of 0.34 mg/l. The ranges of concentrations of cadmium, chromium, cobbolt, copper and zinc in wet and dry seasons were 0.001-0.0l0 & 0.064-0.120, 0.028-0.064 & 0.028-0.064, 0.048-0.144 & 0.048-0.140, 0.013-0.114 & 0.044-0.044, and 0.007-0.070 & 0.008-0.092mg/l respectively, while their mean concentrations were 0.003 & 0.092, 0.04 & 0.04, 0.096 & 0.096, 0.045 & 0.044, and 0.028 & 0.046 mg/l respectively.

## 4. Discussion

### 4.1 Physico-Chemical Parameters

According to the classification based on (Hounslow, 1995), three (3) wells representing 15.8% were moderately acidic (i.e. pH of 4-6.5) while the remaining sixteen (16) wells were neutral (pH 6.5-7.8) in the wet season. On the other hand, during the dry season, four (4) wells, representing 21% were moderately acidic (pH 4-6.5), eleven (11) out of nineteen samples representing 58% were neutral (pH 6.5-7.8) while four (4) out of nineteen (19) were moderately alkaline (pH 7.8-9). However, majority of the wells (79%) and (89.5%) in the dry and wet seasons respectively, had pH values falling within the acceptable range (pH 6.5-8.5) for human consumption (WHO, 2008). Freeze and Chery (1979), classified groundwater on the basis of TDS as fresh when values range 0-1 000 mg/L as fresh, 1 000-10 000 mg/L as brackish, 10 000-100 000 mg/L as saline and greater than 100 000 mg/L as brine. In this study, TDS of all the samples ranged between 34.8-502 mg/l and 23.9-355 mg/l in wet and dry seasons respectively, which obviously were less than 1 000 mg/l implying fresh waters in all the wells. Electrical conductivity (EC) of groundwater ranged from 47.6 µS/cm to 709 µS/cm, with a mean of 356.86 µS/cm for the dry season while it ranged from 51.9 µS/cm to 830 µS/cm with a mean of 413.46 µS/cm in the wet season. Approximately 84.2% and 79% of the analysed samples in the wet and dry seasons respectively had EC values above the (WHO, 2008) maximum allowable limit of 250 µS/cm. Water hardness is the soap consuming property of water caused by the presence of alkali earth metals (calcium and magnesium), and to a lesser extent, the salts of other metals such as iron and manganese. Where these metals combine with bicarbonates they form temporal hardness which can easily be removed by heat, but when in combination with nitrates, they form permanent hardness which is not easily removable by heat (Freeze & Cherry, 1979). According to McGoowan (2000), groundwater with hardness between 0-60 mg/l is considered as soft; between 61-120 mg/l as moderately hard; 121-180 mg/l as hard and greater than 181 mg/l as very hard. In this study, values of hardness of all samples were within permissible limit (< 500 mg/l) (WHO, 2008). Values ranged from 20-214 mg/l with a mean of 93.7 mg/l and 10-218 mg/l with a mean of 87.43 mg/l for samples in wet and dry seasons, respectively.

Generally, hardness in the study area could be described as soft to moderately hard as approximately 69% of the samples had values ranging between 0-120 mg/l while 31% were hard to very hard.

The EC and TDS values along with relatively low-medium total hardness (TH) values suggest a low-medium mineralized soft-moderately hard fresh groundwater system. The total concentration of alkaline earth metal ions, such as calcium and magnesium are determinants of the hardness of water during the wet season as depicted by the relatively strong correlations (0.848 and 0.928) with Total Hardness respectively, as seen from the cross plot of $Ca^{2+}$ + $Mg^{2+}$ (mg/L) versus TH (mg/L) with correlation coefficient of 0.984 (Figure 3a). However, as can be observed from Figure 3b, $Ca^{2+}$ appeared to be the major determinant of total hardness (with a correlation of 0.98), which almost equals the correlation between total hardness and the combined alkali earth metals ($Ca^{2+}$ + $Mg^{2+}$). This must have been due to higher mobility of $Mg^{2+}$ and a possible cationic exchange between $Na^+$ and $Mg^{2+}$ and could be seen from the increased correlation of total hardness with $Na^+$ from 0.107 to 0.463. The increased correlation with $Na^+$ could be a possible cause of the decrease in the mean total hardness of water from 97.33 mg/l to 87.43 mg/l (Table 1). Thus, the possible source of hardness in the groundwaters could be as a result of geogenic introduction of $Ca^{2+}$ and $Mg^{2+}$ ions into water by leaching of rock minerals within the aquifer, which agrees with the findings of Talabi (2013).

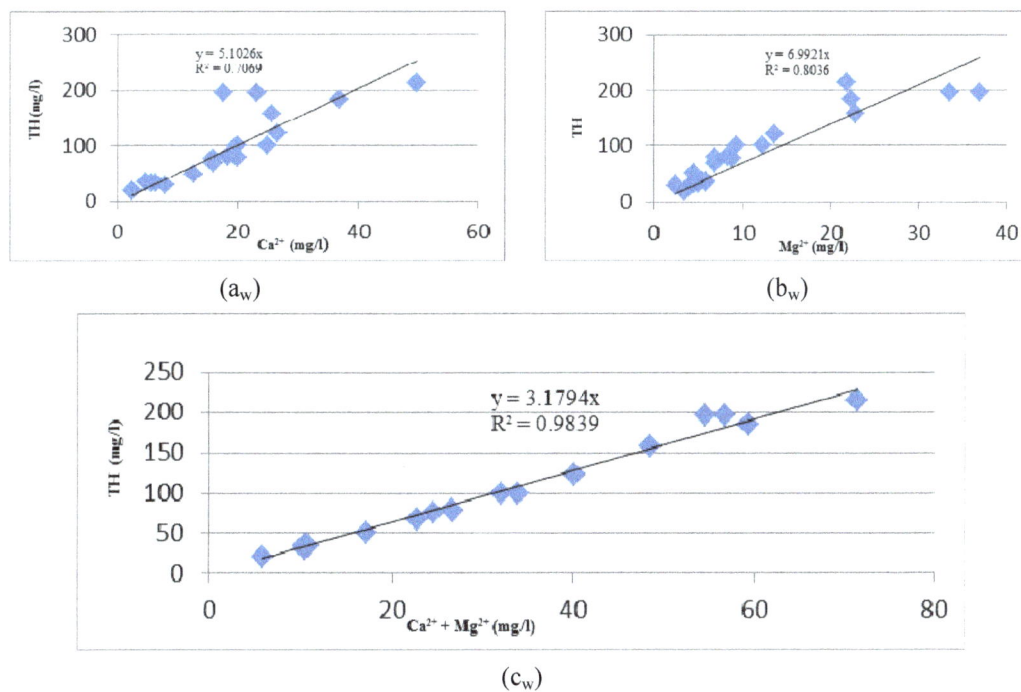

(a_w)

(b_w)

(c_w)

Figure 3a. Plots of Alkali earth metals with total hardness for wet season water samples

(a$_d$)

(b$_d$)

(c$_d$)

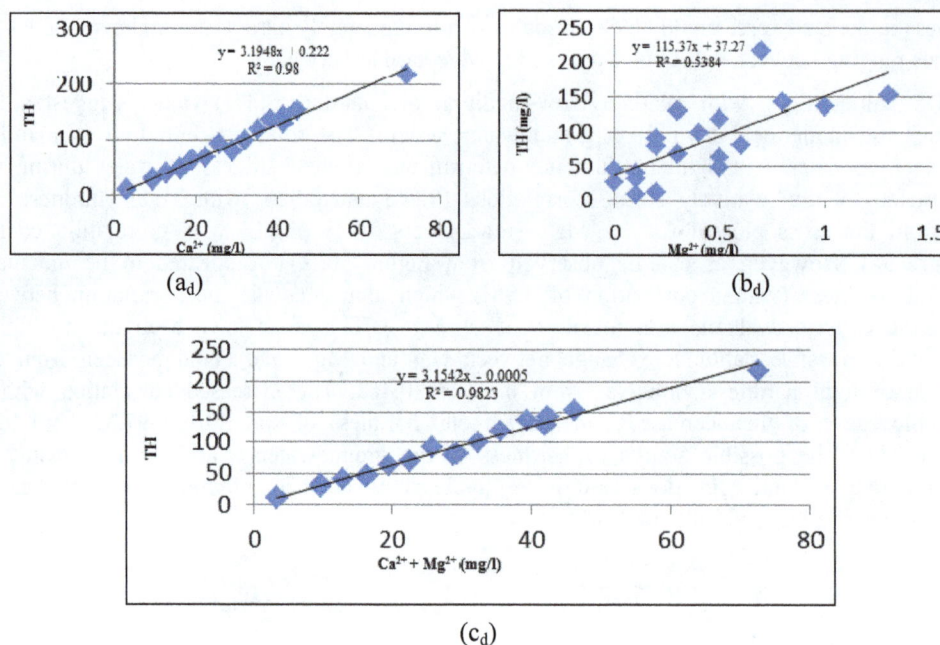

Figure 3b. Plots of Alkali earth metals with total hardness for dry season water samples

On the other hand, with the exception of high nitrate concentrations in three wells, namely Tambing (85.1 mg/l), Naajong-2 (23.04 mg/l) and Yunyo (118.7 mg/l), which exceeded the (WHO, 2008) permissible limit of 10 mg/l, all the anions analysed fell within the acceptable limit for human consumption. The presence of high nitrate concentrations in the three relatively shallow wells (depth of aquifers range between 18-30 m) in the wet period could be an indication of a possible contamination from the extensive use of chemical fertilizers (NPK-fertilizers) for farming during the raining (wet) period, since rain-fed agriculture is the major occupation of the inhabitants in the study area, and this agrees with Yidana & Yidana (2009) who indicated that most nitrate water comes from either agro-chemical or from industrial and also organic sources and findings of Hem ( 2002). This could be very true due to the realization that nitrate concentrations of all water samples in the dry season were found to be extremely very low, ranging from 0-0.2 mg/l with a mean of 0.1 mg/l as there is little or no rain during this period and no farming takes place. This clearly suggests that groundwater contamination of nitrate during wet (rainy) season is due to infiltrating rain water which is contaminated with chemical fertilizers applied to crops by farmers (anthropogenic). Thus, the consumption of groundwater during the wet period by the inhabitants in the three affected communities exposes the inhabitants to such diseases as methaemoglobinemia in children (WHO, 2008).

*4.2 Hydrochemical Facies*

From Table 1, it was observed that all the cations irrespective of the climatic period fell within the acceptable limits (WHO, 2008) for human consumption. Sodium appears to be the dominant cation followed by calcium while bicarbonate is the dominant anion irrespective of the climatic period. The concentrations of sodium varied from 3.9 to 150 mg/l with a mean of 51.44 mg/l while potassium was least ranging from 1.3-8.33 mg/l with an average of 4.1 mg/l with the order of dominance of major cations in the wet and dry periods being Na>Ca>Mg>K and Na>Ca>K>Mg, respectively. On the other hand, bicarbonate appeared to be the major anion irrespective of the climatic regime of the study area. Its concentration in the wet period varied from 19.5 to 368 mg/l with a mean of 199.62 mg/l while in the dry period, the concentrations of bicarbonate ranged from 29.3-429.2 mg/l with a mean of 212.6 mg/l. The orders of dominance of the major anions in the wet and dry periods were $HCO_3$>$NO_3$>Cl>$SO_4$>F>$PO_4$ and $HCO_3$>Cl>$SO_4$>F>$PO_4$>$NO_3$, respectively. The major ionic compositions of groundwaters in the dry and wet seasons of the study area are presented in the Piper trilinear diagram (Piper, 1944) in Figure 4. From Figure 4a and 4b, it can be observed that climate variability has profound influence on the cations composition of groundwaters in the study area. The cationic distribution of water samples obtained during the wet period appeared more mixed (Figure 4b) while water samples obtained during the dry period (Figure 4a) appeared to be depleted in magnesium. The major water types of the groundwaters within the study area were Na-Ca-Mg-$HCO_3$ and Na-Ca-K-$HCO_3$ for the wet and dry seasons, respectively.

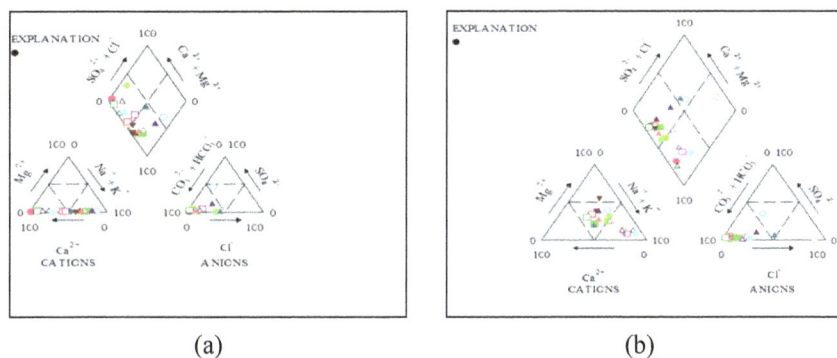

(a)                                      (b)

Figure 4. Piper plot of major ions of groundwaters in the dry (a) and wet (b) seasons

### 4.3 Heavy Metals

From Table 2, the mean concentrations of all the heavy metals analysed in the sampled waters during the raining and dry seasons fell within the acceptable limits of WHO (2008) standards with the exception of cadmium and lead. Generally, the orders of dominance of the concentration of heavy metals in the analysed groundwater samples were Pb>Co>Cd>Zn>Cu>Cr and Co>Cu>Cr>Zn>Cd>Pb in the dry and wet seasons, respectively. According to van Assche (1998) apart from the natural dissolution of cadmium from rocks and soils into groundwater, much could be released into the environment from anthropogenic activities which may include but not limited to the use of agro-chemicals (like phosphate fertilizers), burning of coal, manufacturing of iron, steel and cement, and disposal of waste. Cadmium may enter water and soil in waste from industries or waste disposal plants, or from leaching from landfill sites. The study area is characterised with long periods of bushfires during every dry season as well as the use of agrochemicals such as phosphate fertilizers, weedicides and pesticides which have cadmium as trace constituents. The concentrations of cadmium ranged from 0.001 to 0.12 mg/l with a mean of 0.092 mg/l and 0.001 to 0.01 mg/l with a mean of 0.0033 mg/l for dry and wet seasons, respectively. Analyses of samples derived in the dry season showed that two (2) well had Cd concentration above recommended level while in the wet season; five well had higher Cd concentration. The high Cd concentrations in the identified wells could therefore be due to the intense use of phosphate fertilizers in the study area and possibly from burnt crops and plant which had assimilated Cd which then may be leached during the rainy season to the groundwaters in the aquifers beneath the ground. The ingested cadmium into the human body may expose them to adverse health effects (Forstner & Wittmann, 1983). Ingestions of high doses of cadmium can affect the kidney, lungs and bones of humans. According to Green (2011), the human kidney is the main organ commonly affected by cadmium. Cadmium is known to be able to accumulate in the human up to 20-30 years before producing adverse health effects on the respiratory system as well as the weakening of the bones.

The concentrations of lead in all the sampled waters in the study area during the rainy season were below detection limit, which agrees with the findings of Yidana and Yidana (2009). They concluded that Pb concentrations of groundwaters in the southern part of the voltaian basin of Ghana were generally very low. However, analyses of results of water samples obtained during the dry period showed that concentrations of Pb in all the boreholes exceeded the WHO (2008) permissible limit of 0.01 mg/l. Pb values ranged from 0.016 to 0.77 mg/l with a mean of 0.335 mg/l which clearly contradicts the conclusion established by Yidana and Yidana (2009). The presence of high Pb in groundwaters in the dry periods is rather a disturbing situation as far as exposure to adverse health condition to inhabitants is concerned. This is due to the fact that the period of dry season in the study area is about seven months with groundwater serving as the only source of potable water for the inhabitants. The study area is also located in the hottest parts of Ghana (the three northern regions) and the consumption of water is therefore very intense during the dry period. According to WHO (2008), Green (2011) and Reeves and Vanerppool (1997) the ingestion of water contaminated with Pb at concentrations above the permissible limit may expose humans to cause damage to the brain, kidneys, the central nervous system, the cardiovascular system and the immune system. According to CDC (2000), the most vulnerable group to the effects of Pb ingestion are children and Pb has been shown in several instances to permanently reduce the cognitive capacity of children even at extremely low levels of exposure. Lead is generally known to occur naturally and also exist in old pipes, lead-combining solders, exhaust from motor fumes containing leaded gasoline as well as from industrial sources such as smelters and lead manufacturing and recycling industries and waste sites such as contaminated landfill sites (Jin et al., 2006). According to van Assche (1998), Pb and Cd concentrations in freshwater can be impacted by agricultural activities such as the use of agro-chemicals in farming. Anim-Gyampo et al. (2013), concluded that chemical fertilisers such as phosphate fertilisers, weedicides and pesticides which are massively being utilized in dry-season irrigation farming activities in northern Ghana contain heavy metals such as Pb, Cu, Zn and Cd as trace components, therefore may be leached into the underlying groundwater by infiltrating rainwater and irrigation water. The low concentration levels of Pb in the raining season could be due to dilution by rainwater which readily recharges the groundwater system due to existence of preferred pathways while during the long dry season, no recharging groundwater exist, and coupled with unfavourable climatic conditions such as extremely high evapotranspiration and high temperature, the abstraction frequency is huge. These factors could enhance the concentrations of heavy metals per unit volume of water within the aquifers as could be observed from Table 2. Notwithstanding the above enumerated anthropogenic activities, there is the possibility of increased dissolution of heavy metals into the groundwater system during the dry season by virtue of groundwater-rock interaction as a result of reduced groundwater flow coupled with increased ambient temperatures which can enhance the dissolution processes.

### 4.4 Chemical Quality of Water for Drinking Purposes

Water Quality Index (WQI), which is a quantitative means of evaluating the quality of water, offers a useful representation of overall quality of water. According to Sahu and Sikdar (2008), WQI is defined as a reflection of composite influence of individual quality characteristics on the overall quality of water. WQI is used to assess water quality trends for management purpose. The estimation of WQI requires the selection of parameters of great importance since the selection of many number of parameters widen the water quality index. The importance of selecting parameters depends on the intended use. In this study, ten physico-chemical parameters namely pH, Electrical Conductivity (EC), Zinc, Lead, Cadmium, Fluoride, Chloride, Sulphate, Sodium, and Calcium were used to calculate WQI. According to Yidana and Yiadana (2009), the highest weight of five is assigned to parameters which have the major effects on water quality. In this study, lead, nitrate, and fluoride were assigned the highest weight of 5 because of their importance in the water quality assessment in that order as shown in Table 2. The assessment of the suitability of groundwaters in the study area for human consumption throughout the whole year which is characterised by two climatic seasons (i.e. raining and dry seasons) was achieved by estimating the WQI using Equations (1), (2) and (3) and comparing the results to the criteria defined by Sahu and Sikdar (2008) and using influential parameters defined by Yidana and Yiadana (2009). Weights (W) were assigned to each influential parameter based on their perceived effects on primary health. The Relative Weight ($W_i$) of each influential parameter was estimated using Equation (1);

$$W_i = \frac{w_i}{\sum w_i}$$

(1)

where$\sum w_i$ = sum of the weights of all parameters and the weights are shown in Table 3 below.

Table 3. Assigned weights and estimated relative weights of influencial parameters (wet season)

| Parameters | WHO (2008) | Wet Season | | Dry Season | |
|---|---|---|---|---|---|
| | | Wi | W | Wi | W |
| pH | 7.5 | 4 | 0.13 | 4 | 0.14 |
| SO₄⁻ | 250 | 3 | 0.09 | 3 | 0.1 |
| Cl⁻ | 250 | 3 | 0.09 | 3 | 0.1 |
| F⁻ | 1.5 | 5 | 0.15 | 5 | 0.17 |
| Ca | 75 | 2 | 0.06 | 2 | 0.07 |
| Mg | 30 | 2 | 0.06 | 2 | 0.07 |
| Na | 200 | 2 | 0.06 | 2 | 0.07 |
| Cd | 0.003 | 3 | 0.09 | 3 | - |
| Cu | 2 | 5 | 0.06 | 5 | |
| Zn | 5 | 2 | 0.06 | 2 | 0007 |
| NO₃⁻ | 50 | 5 | 0.15 | 5 | - |
| Pb | 0001 | 5 | - | 5 | 0.17 |
| | Sum | 41 | 1.0 | 41 | 0.89 |

Determination of WQI in this study was done using Equations (2), (3) and (4) in the proceeding steps. In the third step, a quality rating scale, $q_i$, was computed for each parameter using Equation 2 below;

$$qi = (Ci / Si) \times 100 \qquad (2)$$

where $C_i$ and $S_i$ respectively, refer to the concentration and the WHO standard for each parameter, in mg/l.

The water quality sub index for each influential parameter ($SI_i$) was then calculated using Equation 3 below;

$$SI_i = q_i \times W_i \qquad (3)$$

The water quality index (WQI) was estimated using Equation (4) below;

$$WQI = \sum SI_i \qquad (4)$$

Table 4 below shows the results and Figures 4 & 5 show spatial variations of the estimated water quality indices for groundwater in the study area during the rainy and dry seasons;

Table 4. Values of Water Quality Index (WQI) of groundwaters in the Wet Season

| Community | Wet Season | | Dry Season | |
|---|---|---|---|---|
| | WQI | Classification | WQI | Classification |
| Bunkpurugu | 18.85 | Excellent Water | 385.94 | Unsuitable water |
| Poiga | 36.36 | Excellent Water | 104.39 | Poor Water |
| Naanyar | 22.76 | Excellent Water | 100.99 | Poor Water |
| Sakbouk | 21.20 | Excellent Water | 782.37 | Unsuitable water |
| Paknaatik | 20.21 | Excellent Water | 41.30 | Excellent Water |
| Toogeng | 20.87 | Excellent Water | 71.44 | Good Water |
| Gbankpurugu | 22.38 | Excellent Water | 1339.9 | Unsuitable water |
| Naaban | 23.61 | Excellent Water | 609.44 | Unsuitable water |
| Tambing | 13.44 | Excellent Water | 283.74 | Very Poor Water |
| Jilick 1 | 30.18 | Excellent Water | 704.97 | Unsuitable water |
| Naajong 1 | 31.04 | Excellent Water | 553.83 | Unsuitable water |
| Binde | 20.90 | Excellent Water | 820.53 | Unsuitable water |
| Naaniac | 19.56 | Excellent Water | 1000.51 | Unsuitable water |
| Jimbale | 27.85 | Excellent Water | 647.87 | Unsuitable water |
| Mangor | 18.65 | Excellent Water | 646.30 | Unsuitable water |
| Bunbuna | 23.98 | Excellent Water | 813.30 | Unsuitable water |
| Jilick 2 | 31.20 | Excellent Water | 126.31 | Poor Water |
| Naajong 2 | 82.56 | Good Water | 554.82 | Unsuitable water |
| Yunyoo | 202.34 | Very Poor Water | 786.97 | Unsuitable water |

The assessment of water quality based on WQI is usually achieved by using the criteria developed by Yidana and Yiadana (2009), which defines WQI values of less than 50 as excellent water, between 50-100 as good water, 100-200 - poor water, 200-300 - very poor water and WQI values above 300 as unsuitable for human consumption. It can be observed from Table 4 and Figure 5a that groundwater quality in the study area was generally excellent for human consumption during the wet season except one well in Yunyo community that had an estimated WQI of 202.34, far exceeding 100.

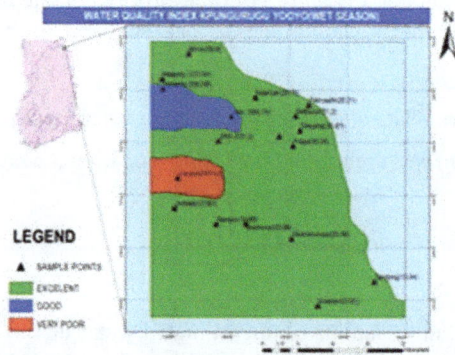

Figure 5a. Variation in water quality (wet)    Figure 5b. Variation in water quality (dry)

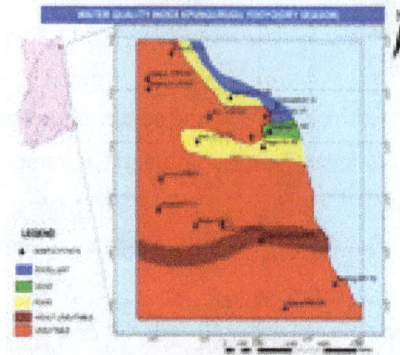

On the contrary, there is a dramatic change in the quality of groundwaters during the dry season (Figure 5b). With the exception of two wells located at Paknaatik and Toogeng with WQI below 100 (41.3 - excellent and 71.4 - good water) the remaining 17 wells had quality being poor water, very poor and unsuitable. It is observed from Table 4 that 13 out of the 19 sampled wells, representing about 68.4% had their water quality changing from excellent water to unsuitable water; 3 wells changed from excellent water to poor water whilst one well located in Yunyo community changed from very poor to unsuitable. This observation clearly shows that among other factors controlling groundwater in the Bunkpurugu-Yunyo District of Ghana, climatic conditions (rainfall) plays a critical role. Only one community (i.e. Paknaatik) has excellent water quality throughout the year with insignificant effect from climate. The seemingly deterioration in the quality of groundwaters in the wells during the dry season could be due to elevations in the concentrations of Pb in almost all the sampled wells from below detection levels in the wet seasons up to 0.776 mg/l in the dry season. The very poor quality of groundwater in Yunyo community could be due to the extremely high levels of nitrate in the wells.

## 5. Conclusion

The water quality of nineteen (19) boreholes in the Bunkpurugu-Yunyo district of north-eastern Ghana had been assessed from the analyses of their hydrochemistry and the use of WQI. Generally, the physico-chemical parameters of most groundwater samples fell within WHO acceptable limits with exception a few parameters, namely Turbidity and Nitrates. Two wells in Tambing and Binde representing 10.5% had turbidity values above WHO (2008) guideline values during the wet season while five wells in Tambing, Binde, Bunkpurugu, Sakbouk and Naaniac representing 21% had turbidity values exceeding acceptable limit for potable drinking water. Hardness of groundwaters in the study area varied significantly with about 36.8% (majority) being soft in both seasons while 5.3% and 21.1% were very hard during the dry and wet seasons (see Table 3). Hardness of waters could be predominantly temporal, due to the dominance of $HCO_3^-$ as the major anion in all samples irrespective of the climatic season while permanent hardness is expected on a minor scale due to the presence of nitrates and sulphates. Majority (84.2%) of the groundwaters are neutral while all samples were fresh (TDS < 1000 mg/l) making the waters generally acceptable for human consumption. The ionic composition and relative dominance varied with prevailing climatic conditions, with $Na^+$ and $HCO_3^-$ being the dominant cation and anion in the raining season $Ca^{2+}$ and $HCO_3^-$ are the dominant ions in the dry period with corresponding water types of Na-Ca-Mg-K-$HCO_3$ and Ca-Mg-(Na-K)-$HCO_3$ for wet season and dry seasons respectively. The concentrations of major cations are in the order of Na>Ca>Mg>K. and Ca>Na>K>Mg while the in the case of the anions, the order was $HCO_3$>$NO_3$>Cl>SO4>F>$PO_4$ and $HCO_3$>Cl>$SO_4$>F>$PO_4$>$NO_3$ respectively, for wet and dry seasons. The orders of heavy metals in the groundwater sampled in the study area were Pb>Co>Cd>Zn>Cu>Cr and Co>Cu>Cr>Zn>Cd>Pb in the dry and wet seasons respectively. The concentrations of heavy metals were generally acceptable for drinking water except Cd in four samples which had value slightly above the WHO (2008) and recommended limit of 0.003 mg/l in the wet season while the concentrations of Pb in all the nineteen samples were above the recommended value of 0.01 mg/l and Cd values in two samples exceeded the recommended of 0.003 mg/l in the dry season. The high Pb concentrations in all the water samples could be due to the intensive use of agro-chemicals such as phosphate fertilizers, weedicides and pesticides which contain Pb and Cd as traces but the low mobility of Pb in groundwater could account for the very high concentrations as compared to Cd which is relatively much mobile. The occurrence of the high Pb in the dry season presents a serious health issues due to the fact that groundwater is the only available potable water for human consumption in the study area with little or no alternative. Thus, the inhabitants stand the risk of kidney damage, weakening of the nervous system and cardiovascular infections. The public health directorate of the Ministry of Health in Ghana must as a matter of necessity carry out monitoring of the health of inhabitant in the study area to ascertain the current health status of the people and also sensitise them on the potential health implications that would likely affect consumers. The quality of groundwaters in the study area and its surroundings must continuously be monitored and if conditions do not improve public awareness must be created to alert inhabitants of the potential health implication associated with the consumption of groundwater in the study area.

## Acknowledgements

The authors would like to greatly acknowledge the contribution of Salamatu Bannib Lambon for undertaking the water sampling for laboratory analysis. We are also indebted to the Environmental Chemistry Department of the Ghana Atomic Energy Commission (GAEC) for offering equipment support and allowing the team to carry out water quality analysis of samples used in this study.

## References

Anim-Gyampo, M., Kumi, M., & Zango, M. S. (2013). Heavy metal concentrations in some selected fish species inTono Irrigation Reservoir in Navrongo, Ghana. *Journal of Environmental and Earth Science, 3*(1), 109-119.

Anim-Gyampo, M., Zango, M. S., & Apori, N. (2012). The Origin of Fluoride in GroundwatersofpaleozoicSedimntary Formations of Ghana- A preliminary Study in Gushiegu District. *Research Journal of Environmental and Earth Science, 4*(5), 546-552.

Anongura, R. S. (1995). Fluorosis survey of Bongo District, Upper East Region.*A report submitted to the Upper Region Community Water Project* (unpublished).

Anongura, R. S., Louw, A. J., & Chikte, U. M. E. (2004). Dental fluorosis in a district of Ghana, West Africa. *J. Dental Res., 83*(Special Issue B): 27 (SA Division IADR), 2003.

Apambire, W. B., Hess, J. W., & Mitchel, F. A. (1997). Geochemistry, Genesis and Health Implications of Fluoriferous Groundwater in the Upper Regions of Ghana. *Environmental Geology, 33*(1), 13-24. http://dx.doi.org/10.1007/s002540050221

Barcelona, M., Gibb, J. B., Helfrich, J. A., & Garske, E. E. (1985). Practical Guide for Groundwater Sampling. *Illinois State Water Survey ISWS*, Contract Report 374.

CDC. (2000). Blood lead levels in young children- United States and Selected States1996-1999. *MMWR, December, 49*(50), 1133-1137.

Claasen, H. C. (1982). Guidelines and Techniques for Obtaining Water Samples that Accurately Represent the Water Quality of an Aquifer. *U.S. Geological Survey Open File Report* (pp. 82-1024).

CWSA, (2009). *The Challenge of fluoride in drinking water- The tragic story of Nayorigu.* Retrieved from http//www.cwsagh.org

Dickson, K. D., & Benneh, G. (1985). *A New Geography of Ghana.* London: Longmans Group Limited.

Forstner, U., & Wittmann, G. T. W. (1983). *Metal pollution in the aquatic environment* (p. 486). Berlin:Springer-Verlag.

Freeze, R. A., & Cherry, J. A. (1979). *Groundwater.* Englewood UK: Prentice-Hall.

Green, N. (2011). *Effects of lead indinking water.* Health and Disability. Retrieved January 10, 2011, from http://www.disabled-world.com/health/lead-water.php

Griffis, J. R., Barning, K., Agezo, F. L., & Akosah, K. F. (2002). *Gold Deposits of Ghana* (p. 362). Ontario: Graphic Evolution in Barrie.

Hem, J. D. (2002). *Study and interpretation of the chemical characteristics of natural water.* US Geochemical Survey Water Supply, Paper 2254.

Hounslow, A. W. (1995). *Water Quality Data: Analysis and Interpretation.* Floriida, U.S.: CRC Press Inc.

Jin, Y. R., Liao, Y. J., Lu, C. W., Li, G. X., Yu, F., & Zhi, X. P. (2006). Health Effects in Children aged 3-6 years induced by environmental lead exposure. *Ecotox. Environ. Safe, 63*(2), 313-317. http://dx.doi.org/10.1016/j.ecoenv.2005.05.011

McGoowan, W. C. (2000). *Water processing residual, commercial, light industrial* (3rd ed.). Lisle, IL, 2000, Water quality Association.

Ministry of Water Resources, Works and Housing-Ghana (MWRWH-NWP). (2007). *National Water Policy* (p. 69).

Obeng, H. (2000). Soil Classification in Ghana. *Centre for Policy Analysis (CEPA): Selected Economic Issues, 3.* Accra, Ghana.

Piper, A. M. A. (1944). Graphic procedure in the geochemical interpretationof water analyses. *Trans. Am. Geophy. Union, 25*, 914-928. http://dx.doi.org/10.1029/TR025i006p00914

Rashed, M. N. (2001). Monitoring of environmental heavy metals in fish from Nasser Lake, Egypt. *Environ. Int., 27*, 27-33. http://dx.doi.org/10.1016/S0160-4120(01)00050-2

Reeves, P. G., & Vanderpool, R. A. (1997). *Cadmium burden of men and women.* WHO report.Regular Consumption of confectionary sunflower kernels containing a natural abundance of cadmium. *Environ. Health Perspective, 105*, 1098-1104. http://dx.doi.org/10.1289/ehp.971051098

Sahu, P., & Sikdar, P. K. (2008). Hydrochemical framework of the aquifer in and around East Kolkata wetlands,West Bengal. *India Environ Geol, 55*, 823-835. http://dx.doi.org/10.1007/s00254-007-1034-x

Talabi, A. O. (2013). Hydrochemistry and Stable Isotopes of ($\delta18O$ and $\delta2H$) Assessment of Ikogosi Spring Waters. *American Journal of Water Resources, 1*(3), 25-33.

Todd, D. (1980). *Groundwater Hydrology* (2nd ed., p. 535). New York: Wiley.

Van Assche, F. J. (1998). A stepwise Model to quantify the relative contribution of different environmental sources to human cadmium exposure. *Paper presented at NiCad. 98, Prague, Czech Republic, September 21-22*, 1998.

WHO. (2008). *Guideline for drinking water quality, 2, Health Criteria and other supporting information.* Geneva; WHO.

Yidana, S. M., & Yidana, A. (2009). Assessing water quality using water quality index and multivariant analysis. *Environ Earth Sci, 59*, 1461-1473. http://dx.doi.org/10.1007/s12665-009-0132-3

# Ground Water Conditions and Spatial Distribution of Lead and Cadmium in the Shallow Aquifer at Effurun- Warri Metropolis, Nigeria

Irwin A. Akpoborie[1], Alex E. Uriri[2] & Oghenevwede Efobo[3]

[1] Department of Geology, Delta State University, Abraka, Nigeria

[2] Department of Geography and Planning, University of Lagos, Lagos, Nigeria

[3] Center for Research in Water and Environment, Abraka, Nigeria

Correspondence: Irwin A. Akpoborie, Department of Geology, Delta State University, Abraka, Nigeria. E-mail: tony.akpoborie@gmail.com

**Abstract**

A water table head distribution map of the shallow Benin Formation aquifer in the Effurun-Warri area has been drawn from dug well data and used to define groundwater gradients as well as identify directions of groundwater movement in this densely populated urban setting. Water samples from forty dug wells were also screened for the presence of lead and cadmium and results showed a variation in concentration from not detectable to 0.04mg/l for each metal. Iso-concentration contours for lead in groundwater suggest that enrichment may be from two sources: wastes from the refinery and petrochemical industrial complex on the northwestern edge of the city and secondly from leachates associated with the many unregulated waste dumpsites. Lead appears to be constrained from spreading eastwards from the industrial complex area by the south and westwards trending groundwater gradient. The city wide prevalence of elevated levels of cadmium is also probably due to leachates from unregulated dumpsites as well as the mixing of groundwater as suggested by existing gradients. Potential implications of the findings for public health, local and regional water quality monitoring are discussed.

**Keywords:** heavy metals, lead, cadmium, groundwater, urban water, leachates, Niger Delta, Benin Formation

## 1. Introduction

Lead and cadmium have been identified as two of several heavy metals that can impair human organ function when ingested even in minute quantities. Sarojam (2011) notes for example, that heavy metals in general are known to cause harmful reproductive effects and draws attention to the non degradability of lead and calcium in nature and hence their tendency to accumulate in the food chain. Ifegwu and Ayankora (2012), WHO (2011, 2010), the American College of Obstetricians and Gynecologists (AMCOG, 2012) provide summary descriptions of the mechanisms through which organ and enzyme dysfunction are associated with their ingestion. Specifically, excessive intake of lead has been associated with multiple problems including cancer, intestinal nephritis, hypospermia, testicular atrophy, learning disorders and in large quantities, death. Indeed, the United States Agency for Toxic Substances and Disease Registry (ATSDR, 2007) states that while the developing nervous system, cardiovascular systems, and the kidney are the most sensitive targets for lead toxicity, lead could potentially affect any system or organs in the body. Cadmium is also toxic to kidneys when ingested in excess.

Several studies including Aremu, Olawuyi, Metshitsuka, Sridhar, & Oluwande (2002), Otobo, Aigbogun, and Ifedili (2007); Akodu, Ozulu and Osagbue (2010); Basorun and Olamiju (2013) Ogbeibu, Chukwurah and Oboh, (2013), Iwegbue, Nwajei, Ogala and Overah (2010) have shown the presence of elevated levels of heavy metals including lead and cadmium in soils, storm runoff and shallow ground water in the Effurun- Warri area. Etchie, Etchie and Adewuyi (2011) use principal component analysis to show that the presence of these heavy metals in parts of the city's groundwater are associated with nearby refinery operations while Abimbola, Oke and Olatunji (2004), Akodu, Ozulu and Osagbue (2010) and Ogbeibu, Chukwurah and Oboh, (2013) show that leachate from unregulated dumpsites in the city are point sources for heavy metal loading of the underlying aquifer. Additional non point sources of these contaminants possibly include open storm water drains that contain sediment with elevated levels of lead and cadmium of up to 1.4 mg/kg and 0.6mg/kg respectively (Egboh, Nwajei & Adaikpoh,

2000) as well as nearby creeks and water courses in the area that are contaminated with industrial effluent (Emoyan, Akporhonor & Akpoborie, 2008; Nduka & Orisakwe, 2009). Unfortunately, a dysfunctional and inadequate public agency water supply system has driven residents, industry and commercial establishments in the Effurun-Warri metropolis to rely on groundwater from shallow dug wells and boreholes drilled into the upper horizon of the Benin Formation aquifer for self supplies. While the chemical quality of water from these sources meets regulatory agency standards for domestic use (Olobaniyi & Owoyemi, 2004; Ejechie, Olobaniyi, Ogban & Ugbe, 2007) the spatial distribution of reported lead and cadmium in the city's shallow groundwater is not well understood. Spatial distribution would be controlled largely by textural as well as lithological variations and by existing groundwater gradients in the aquifer that determine the direction of ground water flow and other mixing phenomena.

The objectives of this study which is part of a comprehensive hydro-geochemical evaluation of the shallow aquifer are twofold: first, to establish for the first time, geo-referenced groundwater gradients in the shallow aquifer; second, determine levels of occurrence of lead and cadmium in groundwater and identify the influence of existing gradients on the spatial distribution of these heavy metals in groundwater in this rapidly expanding urban setting. These are important issues in water supply planning and aquifer protection especially as they relate to the future design and location of solid and liquid waste management systems and public health.

### 1.1 Area of Study

The study area includes the Warri-Effurun metropolis and the adjoining rural fringes that lie   roughly between latitude 5°30'N - 5°45N and longitude 5°15'E - 5°50'E, Figure 1, Figure 2. The small and rural river port of Warri town with a population of a mere 20,000 people in 1933, has expanded to become the present day agglomeration of many towns and communities that include Effurun, Ekpan, Enerhen, Edjeba, Ogunu, Jakpa, Ovian –Aladja. These towns and communities are also the main population centers of four core local government areas (LGAs), namely: Uvwie, Warri South, Warri South West and Udu, Figure 1. The 1991 Nigerian census recorded the population of these LGAs that substantially make up the population of the Warri-Effurun metropolis at 450,362; this increased to 754,931 by the 2006 census and projected to have reached 852,317 by year 2010. Indeed, Babatola and Uriri (2013) projected a population of one million inhabitants by mid- year 2013 for this metropolis that is the hub of the oil and gas industry in the western Niger Delta petroleum province as well as the most populous industrial and commercial center of Delta state, Nigeria.

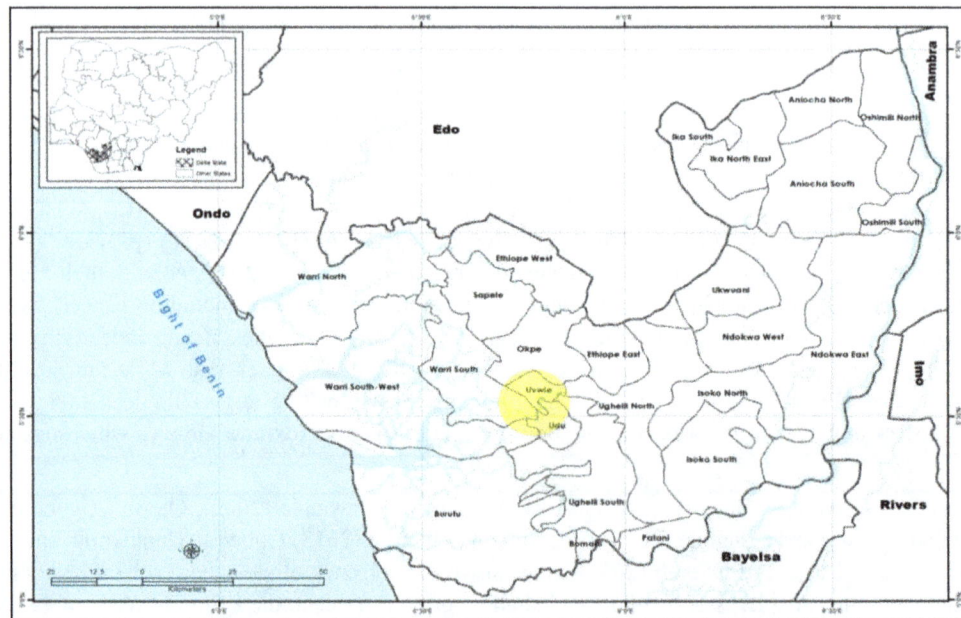

Figure 1. Map of Delta State showing LGA boundaries and the position of the Effurun-Warri Metropolis (marked in yellow)

*1.2 Physiography and Climate*

The Effurun-Warri area is typical Sombreiro- Warri Deltaic Plain (SWDP) terrain which is monotonously lowland and flat with a gentle slope towards existing water courses and swamps. The western edge rests on the equally low lying and extensive wetland, the Brackish Water Mangrove Swamps (BMS) that stretches westwards to the Atlantic coast. The area is drained by the Warri River and its subordinate network of creeks. The swamps on the north bank of the Warri River have largely been reclaimed for urban development and only patches of mangrove vegetation remain on this as well as on the banks of the creeks at Edjeba, Bendel Estate, and Ogunu that flow into the Warri River, Figure 2.

The climate is equatorial, hot (23°C -37°C) and humid (Relative Humidity, 50-70 per cent). There is a dry season from about November to February, and a wet season that begins in March and peaks in July and October. Mean annual rainfall is 2,500 -3000mm (Adejuwon, 2012).

*1.3 Geology and Distribution of Quaternary Recent (Holocene) Deposits*

The geology of the Niger Delta petroleum province is much studied and descriptions may be found in Short and Stauble (1967) among many others. In summary, the Benin Formation is the youngest of the three important formations that constitute the sedimentary fill of the Niger Delta Basin. The formation consists of massive continental/fluvial sands and gravels. The older formations which are encountered only in the subsurface in the Warri area are the Agbada Formation of paralic sands and shales and the basal Akata Formation which consists of holomarine shales, silts and clays. Thus the Benin Formation underlies the Effurun-Warri area but is overlain and masked by deposits of the SDWP and BMS. These deposits are typically successions of fine to medium grained unconsolidated sands interbedded with thin discontinuous layers of clay in the SDWP and thin and dark grey silts and discontinuous clays in case of the BMS (Akpoborie, Ekakite & Adaikpoh, 2000; Akpoborie & Aweto, 2012).

## 2. Methodology

One hundred and twenty nine dug wells dug wells which are evenly distributed in the city were selected for water level monitoring in the wet season. Depth to water level in each of the dug wells were measured with a Solinst model electronic water level indicator. An Ertec model GPS instrument was used to determine coordinates and to locate the well positions on the city map. Because available city maps are devoid of contours as are all maps of this general Niger Delta region, averaged elevation readings from the altimeter module of three GPS instruments at each site were used to approximate the elevation of each well location. Results were employed in generating the depth to water level as well as the water table head distribution using Surfer 8 (Golden Software Inc., 2002).

Furthermore, approximately 40 dug wells located across the city were randomly selected for ground water sampling. Samples were collected from the selected dug wells and screened for the major ions as well as selected heavy metals. In the sampling procedure, replicate water samples were collected from each dug well into sterilized polyethylene bottles. The set of samples designated for heavy metal analysis were immediately stabilized *in situ* with nitric acid, stored in ice boxes and sent to the laboratory within an hour of collection for analysis. At the laboratory, the Pye Unicam Atomic Absorption Spectrophotometer SP 2900 was employed in the determination of levels of lead and cadmium.

## 3. Results

*3.1 Groundwater Conditions*

Data from about forty evenly spread dug well locations out of the one hundred and twenty points measured were selected and employed in constructing water table contour lines of equal head in meters above sea level as shown in Figure 2. The selected locations and associated data are presented in Table 1. Depth to water level in the shallow aquifer ranges from a minimum of 0.75m at Ugboroke through 3m at Effurun Market to a maximum of 5m below ground level at the Airport Junction area. The configuration of the water table reveals the presence of a groundwater mound which is aligned in a northeast to southwest direction in the city and which begins at the Effurun GRA area and continues through Ugborikoko to Ajamimogha. This mound is also somewhat equidistant from the Ogunu Creek to the north and Warri River to the south. Groundwater moves from it southwards from the neighborhood of Okumagba Layout through Igbudu and Agbassa neighbourhoods to Warri River and northwards from Ugborikoko and Ugboroke to feed the Ogunu Creek. At Edjeba and Ajamimogha, movement from it is northwestwards and westwards to join the Ogunu Creek and Warri River respectively. In the vicinity of the NNPC refinery and petrochemical complex located at the northwest corner of the city, groundwater movement is northwest towards the reclaimed wetland area of Ubeji community and also west and southwards.

In the eastern part of the city at Ekete, Oruwhorun and DSC Steel Town, movement is eastwards and away from the Warri River. Therefore, the Warri River and Ogunu Creek appear to have a major influence on groundwater movement in the city. Olobaniyi and Owoyemi's (2004) model has somewhat similar characteristics but has remarkable differences in the configuration of the water table.

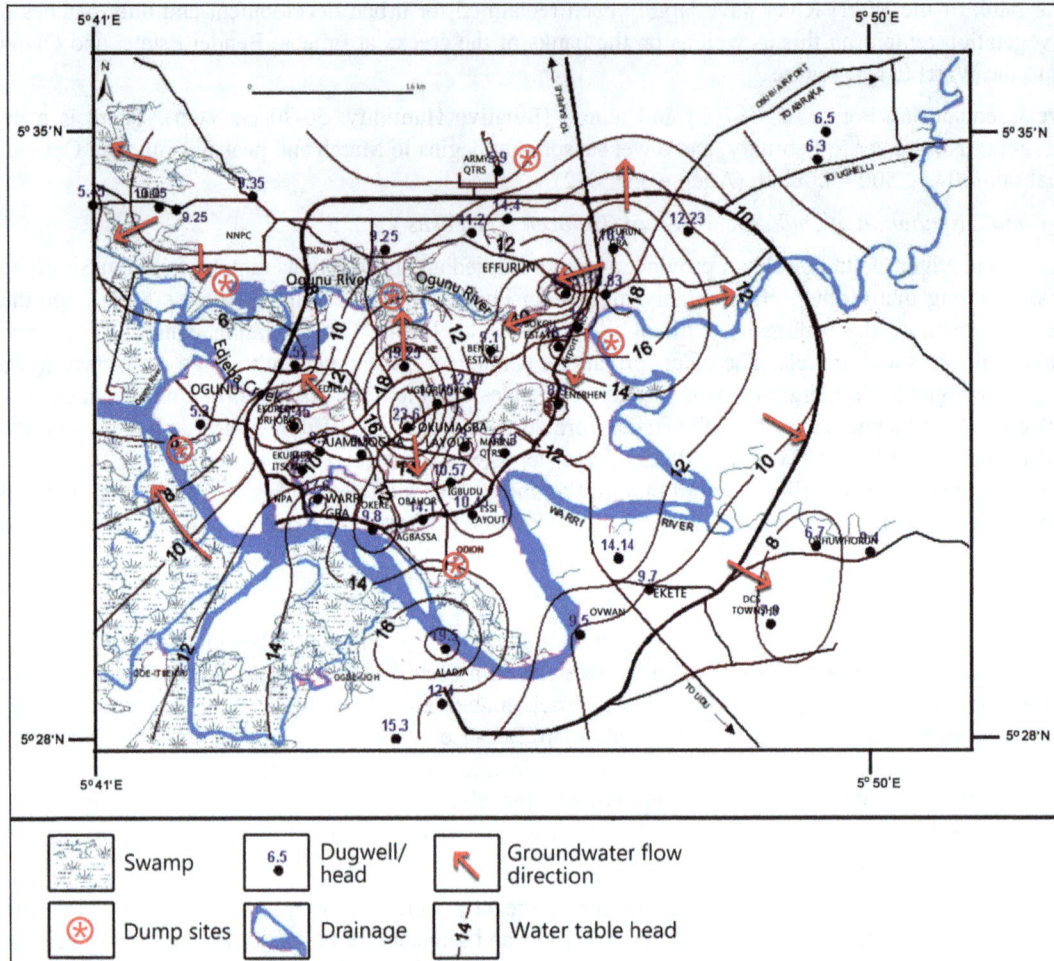

Figure 2. Effurun- Warri metropolis: water table contour lines of equal head in meters above sea level (Base map modified from Urhobo Historical Society, 2014)

Table 1. Depth to water level (DWL) data and calculated head from dug wells in Effurun-Warri

| Easting | Northing | Location Name | DWL | Head |
|---|---|---|---|---|
| 5.704667 | 5.57105 | Ubeji Health Centre | 1.95 | 10.05 |
| 5.707861 | 5.570611 | Town Hall Ubeji | 1.75 | 9.25 |
| 5.692333 | 5.571472 | Seko Est Ubeji Deeper Life Road, Ubeji | 1.55 | 5.45 |
| 5.722167 | 5.573056 | Jeddo/Ubeji Junction | 2.65 | 9.35 |
| 5.746944 | 5.563222 | Ekpan Secondary School | 2.75 | 9.25 |
| 5.729972 | 5.542194 | Opposite Holy Family Catholic Church | 2.45 | 4.55 |
| 5.729694 | 5.531167 | Off Okundolor Street Ekurede Urhobo | 1.55 | 15.45 |
| 5.828472 | 5.58475 | 13, Omoteko Anikopi St. Okuokoko | 3.5 | 6.5 |
| 5.826889 | 5.579694 | 1, James Ovie St. Okuokoko | 2.7 | 6.3 |
| 5.789917 | 5.506889 | 17, Boro St. Owhasa Off. Orhuwhorun Rd | 2.86 | 14.14 |
| 5.795667 | 5.501306 | 3, Visa Link Rd. Off. Orhuwhorun Rd | 2.3 | 9.7 |
| 5.782778 | 5.493056 | Noriel Tueje Compound, Off Ovwian Mrk | 3.5 | 9.5 |
| 5.748722 | 5.474083 | Gbobome Comp. Gramm. Sch. Rd. Aladja | 1.7 | 15.3 |
| 5.757917 | 5.490444 | Tom Compound Off Obierurhu St. Aladja | 1.5 | 19.5 |
| 5.757222 | 5.480417 | Calvary Baptist Church Aladja | 2.6 | 12.4 |
| 5.818028 | 5.495 | 14, White House Lagos Rd DSC | 2.1 | 7.9 |
| 5.836306 | 5.508056 | Ekreku Comp. by Usieffrun Junction Orhuwhorun | 3.6 | 9.4 |
| 5.826472 | 5.509083 | Esyai Close Opp. Itaigho F/S Orhuwhorun | 3.3 | 6.7 |
| 5.789083 | 5.563444 | 3, Onaya St. Off Ovie Palace Rd. Effurun | 1.6 | 18.4 |
| 5.769667 | 5.568861 | 1, Osademe St. (Km2) Off Refinery Rd. | 1.6 | 14.4 |
| 5.768083 | 5.577583 | Army Barrack (Block 23) | 2.1 | 5.9 |
| 5.763 | 5.566306 | 10, Indian Close Off Jakpa Road | 2.8 | 11.2 |
| 5.780389 | 5.55525 | 9, Okito St.Off Jakpa Rd. Sokoh Estate | 2.65 | 4.35 |
| 5.782528 | 5.549111 | BH Conoil, Airport Junction, Effurun | 5 | 15 |
| 5.779028 | 5.545528 | Our Ladys High School Effurun | 3.8 | 18.2 |
| 5.787639 | 5.555139 | Arubaye Comp. by Shrine Effurun Market | 3.17 | 18.83 |
| 5.7615 | 5.527333 | 1, Willy St. Off Deco Road. | 1 | 9 |
| 5.758889 | 5.520778 | 24, Bazunu Rd. Opp. Akajuigo Plaza Igbudu | 2.43 | 10.57 |
| 5.753722 | 5.514022 | 11G, Bazunu Comp. Otowodo Agbassa | 1.9 | 14.1 |
| 5.744472 | 5.512156 | 3, House Rd Main Mkt, Ogbe Joh | 3.2 | 9.8 |
| 5.762917 | 5.514889 | 1, Edesoh St. Off Essi Layout | 2.55 | 10.45 |
| 5.768972 | 5.526194 | 14, Ajuyahs St Marine Quarters | 2.7 | 14.3 |
| 5.778917 | 5.535139 | Urhobo College, Effurun | 2.1 | 8.9 |
| 5.7665 | 5.544806 | Opp. Jnr Staff Club Bendel Estate. | 1.95 | 9.1 |
| 5.750417 | 5.541889 | 3, Akpala St. B/H Holy Family Sch. Ugboroke | 0.75 | 19.25 |
| 5.734556 | 5.526528 | 2, Arimo Close, Off   Olu Palace | 2.1 | 9.9 |
| 5.73125 | 5.523139 | Ogiame Primary School Ekurede Itsekiri | 1.8 | 8.2 |
| 5.712222 | 5.531472 | City Ovwuvwe Comp. Ogunu Village | 2.7 | 5.3 |
| 5.742472 | 5.525972 | 31B, Ekpen St Okere Warri | 3.1 | 9.9 |
| 5.750944 | 5.530889 | 10, Ohwadjeke St Okumagba Layout. | 1.4 | 23.6 |
| 5.756417 | 5.535111 | 11, Grey St Off Poloko Okumagba Layout | 2.25 | 17.75 |
| 5.762194 | 5.537194 | 42, Ohwonigho Comp Ugborikoko | 1.53 | 12.47 |
| 5.803028 | 5.566583 | Ezekiel Close, Jefia Ave. Off   P.T.I Rd | 0.77 | 12.23 |
| 5.734139 | 5.517861 | Central Hospital Warri | 1.4 | 17.6 |

## 3.2 Occurrence and Distribution of Lead and Cadmium in Groundwater

Lead was detected in more than eighty per cent of all samples collected and ranged in concentration from 0-0.04mg/l. The highest value for lead was obtained from a well near the Ubeji Health Center that returned a value of 0.04 mg/l followed by the sample from a dug well at the Ogiame Primary School, Ekurede –Itsekiri area.

Spatial details of the occurrence of lead are presented in Table 2 which consists of data from randomly selected locations and which data has been used in constructing the iso-concentration map shown in Figure 3. The occurrence of groundwater impaired by the presence of lead above the WHO Drinking water guideline value of 0.01mg/l is not limited to these two locations but as suggested by the isocons in Figure 3 spreads to the Ekurede Urhobo, the Warri GRA, NPA and Ogunu neighborhoods. At Ubeji, Figure 3 also suggests the presence of a source of lead contamination around or near the Health Center from which the contamination is spreading westwards.

Cadmium ranged from 0-0.040 mg/l and the highest value was obtained from a well at the Okere neighborhood. Cadmium is predominantly above the WHO limit of 0.003mg/l everywhere in the city except in the isolated areas of Ekete and Ovwian to the east. Iso - concentration contours for cadmium, Figure 4 constructed from data in Table 2 show that the highest concentrations of cadmium in groundwater are in the older and more densely populated neighborhoods of Ekurede Itsekiri, Warri GRA, Ekurede Urhobo, Okere, Edjeba and Agbassa.

Table 2. Levels of occurrence of lead and cadmium at different parts of Effurun-Warri

| Easting | Northing | Location Name | Cadmium(mg/l) | Lead(mg/l) |
|---|---|---|---|---|
| 5.704667 | 5.57105 | Ubeji Health Centre | 0.008 | 0.04 |
| 5.707861 | 5.570611 | Town Hall Ubeji | 0.003 | 0.002 |
| 5.692333 | 5.571472 | Seko Est Ubeji Deeper Life Road, Ubeji | 0.005 | 0.001 |
| 5.722167 | 5.573056 | Jeddo/Ubeji Junction | 0.007 | 0.003 |
| 5.826889 | 5.579694 | 1, James Ovie St. Okuokoko | 0.006 | 0.001 |
| 5.789917 | 5.506889 | 17, Boro St. Owhasa Off. Orhuwhorun Rd | 0.001 | 0.001 |
| 5.795667 | 5.501306 | 3, Visa Link Rd. Off. Orhuwhorun Rd | 0.002 | 0.002 |
| 5.782778 | 5.493056 | Noriel Tueje Compound, Off Ovwian Mrk. | 0 | 0 |
| 5.757222 | 5.480417 | Calvary Baptist Church Aladja | 0.005 | 0.001 |
| 5.818028 | 5.495 | 14, White House Lagos Rd DSC | 0.006 | 0.004 |
| 5.789083 | 5.563444 | 3, Onaya St. Off Ovie Palace Rd. Effurun | 0.005 | 0.003 |
| 5.768083 | 5.577583 | Army Barrack (Block 23) | ND | ND |
| 5.778917 | 5.535139 | Urhobo College, Effurun | 0.009 | 0.001 |
| 5.73125 | 5.523139 | Ogiame Primary School Ekurede Itsekiri | 0.04 | 0.006 |
| 5.742472 | 5.525972 | 31B, Ekpen St Okere Warri | 0.003 | 0.019 |
| 5.750417 | 5.541889 | 3, Akpala St. B/H Holy Family Sch. Ugboroke | 0.009 | 0.001 |
| 5.748722 | 5.474083 | Gbobome Comp. Gramm. Sch. Rd. Aladja | ND | ND |

Note: ND = Not detected.

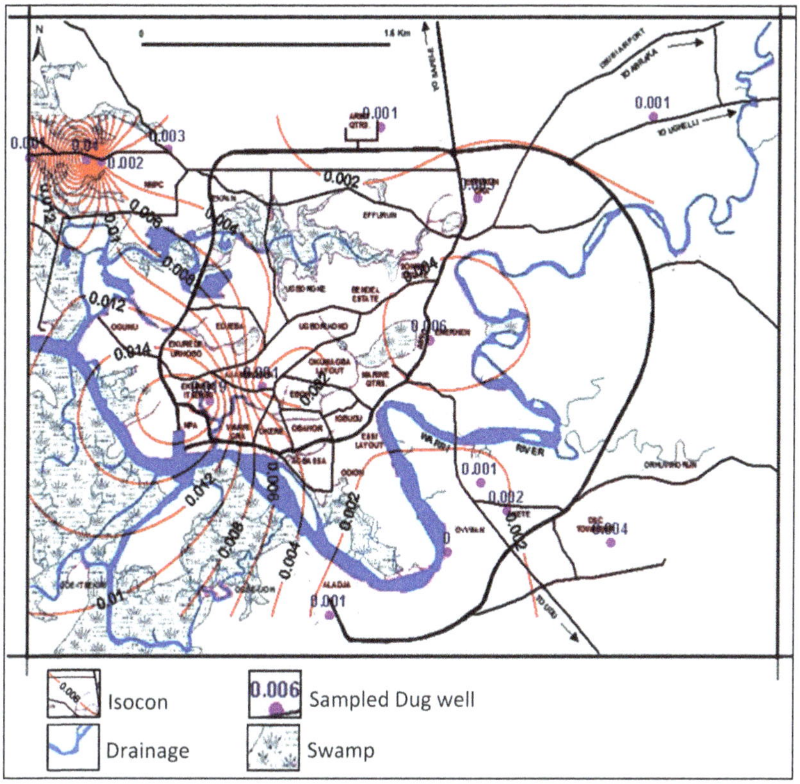

Figure 3. Lead isocons in Effurun –Warri metropolis (Base map modified from Urhobo Historical Society, 2014)

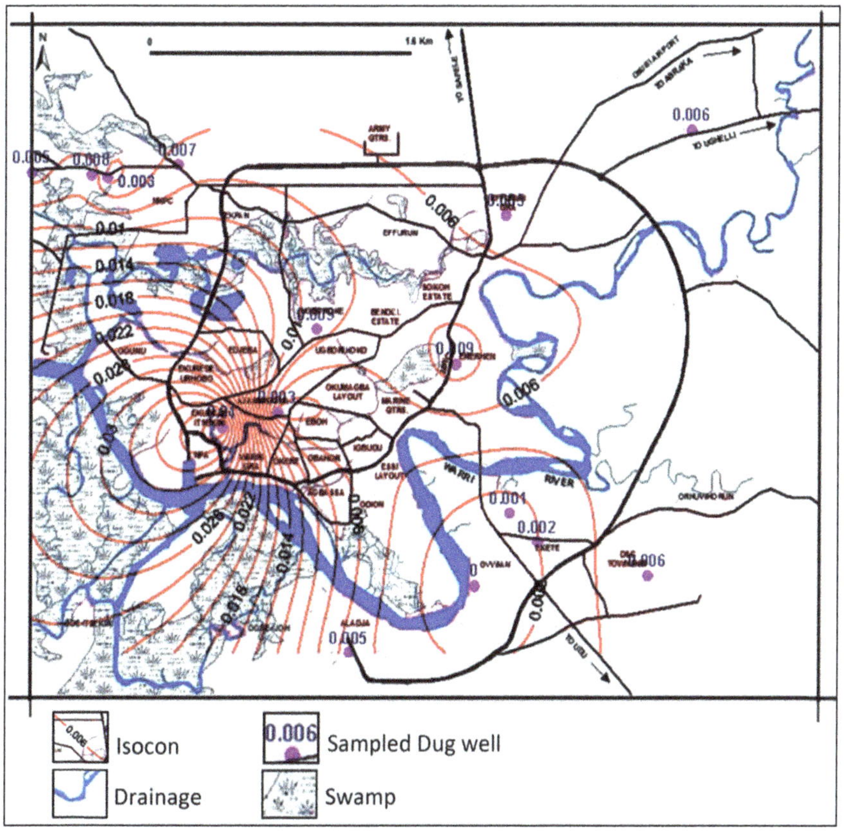

Figure 4. Cadmium isocons in Effurun- Warri metropolis
(Base map modified from Urhobo Historical Society, 2014)

## 4. Discussion

### 4.1 Urbanization, Waste Management and Spatial Distribution of Lead and Cadmium

The distribution of lead and cadmium in groundwater as may be observed from Figure 3and Figure 4 indicate that the western half of the city appears to be more susceptible to heavy metal contamination than the east. A possible explanation for this could be that the western half constitutes the more densely populated older neighborhoods around which other previously rural communities have agglomerated in the past 50 years to form the current metropolis. Westward city expansion has been limited by the BMS on the city's western edge. The NNPC refinery and petrochemical complex located on the northwest corner have also influenced city expansion in that area such that the previously rustic Ubeji community located there has experienced rapid development in the last decade. The implication of this for spatial distribution of contaminants in groundwater relates to the historical location of commerce and industry as well as well as waste disposal sites in the city. Dumpsites and associated leachate in the city have been implicated in the generation of heavy metals and their elevated levels of occurrence in ground water (Adebisi et al., 2013; Iwegbue et al., 2010; Ogbeibu et al., 2013; Akudo et al., 2010; Otobo et al., 2007). These studies all show that leachate from dumpsites is laden with lead and cadmium among others at varying degrees of magnitude. Furthermore, twenty one out of the twenty five large, open and unregulated garbage dump sites identified in a city wide study by Efe, Akosua and Ojoh (2013) are located in the core more densely populated older neighbourhoods of the western section and these dumpsites accounted for approximately 80 per cent of all waste generated in the city. Moreover, many former dumpsites in these older neighbourhoods where land for construction purposes is at a premium have been reclaimed and built up and there is now no evidence of their former existence. However, they would probably continue to possess the potential for pollution depending on the manner and process of reclamation as well as the age of such former dumpsites (Adebisi et al., 2013). Such reclaimed dumpsites essentially fall into the category of sources of heavy metal contamination described as *legacy* sites by the World Health Organization (WHO, 2010).

When viewed together, Figure 2, Figure 3and Figure 4 suggest that the nature of existing groundwater gradients is a likely explanation for the confinement of the highest levels of lead and cadmium to the western sector of the city. It is also responsible for the noticeable presence of lead in groundwater at Ubeji and other communities that are north, west and southwest of the NNPC refinery and petrochemical complex whose waste streams have been identified by Ethcie et al. (2011) and Nduka and Orisakwe (2011) as the source of heavy metal loading to both surface water and groundwater in that area. Ground water to the east of the complex is as a result remarkably and relatively lower in lead content when compared with water from the western sector. This is because ground water movement as shown in Figure 2 is consistently westwards or southwards from the refinery and petrochemical complex area and contaminants from them would move in these directions and would be prevented from moving eastwards by the existing south and northwestward gradient.

Cadmium contamination on the other hand is more widespread and evenly distributed over the city perhaps because it is also associated with soils (Iwegbue et al., 2010) and leachates from existing garbage dumpsites. Furthermore, road side soils, sediment in open drains and storm water runoff have been implicated as continuous sources of cadmium and lead enrichment to ground water by Egboh, Nwajei and Adaikpoh (2000). The groundwater flow regime indicated in Figure 2 ensures hydrogeochemical mixing and the resultant widespread occurrence of cadmium. It therefore appears that the existence of contaminants from point sources has been long term and their spread has over time been governed by existing groundwater gradients and flow directions.

### 4.2 Potential Implications for Public Health and Regional Water Quality Monitoring

The widespread occurrence of lead and cadmium in groundwater is not peculiar to the Warri –Effurun area, but has also been observed in other parts of the western Niger Delta region. West of Warri at Ughoton and Omadino for example, mean lead (0.019mg/l) and cadmium (0.02mg/l) in groundwater from dug wells has been reported by Akpoborie and Aweto (2012). Eriyamremu et al. (2005) have also associated the presence of elevated levels of lead and cadmium in vegetables grown in parts of the Warri area and further west at Forcados to urbanization and industrial pollution.

Furthermore, the linkage between access to clean and acceptable quality water and maternal health in Nigeria has been stressed by Ezenwaji and Otti (2013) who suggest that lack of access is a major factor that contributes to the current high rate of maternal mortality. Indeed, Uriri, Babatola and Akintuyi (2011) and Babatola and Uriri (2013) also forcefully suggest a strong association between poor quality water usage and the abnormally high maternal mortality rates and other negative reproductive health indices they recorded from the Effurun-Warri metropolis. The widespread distribution of lead and cadmium in the city's shallow dug well water, lends support to this argument.

Unfortunately, the expense and equipment requirements for screening for heavy metals preclude their inclusion in the list of parameters that are evaluated by the public water supply agencies in routine water quality certification and monitoring programs (Akpoborie, Ebenuwa & Emonyan, 2011). Thus the possibility exists that groundwater with dangerous heavy mineral content could be and is potentially and routinely being certified by public agencies as fit for drinking and other domestic purposes. Indeed, local and regional water quality surveys that are undertaken in Nigeria (Olobaniyi, Ogban, Ejechi, & Ugbe, 2007; Ince et al., 2010; Edet, Nganje, Ukpong, & Ekwere, 2011) have also ignored the presence of heavy metals in groundwater. This is further compounded by the fact that the regulatory limits for lead and cadmium in drinking water provided by the Standard Organization of Nigeria are contradictory: NIS 554:2007 the Nigerian Standard for Drinking Water Quality specifies a Maximum Permissible Limit (MPL) of 0.003mg/l for Cadmium and 0.01mg/l for lead while NIS 345:2008 the Standard for Mineral Waters that guides the production of packaged water specifies MPLs of 0.01 mg/l and 0.05mg/l for cadmium and lead respectively (SON, 2007; 2008). This inconsistency is indeed the reason the WHO (2006) standard was used as the standard for comparison and discussion in this study.

## 5. Conclusion

A ground water head distribution map of the shallow Benin Formation aquifer in the Effurun-Warri area has been presented and used to identify directions of groundwater movement in the metropolis. The spatial distribution of lead suggest that lead in groundwater exceeds the WHO maximum allowable limits in a large part of the city and that lead enrichment may be from two source areas: the NNPC refinery and petrochemical wastes in the northwestern part of the city and from leachates associated with unregulated dumpsites that have been traditionally used for waste disposal in the city. Existing ground water gradients may be responsible for the confinement of elevated levels of lead in groundwater to the city's western sector. Elevated levels of cadmium in groundwater above WHO maximum allowable limits in drinking water are more widespread and evenly distributed in the area, again because of the prevalent long term use of open and unregulated dumpsites as well as the possible mixing of ground water as suggested by existing gradients. The toxicity of these heavy metals requires that their presence in groundwater be constantly monitored.

## Acknowledgement

This investigation was undertaken with the aid of partial funding and resources provided by the Center for Research in Water and Environment (CREWE), Abraka and for which the authors are grateful.

## References

Abimbola, A. F., Oke, S. A., & Olatunji, A. S. (2002). Environmental Impact assessment of waste Dump sites on geochemical quality of water and soils in Warri metropolis, southern Nigeria. *Water Resources, 13*, 7-11.

Adebisi, N. O., Oluwafemi, O. S., Songca, S. P., & Haruna, I. (2013). Flow system, physical properties and heavy metals concentration of groundwater: A case study of an area within a municipal landfill site. *International Journal of Water Resources and Environmental Engineering, 5*(11), 630-638. http://dx.doi.org/10.5897/IJWREE2013.0427

Adejuwon, O. A. (2012). Rainfall Seasonality in the Niger Delta Belt, Nigeria. *Journal of Geography and Regional Planning, 5*(2), 51-60. http://dx.doi.org/10.5897/JGRP11.096

Adewosu, H. O., Adewuyi, G. O., & Adie, G. G. (2013). Assessment of Heavy Metals in Soil, Leachate and underground water samples collected from the vicinity of Olususun landfill in Ojota, Lagos, Nigeria. *Transnational Journal of Science and Technology, 3*(6), 73-86.

Akoteyon, I. S., Mbata, U. A., & Olalude, G. A. (2011). Investigation of heavy metal contamination in groundwater around landfill site in a typical sub-urban settlement in Alimosho, Lagos-Nigeria. *Journal of Applied Sciences in Environmental Sanitation, 6*(2), 155-163.

Akpoborie, I. A., & Aweto, K. E. (2012). Ground water conditions in the Mangrove Swamps of the Western Niger Delta: Case study of the Ughoton Area, Delta state, Nigeria. *Jour. Env. Hydrology, 20*, 1-14.

Akpoborie, I. A., Ekakite, O. A., & Adaikpoh, E. O. (2000). The Quality of Groundwater from Dug Wells in Parts of the Western Niger Delta. *Knowledge Review, 2*(5), 72-75.

Akpoborie, I. A., Ebenuwa, C. C., & Emonyan, O. O. (2009). *Water Quality Monitoring in Delta State, Nigeria: Status, Rights to Water and the Millennium Development Goals.* Presented at the 2nd Annual Faculty of Social Science International Conference, Delta State University, Abraka, December 10-12.

Akudo, E. O., Ozulu, G. U., & Osogbue, L. C. (2010). Quality Assessment of groundwater in selected waste dump site areas in Warri. *Environmental Research Journal, 4*(4), 281-285. http://dx.doi.org/10.3923/erj.2010.281.285

Allen, J. R. L. (1965). Late Quaternary Niger Delta and Adjacent Areas: Sedimentary Environments and Lithofacies. *Bulletin American Association of Petroleum Geologists, 49*(5), 547-600.

AMCOG. (2012). Committee Opinion No. 533: Lead screening during pregnancy and lactation. *Obtet Gynecol, 120,* 416-20.

Aremu, D. A., Olawuyi, J. F., Metshitsuka, S., Sridhar, M. K., & Oluwande, P. A. (2002). Heavy metal analysis of groundwater from Warri, Nigeria. *International Journal of Environmental Health Research, 12*(3), 261-267. http://dx.doi.org/10.1080/0960312021000001014

Asiwaju-Bello, A. Y. (2007). Contaminant plume migration patterns in ground water around Oke-Odo refuse dump site, Lagos, Nigeria. *Water Resources, 17,* 24-28.

Agency for Toxic Substances and Disease Registry. (2007). *Toxicological Profile for Lead.* ATSDR, Atlanta, Ga., USA.

Babatola, O., & Uriri, A. (2013). Assessment of Maternal Health Intervention Programme of Delta State, Nigeria: Application of the U.N Process Indicators. *Journal of Public Policy and Administration Research, 3*(9), 62-71.

Basorun, J. O., & Olamiju, I. O. (2013). Environmental Pollution and Refinery Operations in an Oil Producing Region of Nigeria: A Focus on Warri Petrochemical Company. *Journal of Environmental Science, Toxicology and Food Technology, 2*(6), 18-23.

Edet, A., Nganje, T. N., Ukpong, A. J., & Ekwere, A. S. (2011). Groundwater chemistry and quality of Nigeria: A status review. *African Journal of Environmental Science and Technology, 5*(13), 1152-1169. http://dx.doi.org/ 10.5897/AJESTX11.011

Efe, S. I. (2005). *Urban effects on precipitation amount and rainwater quality in Warri metropolis* (Ph.D. Thesis). Delta State University, Abraka.

Efe, S. I., Cheke, L. A., & Ojoh, C. O. (2013). Effects of Solid Waste on Urban Warming in Warri Metropolis, Nigeria. *Atmospheric and Climate Sciences, 3,* 6-12. http://dx.doi.org/10.4236/acs.2013.34A002

Egboh, S. H. O., Nwajei, G. E., & Adaikpoh, E. O. (2000). Selected Heavy metals concentration in sediments from major roads and gutters in Warri, Delta State, Nigeria. *Nig. J. Sc. Env., 2,* 105-111.

Ejechi, B. O., Olobaniyi, S. B., Ogban, F. E., & Ugbe, F. C. (2007). Physical and Sanitary Quality of Hand Dug Well Water from Oil producing area of Nigeria. *Environ Monit Assess., 128,* 495-501. http://dx.doi.org/10.1007/s10661-006-9343-1

Emoyan, O. O., Akporhonor, E. E., & Akpoborie, I. A. (2008). Environmental Risk Assessment of River Ijana, Ekpan, Delta State. *Chem. Spec. and Bioavailability, 20*(1), 23-32. http://dx.doi.org/10.3184/095422908X295825

Eriyamvemu, G. E., Asagba, S. O., Akpoborie, I. A., & Ojeaburu, S. I. (2005). Evaluation of Lead and Cadmium Levels in Some Commonly Consumed Vegetables in the Niger Delta Oil Area of Nigeria. *Bull. Env. Contam. Toxicol., 75,* 278-283. http://dx.doi.org/10.1007/s00128-005-0749-1

Etchie, T. O., Etchie, A. T., & Adewuyi, G. O. (2011). Source Identification of Chemical Contaminants in Environmental Media of a Rural Settlement. *Research Journal of Environmental Sciences, 5,* 730-740. http://dx.doi.org/10.3923/rjes.2011.730.740

Ezenwaji, E. E., & Otti, V. I. (2013). Water Related Diseases as a Challenge to the Implementation of Reproductive Health of Pregnant Women in Anambra State, Nigeria. *International Journal of Engineering and Technology, 3,* 2049-3444.

Golden Software Inc. (2002). *Surfer 8.* Golden Software Inc. Co. USA

Ifegwu, C., & Anyakora, C. (2012). Screen for Eight Heavy Metals from Groundwater Samples from a Highly Industrialized Area in Lagos, Nigeria. *African Journal of Pharmaceutical Sciences and Pharmacy, 3*(1), 1-16.

Ince, M., Bashir, D., Oni, O. O. O., Awe, E. O., Ogbechie, V., Korve, K., ... Kehinde, M. (2010). *Rapid assessment of drinking water quality in the federal Republic of Nigeria Country Report of the Pilot Project implementation 2004-2005*. WHO-Unicef.

Iwegbue, C. M., Nwajei, G. E., Ogala, J. E., & Overah, C. L. (2010). Determination of trace metal concentrations in soil profiles of municipal waste dumps in Nigeria. *Environ Geochem Health, 32,* 415-430 http://dx.doi.org/10.1007/s10653-010-9285-y

Nduka, J. K., & Orisakwe, O. E. (2007). Heavy Metal Levels and Physico – Chemical Quality of Potable Water Supply in Warri, Nigeria. *Annali di Chimica, 97,* 867-874. http://dx.doi.org/10.1002/adic.200790071

Nduka, J. K., & Orisakwe, O. E. (2009). Effect of Effluents from Warri Refinery Petrochemical Company WRPC on Water and Soil Qualities of "Contiguous Host" and "Impacted on Communities" of Delta State, Nigeria. *The Open Environmental Pollution & Toxicology Journal, 1,* 11-17. http://dx.doi.org/10.2174/1876397900901010011

Ogbeibu, A. E., Chukwurah, A. E., & Oboh, I. P. (2013). Effects of Open Waste Dump-site on its Surrounding Surface Water Quality in Ekurede-urhobo, Warri, Delta State, Nigeria. *Natural Environment, 1*(1), 1-16. http://dx.doi.org/ 10.12966/ne.06.01.2013

Olobaniyi, S. B., & Owoyemi, F. B. (2004). Quality of Groundwater in Deltaic Plain Sands Aquifer of Warri and Environs, Delta State, Nigeria. *Water Resources, 15,* 38-45.

Olobaniyi, S. B., Ogban, F. E., Ejechi, B. O., & Ugbe, F. C. (2007). Quality of groundwater in Delta State, Nigeria. *Jour. Env. Hydrology, 15,* 1-11.

Otobo, E., Aigbogun, C. O., & Ifedili, S. O. (2007). Geoelectrical Evaluation of Waste Dump Sites at Warri and its Environ, Delta State, Nigeria. *J. Appl. Sci. Environ. Manage, 11*(2), 61-64.

Sarojam, P. (2011). Analysis of Pb, Cd and As in Tea Leaves Using Graphite Furnace Atomic Absorption Spectrophotometry, Application Note, PerkinElmer, Inc., USA. Retrieved from http://www.perkinelmer.com/CMSResources/Images/APP_Metals_In_Tea_Leaves_PinAAcle_GFAA.pdf

Short, K. C., & Stauble, A. J. (1967). Outline of geology of Niger delta. *Bull. Amer. Assoc. Petr. Geol., 54*(5), 761-779.

Standards Organization of Nigeria. (2007). *Nigerian Standard for Drinking Water Quality* (p. 30). *NIS 554:2007*. Abuja.

Standards Organization of Nigeria. (2008). *Standard for Natural Mineral Waters* (p. 10). *NIS 345:2008,* Abuja.

Urhobo Historical Society. (2014). *A map of Effurun-Warri Area, 1986*. Retrieved from www.waado.org/warri/maps/effurun_warri_1986.html

Uriri, A., Babatola, O., & Akintuyi, A. (2011). *Patterns of Antenatal, Postnatal & Maternal Deaths in Delta State, Nigeria: A Spatio-Temporal Analysis: 2007-2010*. University of Lagos 7th Annual Research Conference & Fair: 19th -21st, October.

World Health Organization. (2006). *Guidelines and Standards for Drinking Water* (3rd ed). WHO, Geneva.

WHO. (2011). *Cadmium in Drinking-water Background document for development of WHO Guidelines for Drinking-water Quality* WHO/SDE/WSH/03.04/80/Rev/1

WHO. (2010). *Childhood Lead Poisoning*. WHO, Geneva, Switzerland.

# Permissions

All chapters in this book were first published in EP, by Canadian Center of Science and Education; hereby published with permission under the Creative Commons Attribution License or equivalent. Every chapter published in this book has been scrutinized by our experts. Their significance has been extensively debated. The topics covered herein carry significant findings which will fuel the growth of the discipline. They may even be implemented as practical applications or may be referred to as a beginning point for another development.

The contributors of this book come from diverse backgrounds, making this book a truly international effort. This book will bring forth new frontiers with its revolutionizing research information and detailed analysis of the nascent developments around the world.

We would like to thank all the contributing authors for lending their expertise to make the book truly unique. They have played a crucial role in the development of this book. Without their invaluable contributions this book wouldn't have been possible. They have made vital efforts to compile up to date information on the varied aspects of this subject to make this book a valuable addition to the collection of many professionals and students.

This book was conceptualized with the vision of imparting up-to-date information and advanced data in this field. To ensure the same, a matchless editorial board was set up. Every individual on the board went through rigorous rounds of assessment to prove their worth. After which they invested a large part of their time researching and compiling the most relevant data for our readers.

The editorial board has been involved in producing this book since its inception. They have spent rigorous hours researching and exploring the diverse topics which have resulted in the successful publishing of this book. They have passed on their knowledge of decades through this book. To expedite this challenging task, the publisher supported the team at every step. A small team of assistant editors was also appointed to further simplify the editing procedure and attain best results for the readers.

Apart from the editorial board, the designing team has also invested a significant amount of their time in understanding the subject and creating the most relevant covers. They scrutinized every image to scout for the most suitable representation of the subject and create an appropriate cover for the book.

The publishing team has been an ardent support to the editorial, designing and production team. Their endless efforts to recruit the best for this project, has resulted in the accomplishment of this book. They are a veteran in the field of academics and their pool of knowledge is as vast as their experience in printing. Their expertise and guidance has proved useful at every step. Their uncompromising quality standards have made this book an exceptional effort. Their encouragement from time to time has been an inspiration for everyone.

The publisher and the editorial board hope that this book will prove to be a valuable piece of knowledge for researchers, students, practitioners and scholars across the globe.

# List of Contributors

**Jeroen Provoost**
Independent researcher, Finland

**Lucas Reijnders**
Open University Netherlands (OUNL), Department of Science, Valkenburgerweg 177, 6419 AT Heerlen, Netherlands

**Jan Bronders**
Flemish Institute for Technological Research (VITO), Boeretang 200, 2400 Mol, Belgium

**Ilse Van Keer**
Flemish Institute for Technological Research (VITO), Boeretang 200, 2400 Mol, Belgium

**Steven Govaerts**
Flemish Institute for Technological Research (VITO), Boeretang 200, 2400 Mol, Belgium

**Ayoola O. Oluwajobi**
Department of Biological Sciences, School of Natural and Applied Sciences, Federal University of Technology, Minna, Nigeria
Department of Biology, Institute of Applied Sciences, Kwara State Polytechnic, Ilorin, Nigeria

**Olamide A. Falusi**
Department of Biological Sciences, School of Natural and Applied Sciences, Federal University of Technology, Minna, Nigeria

**Nuha A. Zubbair**
Department of Biology, Institute of Applied Sciences, Kwara State Polytechnic, Ilorin, Nigeria

**Orok E. Oyo-Ita**
Environmental & Petroleum geochemistry Research Group (EPGRG), Department of Pure and Applied Chemistry, University of Calabar, Nigeria

**Bassey O. Ekpo**
Environmental & Petroleum geochemistry Research Group (EPGRG), Department of Pure and Applied Chemistry, University of Calabar, Nigeria

**Peter A. Adie**
Department of Chemistry, Benue State University, Makurdi, Nigeria

**John O. Offem**
Environmental & Petroleum geochemistry Research Group (EPGRG), Department of Pure and Applied Chemistry, University of Calabar, Nigeria

**E. N. Vaikosen**
Department of Pharmaceutical and Medicinal Chemistry, Faculty of Pharmacy, Niger Delta University, Wilberforce Island, Nigeria

**B. U. Ebeshi**
Department of Pharmaceutical and Medicinal Chemistry, Faculty of Pharmacy, Niger Delta University, Wilberforce Island, Nigeria

**B. Airhihen**
Department of Pharmacology and Toxicology, Faculty of Pharmacy, Niger Delta University, Wilberforce Island, Nigeria

**Mamiseheno Rasolofonirina**
Department of Physics, Faculty of Sciences, University of Antananarivo, Madagascar

**Voahirana Ramaroson**
Department of Physics, Faculty of Sciences, University of Antananarivo, Madagascar
Department of Isotope Hydrology, Institut National des Sciences et Techniques Nucléaires, Madagascar

**Raoelina Andriambololona**
Institut National des Sciences et Techniques Nucléaires, Madagascar

**Ahmed Th. Ibrahim**
Department of Zoology, Faculty of Science, Assiut University, New Valley branch Assiut, Egypt

**Marwa A. Magdy**
Department of Zoology, Faculty of Science, Assiut University, New Valley branch Assiut, Egypt

**Emad A. Ahmed**
Department of Zoology, Faculty of Science, Assiut University, Assiut, Egypt

**Hossam M. Omar**
Department of Zoology, Faculty of Science, Assiut University, Assiut, Egypt

**Jing Chen**
Radiation Protection Bureau, Health Canada, Ottawa, Canada

**Irwin Anthony Akpoborie**
Department of Geology, Delta State University, Abraka, Nigeria

**Kizito Ejiro Aweto**
Department of Geology, Delta State University, Abraka, Nigeria

**Oghenero Ohwoghere-Asuma**
Department of Geology, Delta State University, Abraka, Nigeria

**T. Avellan**
Ludwig-Maximilians-Universität Munich, Dept. for Geography and Remote Sensing, Munich, Germany

**F. Zabel**
Ludwig-Maximilians-Universität Munich, Dept. for Geography and Remote Sensing, Munich, Germany

**B. Putzenlechner**
Ludwig-Maximilians-Universität Munich, Dept. for Geography and Remote Sensing, Munich, Germany

**W. Mauser**
Ludwig-Maximilians-Universität Munich, Dept. for Geography and Remote Sensing, Munich, Germany

**Mark P. Wachowiak**
Department of Computer Science and Mathematics, Nipissing University, Canada

**Renata Wachowiak-Smolíková**
Department of Computer Science and Mathematics, Nipissing University, Canada

**Brandon T Dobbs**
Department of Computer Science and Mathematics, Nipissing University, Canada
Department of Geography, Nipissing University, Canada

**James Abbott**
Department of Geography, Nipissing University, Canada

**Daniel Walters**
Department of Geography, Nipissing University, Canada

**Femi Francis Oloye**
Department of Chemistry, Adekunle Ajasin University, Akungba Akoko, Nigeria
Department of Chemistry, University of Aberdeen, United Kingdom

**Isaac Ayodele Ololade**
Department of Chemistry, Adekunle Ajasin University, Akungba Akoko, Nigeria

**Oluyinka David Oluwole**
Department of Chemistry, Adekunle Ajasin University, Akungba Akoko, Nigeria
Department of Chemistry, Rhodes University, South Africa

**Marcus Oluyemi Bello**
Department of Microbiology, Adekunle Ajasin University, Akungba Akoko, Nigeria

**Oluwabunmi Peace Oluyede**
Department of Environmental Biology and Fisheries, Adekunle Ajasin University, Akungba Akoko, Nigeria

**Oluwaranti Ololade**
Department of Chemistry, Federal University of Technology, Akure, Nigeria

**Xing Li**
Graduate School of Horticulture, Chiba University, Matsudo, Japan

**Changyuan Tang**
Graduate School of Horticulture, Chiba University, Matsudo, Japan

**Zhiwei Han**
Graduate School of Horticulture, Chiba University, Matsudo, Japan

**Piao Jingqiu**
Graduate School of Horticulture, Chiba University, Matsudo, Japan

**Cao Yingjie**
Graduate School of Horticulture, Chiba University, Matsudo, Japan

**Zhang Chipeng**
Graduate School of Horticulture, Chiba University, Matsudo, Japan

**A. Sridhar Kumar**
Department of Environmental Science, Osmania University, Hyderabad, Telangana State, India

**A. Madhava Reddy**
Environmental Specialist, Hyderabad, Telangana State, India

**L. Srinivas**
Research Scholar, Department of Botany, Osmania University of Hyderabad, India

**P. Manikya Reddy**
Professor, Department of Botany, Osmania University-Hyderabad, India

**Ahmed Usman**
Department of Pure and Industrial Chemistry, Bayero University, Kano, Nigeria

**Umar Ibrahim Gaya**
Department of Pure and Industrial Chemistry, Bayero University, Kano, Nigeria

**Maxwell Anim-Gyampo**
Department of Earth and Environmental Science, University for Development Studies, Navrongo, Ghana

**Musah Saeed Zango**
Department of Earth and Environmental Science, University for Development Studies, Navrongo, Ghana

**Boateng Ampadu**
Department of Earth and Environmental Science, University for Development Studies, Navrongo, Ghana

**Irwin A. Akpoborie**
Department of Geology, Delta State University, Abraka, Nigeria

**Alex E. Uriri**
Department of Geography and Planning, University of Lagos, Lagos, Nigeria

**Oghenevwede Efobo**
Center for Research in Water and Environment, Abraka, Nigeria